AF361942

Advances in Food Processing (Food Preservation, Food Safety, Quality and Manufacturing Processes)

Advances in Food Processing (Food Preservation, Food Safety, Quality and Manufacturing Processes)

Editors

Theodoros Varzakas
Panagiotis Tsarouhas

MDPI • Basel • Beijing • Wuhan • Barcelona • Belgrade • Manchester • Tokyo • Cluj • Tianjin

Editors

Theodoros Varzakas
Food Science and Technology
University of the Peloponnese
Kalamata
Greece

Panagiotis Tsarouhas
Department of Supply Chain
Management
International Hellenic University
Katerini
Greece

Editorial Office
MDPI
St. Alban-Anlage 66
4052 Basel, Switzerland

This is a reprint of articles from the Special Issue published online in the open access journal *Applied Sciences* (ISSN 2076-3417) (available at: www.mdpi.com/journal/applsci/special_issues/ Preservation_Safety_Quality).

For citation purposes, cite each article independently as indicated on the article page online and as indicated below:

LastName, A.A.; LastName, B.B.; LastName, C.C. Article Title. *Journal Name* **Year**, *Volume Number*, Page Range.

ISBN 978-3-0365-1842-8 (Hbk)
ISBN 978-3-0365-1841-1 (PDF)

Contents

About the Editors

Theodoros Varzakas

Theodoros Varzakas is a senior full professor at the Department of Food Science and Technology, University of Peloponnese, Greece, specializing in issues of food technology, food quality, and safety; a section editor in chief (2020–) for Food Security and Sustainability in the journal *Foods*; a former editor in chief for the *Current Research in Nutrition and Food Science Journal* (2015–2019); and a reviewer and member of the editorial board in many international journals. Varzakas has written more than 200 research papers and reviews and has presented more than 160 papers and posters at national and international conferences. He has written and edited six books in Greek and six in English.

Panagiotis Tsarouhas

Panagiotis Tsarouhas has a Diploma in Engineering and a PhD. He is an associate professor at International Hellenic University (Greece), Department of Supply Chain Management (Logistics). He worked for about ten years in the Technical Department at 'Chipita International SA' in Greece as the head of production and maintenance operations in plants for the food industry. He has about 30 years of research/teaching experience and has published more than 90 research papers in international journals, book chapters, and conference proceedings. He has been involved in many research and practical projects in the field of reliability analysis and maintenance engineering. His areas of interest are reliability and maintenance engineering, quality engineering, and supply chain management.

Preface to "Advances in Food Processing (Food Preservation, Food Safety, Quality and Manufacturing Processes)"

The aim of this e-book and Special Issue is to compile advances in the area of food manufacturing including packaging to address issues of food safety, quality, fraud, and how these processes (new or old) could affect the organoleptic characteristics of foods, with the aim to promote consumers' satisfaction. Moreover, food supply issues are explored. This Special Issue shows that new and improved technologies have been employed in the area of food manufacturing to address consumer needs in terms of quality and safety. The issues of research and development should be taken into account seriously before launching a new product onto the market. Finally, food fraud and authenticity are very important issues, and the food industry should focus on addresssing them.

Theodoros Varzakas, Panagiotis Tsarouhas

Editors

applied sciences

MDPI

Editorial

Advances in Food Processing (Food Preservation, Food Safety, Quality and Manufacturing Processes)

Theodoros Varzakas [1],* and Panagiotis Tsarouhas [2],*

[1] Department of Food Science and Technology, University of Peloponnese, Antikalamos, 24100 Kalamata, Greece
[2] Department of Supply Chain Management (Logistics), International Hellenic University, Kanellopoulou 2, 60100 Katerini, Greece
* Correspondence: theovarzakas@yahoo.gr (T.V.); ptsarouhas@ihu.gr (P.T.)

Citation: Varzakas, T.; Tsarouhas, P. Advances in Food Processing (Food Preservation, Food Safety, Quality and Manufacturing Processes). *Appl. Sci.* **2021**, *11*, 5417. https://doi.org/10.3390/app11125417

Received: 31 May 2021
Accepted: 10 June 2021
Published: 10 June 2021

Publisher's Note: MDPI stays neutral with regard to jurisdictional claims in published maps and institutional affiliations.

The aim of this special issue was to bring about advances in the area of food manufacturing, including packaging, addressing issues of food safety, quality, fraud and how these processes (new and old) could affect the organoleptic characteristics of foods, with the aim of promoting consumer satisfaction. Moreover, food supply issues have also been explored.

In this direction, packaging is the last step in the manufacturing process and makes the product attractive. In addition, it can extend the shelf life of the product. Hence, edible coatings have been used extensively. Acevedo-Correa et al. [1] report on the effect on cassava chips of coatings consisting of pectin and whey protein films. The coating type affected all sensory parameters except crispness, and temperature only influenced the color of the control chips.

Ishkeh et al. [2] employed chitosan as an edible coating to increase the storage of raspberries, and nanoparticles of chitosan were used to increase chitosan efficiency. This methodology was considered effective, safe and environmentally friendly, with health-promoting effects.

In regard to the development of new technologies and the advancement of existing technologies, Negara et al. [3] developed high-frequency defrosting, superheated steam and quick-freezing treatments to improve the quality of seafood home meal replacement products. They examined the chemical, microbiological and organoleptic properties of the products and saw no statistical changes over 90 days. The optimal temperature and heating time for this technology were also determined.

Rana et al. [4] also employed superheated steam roasting (270 °C for 4 min) and hot smoking (70 °C) of chub mackerel fillets to extend their shelf-life at the market level. Oak sawdust for 25 min of smoking time offered the best organoleptic characteristics, along with the optimal physiochemical and microbiological parameters.

Ultrasonic pretreatment was employed in the work of Taghinezhad et al. [5] for drying kiwi fruits under hybrid hot air–infrared conditions. The effective moisture diffusivity coefficient (Deff) was estimated, along with other parameters, and different models were used. The results revealed the effect of temperature and ultrasonic pretreatment time, along with sample thickness, on these parameters.

The study by Liu et al. [6] enabled calcium alginate ball encapsulation in fruit-flavored drinks (Boba milk tea made of tapioca). This technology showed that the organoleptic control and microbiological quality of the products were not affected if the product was stored at a temperature less than 10 °C.

Diaz-Bustamante et al. [7] contributed to the design of costeño-type artisan cheeses with specific functionalities through multiscale modeling by means of the substitution and reduction of NaCl and an increase in cooking temperature. The microstructure was not affected by the reduction in salt or by modifications in the cheese-making process.

The next two studies, which were carried out in Greece, investigated safety and quality aspects. Voidarou et al. [8] screened raw unprocessed honeycombs filled with oregano honey from Epirus, Greece, for bacteriocinogenic lactic acid bacteria and *Bifidobacterium* spp. exerting inhibitory action towards some pathogens and spoilage microorganisms isolated from fresh fruits and vegetables. The results suggested that bacteriocin-like substances are involved in this process.

Skiada et al. [9] evaluated and discriminated between monovarietal extra virgin olive oils obtained from olive cultivars cv. Lianolia Kerkyras and cv. Koroneiki, produced in Greece, based on their chemical characteristics, followed by statistical and chemometric analysis. They showed that possible authenticity tools could include sterol and fatty acid composition markers.

The work of Agustinisari et al. [10] deals with research and development, and more specifically with the formulation of protein and polysaccharides (whey protein–maltodextrin) in eugenol encapsulation and the determination of the effect of eugenol and chitosan concentrations on the characteristics of emulsions and spray-dried powders. The presence of chitosan resulted in more stable emulsions.

The next study, by Cucci et al. [11], reports on a new methodology for the rapid measurement of the redox potential as an indicator of color changes for meat juice in carcasses without slowing down the slaughter line to avoid food waste. A symmetry was highlighted between these two parameters.

Food fraud is very frequent nowadays and could become very complex. However, analytical tools have been employed to address this phenomenon. MALDI-TOF MS has been used in different matrices and has a lot of advantages; hence, a review by Zambonin [12] demonstrates all the existing applications for the detection of food fraud.

Supply chain management needs to take into account issues of corporate social responsibility. A case study was employed here, representing the dairy sector in Vietnam. The study shows the close relationship between the company and consumers, which is reflected in its financial performance [13].

Xu et al. [14] proposed an advanced online-to-offline (O2O) strategy for a single-vendor single-retailer integrated system with coordination mechanisms consisting of revenue-sharing, buy-back and quantity flexibility.

This special issue has shown that new and improved technologies have been employed in the area of food manufacturing, addressing consumer needs in terms of quality and safety. The issues of research and development should be taken into account before launching a new product into the market. Finally, food fraud and authenticity are very important issues and the food industry should focus on these.

Conflicts of Interest: The authors declare no conflict of interest.

References

1. Acevedo-Correa, D.; Jaimes-Morales, J.; Montero-Castillo, P. The Effect of Edible Coatings on Selected Physicochemical Properties of Cassava Chips. *Appl. Sci.* **2021**, *11*, 3265. [CrossRef]
2. Ishkeh, S.; Shirzad, H.; Asghari, M.; Alirezalu, A.; Pateiro, M.; Lorenzo, J. Effect of Chitosan Nanoemulsion on Enhancing the Phytochemical Contents, Health-Promoting Components, and Shelf Life of Raspberry (*Rubus sanctus* Schreber). *Appl. Sci.* **2021**, *11*, 2224. [CrossRef]
3. Negara, B.; Kim, S.; Sohn, J.; Kim, J.-S.; Choi, J.-S. Application of High-Frequency Defrosting, Superheated Steam, and Quick-Freezing Treatments to Improve the Quality of Seafood Home Meal Replacement Products Consisting of the Adductor Muscle of Pen Shells and Common Squid Meat. *Appl. Sci.* **2021**, *11*, 2926. [CrossRef]
4. Rana, M.; Mohibbullah, M.; Won, N.; Baten, A.; Sohn, J.; Kim, J.-S.; Choi, J.-S. Improved Hot Smoke Processing of Chub Mackerel (*Scomber japonicus*) Promotes Sensorial, Physicochemical and Microbiological Characteristics. *Appl. Sci.* **2021**, *11*, 2629. [CrossRef]
5. Taghinezhad, E.; Kaveh, M.; Szumny, A. Thermodynamic and Quality Performance Studies for Drying Kiwi in Hybrid Hot Air-Infrared Drying with Ultrasound Pretreatment. *Appl. Sci.* **2021**, *11*, 1297. [CrossRef]
6. Liu, Y.; Cheng, H.; Wu, D. Preparation of the Orange Flavoured "Boba" Ball in Milk Tea and Its Shelf-Life. *Appl. Sci.* **2020**, *11*, 200. [CrossRef]
7. Diaz-Bustamante, M.L.; Reyes, L.H.; Solano, O.A.A. Application of a Multiscale Approach in the Substitution and Reduction of NaCl in Costeño-Type Artisan Cheese. *Appl. Sci.* **2020**, *10*, 9008. [CrossRef]

8. Voidarou, C.; Alexopoulos, A.; Tsinas, A.; Rozos, G.; Tzora, A.; Skoufos, I.; Varzakas, T.; Bezirtzoglou, E. Effectiveness of Bacteriocin-Producing Lactic Acid Bacteria and *Bifidobacterium* Isolated from Honeycombs against Spoilage Microorganisms and Pathogens Isolated from Fruits and Vegetables. *Appl. Sci.* **2020**, *10*, 7309. [CrossRef]
9. Skiada, V.; Agriopoulou, S.; Tsarouhas, P.; Katsaris, P.; Stamatelopoulou, E.; Varzakas, T. Evaluation and Origin Discrimination of Two Monocultivar Extra Virgin Olive Oils, Cultivated in the Coastline Part of North-Western Greece. *Appl. Sci.* **2020**, *10*, 6733. [CrossRef]
10. Agustinisari, I.; Mulia, K.; Nasikin, M. The Effect of Eugenol and Chitosan Concentration on the Encapsulation of Eugenol Using Whey Protein–Maltodextrin Conjugates. *Appl. Sci.* **2020**, *10*, 3205. [CrossRef]
11. Cucci, P.; N'Gatta, A.C.K.; Sanguansuk, S.; Lebert, A.; Audonnet, F. Relationship between Color and Redox Potential (Eh) in Beef Meat Juice. Validation on Beef Meat. *Appl. Sci.* **2020**, *10*, 3164. [CrossRef]
12. Zambonin, C. MALDI-TOF Mass Spectrometry Applications for Food Fraud Detection. *Appl. Sci.* **2021**, *11*, 3374. [CrossRef]
13. Chen, T.-S.; Huang, Y.-F.; Weng, M.-W.; Do, M.-H. Two-Stage Production System Pondering upon Corporate Social Responsibility in Food Supply Chain: A Case Study. *Appl. Sci.* **2021**, *11*, 1088. [CrossRef]
14. Xu, N.; Huang, Y.-F.; Weng, M.-W.; Do, M.-H. New Retailing Problem for an Integrated Food Supply Chain in the Baking Industry. *Appl. Sci.* **2021**, *11*, 946. [CrossRef]

 *applied sciences*

Review

MALDI-TOF Mass Spectrometry Applications for Food Fraud Detection

Carlo Zambonin 

Dipartimento di Chimica, Università degli Studi di Bari "Aldo Moro", Via E. Orabona 4, 70126 Bari, Italy; carlo.zambonin@uniba.it

Abstract: Chemical analysis of food products relating to the detection of the most common frauds is a complex task due to the complexity of the matrices and the unknown nature of most processes. Moreover, frauds are becoming more and more sophisticated, making the development of reliable, rapid, cost-effective new analytical methods for food control even more pressing. Over the years, MALDI-TOF MS has demonstrated the potential to meet this need, also due to a series of undeniable intrinsic advantages including ease of use, fast data collection, and capability to obtain valuable information even from complex samples subjected to simple pre-treatment procedures. These features have been conveniently exploited in the field of food frauds in several matrices, including milk and dairy products, oils, fish and seafood, meat, fruit, vegetables, and a few other categories. The present review provides a comprehensive overview of the existing MALDI-based applications for food quality assessment and detection of adulterations.

Keywords: MALDI-TOF; applications; food; fraud; adulteration; quality

 check for **updates**

Citation: Zambonin, C. MALDI-TOF Mass Spectrometry Applications for Food Fraud Detection. *Appl. Sci.* **2021**, *11*, 3374. https://doi.org/10.3390/app11083374

Academic Editors: Theodoros Varzakas and Tsarouhas Panagiotis

Received: 16 March 2021
Accepted: 7 April 2021
Published: 9 April 2021

Publisher's Note: MDPI stays neutral with regard to jurisdictional claims in published maps and institutional affiliations.

1. Introduction

Defining and identifying the types of food fraud is still an open issue since the relevant literature is characterized by a considerable variability of the related terms [1]. The European Commission defines food frauds as "intentional actions by businesses or individuals for the purpose of deceiving purchasers and gaining undue advantage therefrom, in violation of the rules referred to in Article 1(2) of Regulation (EU) 2017/625 (the agri-food chain legislation)" [2]. The most common foods subjected to illegal manipulations [3] include oil, fish, honey, milk and dairy products, meat, grain-based foods, fruit juices, wine and alcoholic beverages, organic foods, spices, coffee, tea, and many others.

According to the European Commission [2,4], the most common fraud is adulteration, which can occur in food products in different ways: "replacing a nutrient, an ingredient, a food or part of a food with another one with lower value" (substitution), "mixing an ingredient with high value with an ingredient with a lower value" (dilution), "adding unknown and undeclared compounds to food products in order to enhance their quality attributes" (unapproved enhancement), and "hiding the low quality of food ingredients or products" (concealment). Food frauds can also take others forms, including mislabeling, the process of putting false claims on packaging for economic gain, counterfeiting, when intellectual property rights are infringed, and grey market, referring to unauthorized sales channels for products.

Based on the above consideration, a chemical analysis on a given food can be carried out for many reasons, such as individuation of adulteration, authentication and traceability, confirmation of geographical origin assessment of toxicity, and many others. Food sample analysis, then, constitutes a crucial challenge for analytical chemistry and numerous scientists are focused on the development of reliable, fast, cost-effective analytical processes to solve analytical problems related to food frauds. Food-based matrices are complex and characterized by a wide range of chemical composition that influence the performance of chemical analytical measurements, while the nature of the manipulation is often unknown, making the analysis even more complicated. Moreover, as technologies develop to detect deceptions, they become more sophisticated since fraudulent suppliers also adapt to finding new ways to circumvent the controls.

Targeted and non-targeted analytical approaches are mainly used for food controls [5–7]. A targeted analysis, generally laborious and time-consuming, is based on the knowledge of the contaminants and focused on the detection of one or a few classes of compounds that represent the markers of the fraud. Since in most cases the type of fraud is unknown, a non-targeted analysis is often necessary. This approach relies on the fast instrumental acquisition of the chemical profile of the whole foodstuff sample, usually represented by spectra, which should provide a unique fingerprint as a reference for suspect samples, often with the help of chemometric data handling [8].

Spectroscopic techniques [5,9–16], namely nuclear magnetic resonance (NMR), near- and mid-infrared (NIR, MIR), Raman, ultraviolet–visible (UV-VIS), and X-ray fluorescence spectroscopy (XRF), have been largely used to successfully carry out non-targeted analysis. Different mass spectrometry (MS) techniques and ionization approaches, including matrix-assisted laser desorption ionization time-of-flight (MALDI-TOF), electrospray (ESI), ambient mass spectrometry (AMS), and high-resolution mass spectrometry (HRMS), have been also widely employed for the same purpose [6,17–23], and their constant advances have enabled the development of numerous new methodologies for food quality and safety controls.

Among them, MALDI-TOF represents an ideal option for fast, reliable, and accurate detection of food frauds due to some intrinsic characteristics such as ease of use and speed with which data can be collected even from complex samples. MALDI MS is based on the use of a pulsed laser beam that hits the sample, represented by co-crystals of the so-called "matrix," present in a vast excess, and the analyte. The laser energy is absorbed by the matrix that vaporized carrying intact analyte molecules into the vapor phase. During this process, ions (mainly H^+ and Na^+) are released by the matrix leading to the formation of charged analyte molecules. Moreover, anions can also be generated by abstracting H^+ or Na^+ from the analyte. After being accelerated in an electric field and passing a charged grid, the ions are separated in a TOF mass analyzer based on their m/z ratios (low mass ions arrive at the detector in a shorter time than high mass ions). A "linear" geometry of TOF analyzers is used for the analysis of high molecular weight molecules. Smaller molecules can be analyzed using a "reflectron" configuration able to balance the different initial velocities of the ions during the vaporization process, leading to a consistent improvement of resolution. MALDI mass spectrometers are capable to rapidly generate spectra, profiles, and/or fingerprints from food matrices, often with simple preparation or no sample preparation at all. The potential of the MALDI approach applied to the fight against food fraud, sometimes coupled to multivariate statistical methods to extract information from complex analytical data such as mass spectra, has been clearly understood, as demonstrated by a considerable number of related publications. The purpose of the present review is to provide a comprehensive overview of the relevant literature. Based on the research carried out, most of the existing MALDI applications dedicated to the field of food fraud have concerned the analysis of milk and dairy products, oils, fish and seafood, vegetables, fruit, meat, and, to a minor extent, a few other categories. Most of the works have been focused on the evaluation of food quality and detection of adulterations, while only few applications have been devoted to the investigation of the geographical origin.

2. MALDI Applications

2.1. Milk and Dairy Products

Due to their high nutritional value, milk and derived products are largely consumed all over the world. According to the Food and Agriculture Organization of the United Nations (FAO), worldwide milk production has grown from 522 to 798 million tons in three decades (1986–2016) [24] and was expected to grow in 2020 to 859 million tons [25]. The increase in consumption also increases the number of dishonest producers willing to commit fraud to increase their earnings, making the safety and authenticity of milk and derived products an area of growing attention and concern, as demonstrated by several regulations and governing bodies [26,27]. Traditionally, milk dilution by water has been the most common illegal practice, together with the selling of skimmed milk and semi-skimmed milk instead of whole milk. As detection methods technologies has improved, fraudulent suppliers evolved [28] towards most sophisticated fraudulent approaches, such as the addition of milk anhydrous products (caseins and caseinate, milk protein concentrate, whey proteins) to liquid milk and the mixing of high-quality milk, such as buffalo, sheep, and goat, with less expensive products, such as cow milk. Many MALDI-based analytical methods have been developed and improved in the last years to assess the authenticity of milk and dairy products.

Buffalo milk has been extensively investigated by different MALDI proteomic approaches searching for possible adulteration. Cozzolino and co-workers have developed two different methods for the identification of adulteration of water buffalo and sheep milk and water buffalo mozzarella, respectively. In the first work [29], the investigation was performed on raw buffalo and sheep milk samples. The determination of the presence of bovine milk or the addition of powdered milk to fresh raw milk was accomplished by evaluating the protein profiles coming from the most abundant whey proteins, lactalbumin, and lactoglobulins. In the second work [30], patterns of the same whey proteins were obtained from the direct analysis of water buffalo mozzarella cheese, which was able to differentiate between mozzarella cheeses made from pure water buffalo milk or from mixtures of bovine-sheep-buffalo milks. Interesting quality information on the partial resistance of the detected proteins to the thermal and enzymatic processes involved in the cheese production was also inferred. In both methods, mixtures with different percentages of less expensive milk added to water buffalo milk were prepared, and detection limits below 5% were always obtained in adulterated milk or cheese, respectively. The strategies were also fast, accurate, and practically did not require sample pre-treatment.

Sassi at al. have successfully optimized an integrated MALDI-TOF platform [31] for obtaining milk peptidomic/proteomic profiles to detect the addition of either nondeclared bovine material to water buffalo, goat, and ovine milks, or of powdered bovine milk to the fresh bovine milk. Milk samples were directly analyzed permitting a fast individuation of illegal adulterations at protein and peptide level and allowing the identification of unique diagnostic ions of thermal treatment in different types of commercial milks.

A further MALDI proteomic study for the detection of ricotta buffalo cheese adulteration with bovine milk by obtaining peptide profiles for both matrices was developed [32], followed by the search for signals that could represent specific markers for each type of dairy product. A peptide marker corresponding to the region 149–162 of β-lactoglobulin was correlated to the univocal presence of bovine milk in ricotta buffalo cheese at a 5% level.

According to the European reference method, the fraudulent addition of bovine milk in water buffalo milk can be individuated by concomitant isoelectric focusing detection of bovine γ_2- and γ_3-casein fragments after plasminolysis. However, this approach can produce false positive results due to a water buffalo β-casein peptide, which is also formed after plasminolysis of water buffalo milk and comigrates in isoelectric focusing with bovine γ_2-casein. Thus, Caira at al. have developed a proteomic approach [33] to obtain the unambiguous detection of bovine milk in water buffalo milk and derived products (LOD of 0.8% *v/v*) based on the MALDI determination of specific bovine and water buffalo β-casein phosphopeptide markers. The procedure was proposed as an integrative/alternative to the reference method.

Other studies were focused on the development of MALDI methods for the detection of the addition of sheep and goat milk with the less expensive bovine milk. Contrary to some of the above-mentioned works, which were focused on the identification of specific signals that can work as adulteration markers, one study, divided into two parts [34,35], proposed a different approach. The authors suggested that, in the analysis of a complex matrix such as milk, the simultaneous use of many variables reveals more information compared to the use of single or few variables and proposed a method for quantitative and multivariate use of the whole MALDI mass spectra of milk samples to determine adulteration. The first part of the work [34] was focused on the optimization of sample preparation and instrument setup, the second on the quantitative determination of cow, goat, and sheep milk in mixed milk samples [35]. The results obtained seemed to confirm the original hypothesis since, with optimal parameter settings, multivariate regression on whole spectra allowed to determine the concentrations of milk in mixtures with good accuracy. A further study to detect the presence of cow milk in goat and sheep milk using whole MALDI mass spectra in combination with multivariate techniques was reported by Nicolau at al. [36]. Binary and tertiary mixtures of cow, sheep, and goat milk were easily and rapidly analyzed and accurate information on the amount and type of adulteration were obtained, with typical errors in the range 2–10% for cow milk.

The potential of proteomic approaches to fight illegal manipulations of milk was again demonstrated [37] by the detection of cow milk-specific peptide markers in sheep and goat milk and goat milk-specific peptide markers in sheep milk, by analyzing whole milk samples mixtures subjected to in-solution tryptic digestion. Seven peptide markers of cow milk and two peptide markers of goat milk were identified by MALDI-TOF analysis and the approach was able to detect adulteration up to a 5% level. Moreover, the same markers were found in cheese samples, demonstrating the applicability of the procedure even to milk-derived products. The authors stated that the use of the α-cyano-4-chlorocinnamic acid matrix was essential to reach the high sensitivity observed for the marker peptides. Moreover, the results were found in good agreement with those obtained with a traditional approach based on SDS-PAGE/in-gel digestion. The same research group reported another application [38] for the extraction and MALDI determination of phospholipids in milk, using the same matrix. The method was able to provide peculiar milk phospholipid profiles that were used for the detection of cow milk in sheep and goat milk. The abundance ratio of specific ions (m/z 703 and 706) was found to be species-specific and was used for the identification of the adulteration.

Very recently, liquid atmospheric pressure—MALDI mass spectrometry—was used [39] to optimize a very accurate approach to classify goat and sheep milk and sheep milk containing 10% of goat milk, to evaluate colostrum quality and postnatal stages, through the recording of the milk lipid/protein profiles or the detected orthologs of single proteins.

Donkey milk is a safe alternative for individuals which are allergic to cow milk and is often the object of adulteration, the detection of which is of critical importance for the safety of the consumers. The MALDI mass spectra profiles of α-lactalbumin and β-lactoglobulin were used to detect the fraudulent addition of goat or cow milk in donkey milk [40] in a protocol developed for routine analysis and potentially extendable to other milk species. Detection limits for the analysis of defatted milk samples were in the range 0.5–2.0%, comparable to those obtainable with traditional complex approaches. Despite the typical MALDI quantitative issues, the measured values were in good agreement with the actual composition of the analyzed mixtures. Another study [41] investigated donkey and goat milk adulteration by cow and sheep milk using MALDI in combination with unsupervised hierarchical clustering, principal component (PCA), and Pearson's correlation, focusing on the mass spectra profiles in the m/z mass range 2000–25,000 Da. The approach was shown to be rapid, robust, and sensitive, with detection limits of 0.5%.

The adulteration of fresh milk by reconstituted milk and the selling of reconstituted milk as fresh product is economically advantageous when either a surplus of milk powder exists, or when the importation of dried milk powder is subsidized. Moreover, the addition of powdered derivatives is difficult to detect because the adulterant materials have almost the same chemical composition of liquid milk. Specific peptide markers, attributable to modified whey proteins and/or caseins, formed by thermal degradation during milk powder production, were identified and used to detect the presence of powdered milk in liquid milk samples [42]. Whey and casein fractions of milk samples were directly digested in solution and analyzed by MALDI using α-cyano-4-chlorocinnamic acid as a matrix, which enhanced the detection of more acid peptides. The approach was able to detect the peptides diagnostic for the presence of powder in liquid milk even at a 1% level. The results were in good agreement with those obtained with a reliable but time-consuming 2D gel approach.

On the contrary, milk powder itself could be the object of adulteration when non-milk fat such as vegetable oils and fats are added. Garcia and co-workers [43] successfully developed a MALDI-qTOF method for the fast individuation of non-milk fats and oils in milk powder. Samples were subjected to a simple n-hexane extraction prior to MS analysis, which provided rapid and unambiguous profiles of the triacylglycerols (TAGs) composition capable to characterize the adulterant and estimate the adulteration level.

Other MALDI applications on the detection of frauds were focused on different kind of milks and adulterations, as well as to the analysis of cheese samples. Hinz at al. [44] compared the principal proteins in bovine, caprine, buffalo, equine, and camel milk, finding interesting differences between the species that could be used to identify sources of hypoallergenic alternatives to bovine milk and detect adulteration of milk samples and derived products. A proteomic approach for the MALDI-TOF analysis of commercial bovine milk, based on in-solution digestion of the whole samples, was also suggested [45] for the routine analysis of raw and processed foods and to detect adulterations.

Due to similar properties to bovine milk, soya milk is added to bovine milk for revenue maximization. In a recent work, England and colleagues [46] used MALDI to develop a solventless, sensitive, and cost-effective method for the discrimination of bovine milk from soya and coconut milk as well as for the individuation of milk adulteration. Samples were simply diluted in water, combined with the matrix, and subjected to instrumental analysis, which allowed to obtain unique lipid profiles (mainly phosphatidylcholine and TAGs).

Magenis at al. reported a qualitative method to check the authenticity of a typical Brazilian cheese using β-lactoglobulin as an adulteration marker [47], showing good precision and sensitivity (7 mg/g). The procedure involved SDS–PAGE and in-gel trypsin digestion followed by MALDI analysis. The applicability to real samples was demonstrated by the analysis of 42 commercial samples, 18 of which were found to be adulterated.

Direct Imprinting in Glass Surface Mass Spectrometry (DIGS-MS) for qualitative cheese analysis in a MALDI instrument was also successfully used [48] to identify complex lipids to be used as quality and/or adulteration marker in different cheese samples. The integration of analytical and statistical data could also be employed for the control of productive stages.

In another recent work [49], Rau and co-workers developed an original method to identify the dairy animal species of mozzarella and white brined cheese by MALDI-TOF in combination with direct protein extraction without tryptic digestion and a small in-house reference spectra database.

Some further MALDI works were oriented towards the quality evaluation of milk and dairy products. The selective extraction of phospholipids from dairy products such as milk, chocolate milk, and butter by micro-solid phase extraction (μ-SPE), based on homemade titanium dioxide (TiO_2) microcolumns, followed by MALDI-TOF MS, was developed by Calvano at al. [50]. Since α-lactalbumin and β-lactoglobulin are among the main cow milk allergens, Gasilova at al. [51] developed a sensitive quantitative method for their determination by immunoaffinity capillary electrophoresis, using magnetic beads functionalized with opportune antibodies, coupled to MALDI. A transient isotachophoresis preconcentration step allowed to obtain LOD values of 0.02 and 0.03 μg/mL for β-lactoglobulin and α-lactalbumin, respectively, suitable for allergen detection. The method was then successfully tested on cow milk and fortified soy milk, proving the capability to determine the analytes at both high and low concentration levels.

To identify and characterize oxidized and glycated phospholipids in heat-treated food, namely milk powders, pasteurized milk, ultra-high-temperature milk, and soy flour, an extraction protocol [52] was developed by means of a methanolic solution of 1,8-bis(dimethylamino) naphthalene (DMAN), used as both extraction medium and matrix for the successive MALDI detection, performed in negative ion mode. Thermally modified lipid products were first characterized by heating representative standards and eventually determined in real samples. MALDI-TOF, combined with C18-stage tip extraction, was also used [53] to rapidly obtain peptide profiles from different commercial milk products in view of the identification of possible changes induced by different heating regimens and storage conditions. Several peptide ions showed relative abundance variations following specific treatments, such as heating or proteolytic activity of enzymes during storage, suggesting their potential use as markers to draw information on milk types and freshness. A similar approach permitted [54] the generation of polypeptide profiles from buffalo milk samples subjected to different freezer storage times, to assess their freshness through the identification of specific markers. The statistical evaluation of data relevant to the analysis of several fresh and frozen samples allowed to identify 28 polypeptide markers of freezing storage originated from the breakdown of buffalo proteins, being α-lactalbumin, β-lactoglobulin, γ2-, γ3-, and γ4-caseins, β-casein-derived phosphopeptides, and GLY-CAM1 phosphorylated peptides as the most significant components. Their progressive formation even in freezing conditions was attributed to an unknown protease stable at low temperatures. MALDI profiling of milk proteins, in combination with multivariate data analyses, also proved to be an optimal mean to discriminate between different milk types [55]. Protein fingerprints of mammalian (various species), human (at different lactation stages), and formula (different brands) milks were rapidly obtained, permitting to obtain key information on the matrices under study.

2.2. Oils

Olive oil is one of the main ingredients in the Mediterranean diet [56]. The International Olive Council defines two categories of olive oil [57], the first comprising oils suitable for consumption, such as extra-virgin, virgin, and ordinary olive oil, the second comprising oils that must be further processed prior to ingestion. Extra-virgin olive oil (EVOO) is the most valuable, since it is obtained from olive fruits using mechanical processes or other physical means [58] that lead to a product of unmatched value, characterized by unique features, such as nutritional quality, health benefits, and pleasant flavor. These features make EVOOs expensive products that are continuously counterfeited in many countries [59] with oils produced from cheaper fruits/seeds or with other lower quality olive oils or even mislabeling virgin/refined olive oils. Of course, analytical methods are crucial for detecting EVOO frauds [60] and many scientists are committed to the development of MALDI-TOF MS applications.

A matrix-less laser desorption/ionization (LDI) approach using a stainless-steel target plate was proposed [61] for the fast analysis of diluted soy, sunflower, and extra-virgin olive oil samples. MS spectra characterized by oil-specific profiles and free of MALDI-matrix peaks were obtained, allowing the easy discrimination of the oil under study. Most of the m/z ions present in the spectrum were easily attributed to tri/diacylglycerols compounds, some of which were found to be diagnostic ions for olive (m/z 907.77) and sunflower oil (903.79 and 901.78), thus permitting the detection of adulteration of olive with sunflower oil. All the diagnostic ions were potentially attributable to different compounds, namely OOO, SOL, SLO, LSO (907.77), OLL, LOL, SLLn, LSLn, SLnL, OOLn, OLnO (903.79), LLL, OLLn, OLnL, and LOLn (901.78), with O being oleic acid, S, stearic acid, L, linoleic acid, and Ln, linolenic acid. Over the following years, the same research group published three more papers focused on the development of MALDI methods for the challenging detection of hidden hazelnut oil in extra virgin olive oil, each time using different classes of compounds as markers for the adulteration. In the first work [62], the polar fraction of oils was enriched and characterized using hydrophilic liquid chromatography micro-columns coupled with MALDI. Lysophosphatidylcholine (LPC) (16:0/0:0), LPC (18:1/0:0), and LPC (18:2/0:0) were identified and used as diagnostic compounds for the presence of hazelnut oil in EVOO to a level of 5%. The second application [63] exploited a modified Bligh–Dyer method for the selective extraction of phospholipids, present in seed oils at much higher concentration levels and then used as markers for hazelnut in olive oil. The solution resulting from the combination of α-Cyano-4-hydroxycinnamic acid (176 mM) and tributylamine (equimolar) was used as both extraction solvent and MALDI matrix. The method was capable to detect adulterations at a 1% contamination level. The last work [64] used cold acetone precipitation followed by in-solution tryptic digestion and final MALDI analysis for the revealing of hazelnut peptide markers arising from the main hazelnut proteins Cor a 9, Cor a 11, and Cor a 1. These markers can potentially be used for the determination of hazelnut traces in oils and processed foods. SDS–PAGE analysis confirmed the presence of hazelnut proteins in hazelnut extracts with molecular masses in the range of 10–60 kDa.

Arlorio and colleagues [65] reported that the adulteration of extra virgin olive oil with solvent-extracted hazelnut oil can be directly traced down to a 1% level by SDS–PAGE analysis of hazelnut proteins, even if a MALDI analysis was indeed necessary to confirm the identity of the alleged allergens, i.e., two oleosin isoforms and Cor a 9. MALDI-TOF, combined with unsupervised hierarchical clustering, principal component analysis, and Pearson's correlation analysis, was also successfully adopted [66] to detect very low amounts (0.5%) of corn oil in EVOO.

As described in the work of Calvano at al. [61], when the fraud to be revealed concerns the presence of seed oils in EVOO, it may be sufficient to target their TAGs profile to characterize the samples and eventually determine the presence of the adulterant product. In fact, two more applications based on MALDI in conjunction with statistical approaches were successfully developed for the detection of sunflower (or refined) [67] and canola [68] oils in EVOO, respectively, using the TAGs-profiles present in the relevant mass spectra.

A hardly detectable adulteration is represented by the addition of sunflower oil to poppy seed oil because of the similar fatty acid ratios. In fact, a further application [69] reported the successful MALDI-TOF MS detection of mixtures of sunflower oil with high levels of triolein (high-oleic acid type) down to the 5–10% level, but the same approach failed to discover adulteration of pure poppy seed oil by sunflower oil.

Several works were focused on the quality evaluation of edible oils (mainly EVOO). Taking into account that squalene concentrations in olive oil vary considerably depending on the cultivar, its determination in this matrix should be able to discriminate between varieties. Thus, Zambonin at al. [70] reported on the LDI-TOF MS determination of squalene and derived oxides involved in the biosynthetic pathway of cholesterol in extra virgin olive oil. The same research group used the same matrix-less approach to characterize olive and sunflower oils before and after thermally assisted oxidation [71]. The obtained MS profiles provided information about the identity of thermally induced oxidation products, such as TAGs epoxy/hydroxy, hydroperoxy derivatives, and β-scission products, making the method a tool to rapidly check the quality of cooking oil. TAGs thermal oxidation of sunflower and olive oils was also studied by Picariello at al. two years later [72]. To increase the detection of oxidized components, a chromatographic separation of polar and non-polar compounds was performed on silica gel before MALDI analysis. Tri- and diacylglycerols, TAGs oxidative dimers, oxidized TAGs, and TAGs fragments arising from the β-scission of linoleyl, peroxy, and alkoxy radicals were observed in the relevant spectra.

For authentication and characterization purposes, detailed MALDI TAG profiles of olive oils were also successfully obtained from samples coming from six different cultivars [73] and of two different varieties grown in the same area at different olive ripening stages [74].

Shen and co-workers [75] optimized a matrix solid-phase dispersion (MSPD) procedure to extract several phospholipids from olive fruit and oil samples, exploiting the ability of the sorbent (TiO$_2$ nanoparticles) to selectively interact with the phosphate group of the analyte by a chelating bidentate bond. After elution, MALDI-TOF analysis in both positive and negative ion modes was performed. The method proved to possess a great potential in lipidomic fingerprinting of olive samples for quality control.

Different analytical instrument, namely MALDI, GC, and LC-MS, were synergically used to perform a study of the lipid composition of commercial extra virgin and virgin olive oils [76], in an attempt to define specific lipid patterns and/or establish a set of lipid markers representing the identity of a specific oil. Significant differences between EVOOs and VOOs were found, and five classes of phospholipids were identified in the polar lipid fraction, with remarkable variations in phosphatidylcholines.

MALDI-SpiralTOF was employed [77] to obtain TAG fingerprints relevant to sesame, sunflower, and olive oils to draw traceability and authenticity information. The application of PCA allowed the discrimination of Istrian olive oils from those coming from other Croatian coastal regions. Furthermore, high energy collision-induced dissociation (CID) MALDI-TOF/TOF analysis of TAGs was suggested as a mean that could correlate oil TAGs to the geographical origin, analyzing a high number of samples that can provide statistically significant data. A multi-instrumental analysis (MALDI for TAGs, GC-MS for fatty acids, and NIRS for non-selective analysis) coupled to PCA and partial least square-discriminant analysis (PLS-DA) was also used [78] for geographical origin sample grouping evaluation and for testing of predictive capabilities of measured variables on accurate classification of EVOO regional category.

A synergic MALDI-TOF/GC-MS approach was proposed to study the different components of pomegranate oil [79] and of several nut oil varieties [80], respectively. In particular, MALDI confirmed its ability and reliability to profile the TAG component, showing unique fingerprints that allows to differentiate pomegranate from most edible oils and nut oils exhibiting quite similar fatty acid composition (hazelnut, pistachio, and beech oil) between each other.

Unique TAGs profiles were also displayed [81] by different Amazonian oils and fats, characterized by MALDI-TOF without prior separation. The variable combinations of fatty acids provided detailed information that could permit the development of a database from spectra to be used for fast and reliable typification, screening, and quality control. The triacylglycerol content of shea butter fat, palm kernel oil, and peanut oil was also rapidly characterized [82]. The potential of the method as a tool for quality control of the matrices under study was demonstrated by the detection, in addition to intact and specific triacylglycerols, of oxygenated and fragmented TAGs, likely arising from poor handling and production processes.

2.3. Fish and Meat

Fish represents and is perceived by consumers as a healthy and nutritious food resource. However, a common problem in the fish processing industry is adding or replacing cheap fish instead of expensive ones, as an accidental event due to the lack of experience or as a deliberate fraud. An adequate level of protection is then required to ensure seafood quality and safety, making the development of rapid, effective, accurate, and reliable analytical methods the key to effectively supervising the market. This need has prompted the issue of directives and regulations for quality control and encouraged the development of innovative MALDI analytical methodologies, almost always based on proteomic techniques.

Highly specific mass spectrometric profiles from 25 different fish species were obtained using a new MALDI method [83] meant for the fast assessment of authenticity and fraudulent substitutions. Specific protein signals at m/z values around 11 kDa were selected as markers to discriminate the various species, while a structural characterization permitted to identify some major fish allergens arising from parvalbumin.

In a pilot study [84], three freshwater fish species were discriminated based on the cluster analyses of the overall profiles generated by the MALDI analyses of muscle and liver tissues, after a simple single-step extraction procedure. Each tissue provided species-specific profiles, even if muscle is more suitable for routine controls since fillets are more commonly commercialized. The authors pointed out that the clear discrimination of the species was possible even though the settings and statistical algorithms used were originally designed for the analysis and identification of bacteria and fungi. Then, they suggested that performing an optimization of mass spectral analysis and data processing targeted on fish species could lead to exhaustive evaluation of the potential of MALDI analysis for studies on freshwater fishes.

The analysis of muscles taken from two closely related fish species was undertaken by two-dimensional gel electrophoresis (2-DE) [85] and the interspecies differences between the protein spots, which could be useful for their differentiation, were visually identified. MALDI-TOF MS and/or LC-MS were then necessary to identify 19 proteins, some of which resulted to be specific for each species. Another MALDI method [86] for the profiling of protein extracted from fish muscle was reported in a book chapter by Siciliano at al. for fish authentication purposes and to detect fraudulent substitutions.

Since universal sample preparation protocols prior to MALDI analysis were available to other species but not to fish, Spielmann at al. [87] tested the performances of five preparation protocols to verify the capability to produce reproducible and high-quality spectra dependent on storage conditions and food processing, and eventually differentiate between different fish species. After the optimization of the best protocol, it was concluded that MALDI could be used for the purpose as soon as a valid species database is available. This need was evidently shared by Stahl and co-workers, who almost simultaneously used MALDI to establish a database [88] of protein patterns from 54 fish species susceptible to fraud, in order to detect and prevent substitution, characterized by low intraspecies but high interspecies variability.

MALDI-TOF combined with multivariate analysis was exploited [89] for the development of a method to directly analyze fish skin or muscle tissues surface, for authentication purposes. Samples were quickly discriminated, hinting at potential future developments for the authentication of seafood in general and/or other protein food products.

Another study centered on the comparison of several one-step sample pre-treatment protocols of fish muscle tissues, prior to MALDI determination of the relevant extracts, was recently carried out [90] by Wang and colleagues. The best results in terms of repeatability and spectral resolution were obtained by boiling samples in 0.1 M trifluoroacetic acid (TFA) for 5 min, permitting to discriminate between different fish samples, when combined with similarity coefficient-based analysis for their mass spectra.

MALDI also turned out to be a good system for assessing fish freshness. For instance, the effect of storage time on fish muscle proteins was investigated using 2-DE coupled to MALDI [91], permitting the identification of three altered proteins. More recently [92], the same approach was involved in the analysis of fish samples, both fresh and stored in a refrigerator for different days, looking for protein markers for its freshness. It was found that the comparison of three specific proteins, l-lactate dehydrogenase, adenylate kinase isoenzyme 1, and myosin heavy chain, permitted to trace the freshness of the products. In a different study, the vitreous fluid of the eyes was taken from fishes subjected to different days of post-mortem storage, which was immediately analyzed by MALDI-TOF to evaluate fish spoilage [93]. Software spectra processing allowed the identification of four m/z ions able to differentiate between the tested days of storage, although a limited applicability was shown within the end of the tested period.

A further study [94] of a comparison of different protein extraction protocols from seafood (marine mussel) prior to 2-DE was conducted by Campos and co-workers. Then, MALDI-TOF/TOF analysis of the gels was performed and proteins with several functions, such as energy metabolism, cell signaling and regulation, and stress response, were identified. The method was a precious tool to investigate the protein expression in the species under investigation and potentially in other affiliated species.

Several analytical methods are used in the meat industry for detecting contamination, adulteration, and authenticity, to safeguard safety, and to reassure consumers, who are increasingly aware and attentive to the quality of the food they consume. Two MALDI methods to determine the origin of raw and processed meat (pork, beef, horse, veal, and chicken) and of gelatin (pork or beef), respectively, were developed by Flaudrops and colleagues [95]. In the first method, intact proteins were determined in linear mode prior to cluster analysis, allowing the separation of meat into distinct mass spectra clusters depending on the origin. In the second method, gelatin was digested with trypsin and analyzed in reflectron mode. The relevant spectra showed specific profiles that permitted to distinguish pork from bovine gelatin (1% of gelatin in spiked candies and 20% of pork gelatin in beef gelatin). Although less sensitive compared to other approaches, these methods are fast and easy to perform and can be conveniently used for fast screening controls.

An approach to reveal the differences in protein expression between young and old buffalo meat looking for biological markers of tenderness was performed by 2-DE coupled to MALDI-TOF/TOF [96]. Structural proteins with expression levels associated with meat tenderness were successfully identified through digestion and MS/MS analysis of selected gel spots. Moreover, the study demonstrated that through the ageing process, it was possible to reduce the variation in tenderness between young and old buffalo meat that, consequently, requires different processing strategies for their effective utilization. The 2-DE-MALDI combination was also exploited [97] to compare meat quality traits and to identify different protein expression between low and high pH pig muscles. Fourteen proteins that differed in spot density between the two groups were identified and nine of them involved in meat quality attributes significantly increased in high-pH muscles. The whole results obtained in the work provided useful information to understand the molecular mechanism at the base of meat quality.

Furthermore, the 2-DE-MALDI-TOF/TOF approach was again used [98] to assess the protein modifications occurring in the duck breast muscle collected form three breeds during the early post-mortem storage period. Evidence of changes in the protein expressions for each breed were found in several spots, some of which (10 for each breed) were subjected to MS/MS analysis, which allowed the identification of at total of 22 proteins.

2.4. Fruits and Vegetables

Intake of fruits and vegetables is highly recommended to promote good health due to their concentrations of bioactive compounds, including vitamins, minerals, antioxidants, and fiber. Indeed, the quality of these precious foods must be safeguarded and monitored using suitable analytical methods. MALDI-TOF was adopted for the development of many related applications. For instance, fruit skins were analyzed in two different works. It was demonstrated [99] that anthocyanins profiles of the berry skins of 23 red grape varieties can be easily and rapidly obtained by MALDI-TOF, enabling the identification of several varietal traits useful for the differentiation of authentic cultivars from hybrid ones on a molecular basis. Very recently, proanthocyanidin profiles of different fruit skins were obtained [100] by MALDI-TOF coupled to multivariate analysis, which proved to be a useful tool to evaluate authenticity in mixtures of different proanthocyanidin.

Three different MALDI studies were focused on the assessment of the authenticity of saffron, an expensive spice made from *Crocus sativus* L. dried stigmas, which is often the object of frauds. MALDI MS imaging (MSI) was successfully applied [101] to the determination of different crocins species in saffron. This original approach was also compared to a traditional one, i.e., MALDI analysis after solvent extraction, and the relevant results were found in good agreement. The other two studies were both performed by Aiello at al., using MS or MS/MS as mass analyzer and curcumin as the non-isotopic isobaric internal standard. In the first one [102], powdered saffron was subjected to a one-step sample pre-treatment and subjected to MALDI analysis. Crocins C-1–C-6, flavonols, unknown highly glycosylated crocins C-7, C-8, and C-9, and carotenoid-derived metabolites were detected. The method was shown to be fast, sensitive, reproducible, and suitable for routine quality controls, including the assessment of adulterations, as ascertained through the analysis of commercial samples. In the second one [103], picrocrocin was used as a saffron authenticity marker. Again, a rapid and simple method was developed, and very good quantitation parameters were obtained, with LOD (47.63 ppm) comparable or even lower than those obtained with other approaches.

The high demand for cereals, a major part of the diet in most countries providing nutrients such as vitamins, minerals, carbohydrates, fats, and proteins, makes relevant frauds a severe global problem. Sequenom® MassARRAY® (Agena Bioscience, San Diego, CA, USA) MALDI-TOF MS, a reliable platform for detection and validation of single nucleotide polymorphisms (SNPs) for varietal analysis, was used to target specific genes in order to genotype and differentiate 35 barley varieties [104]. Information was drawn from the majority (33) of the tested loci (45), permitting the generation of a unique barcode of SNPs for each barley cultivar. This reliable and fast approach can potentially identify more than 150,000 SNPs per day, making the platform a suitable alternative for cereal varieta identification.

Jin at al. [105] used 2-DE followed by MALDI-TOF to compare the proteome extracted from barley malts of two selected cultivars, to find correlations between the protein content and malt qualities. Almost 700 total spots were detected; most were shared by the two cultivars, 377 of which were attributed to 192 proteins, mainly enzymes and enzyme inhibitors. The same research group performed another comparative study on green malts from two cultivars using two-dimensional fluorescent difference gel electrophoresis (2D-DIGE) [106]. Several metabolic proteins were identified by MS/MS and significant differences between cultivars were detected, suggesting potential applications for malt quality assessment. MALDI-TOF and two-dimensional difference gel electrophoresis (2D-DIGE) were also used by Fernando at al. [107] in a differential proteomic study on wheat grain, grown under normal CO_2 concentration and Free Air CO_2 Enrichment, respectively. It was found that a decrease of 9% in the protein content, in particular glutenin high molecular weight subunits associated with lower flour rheological properties and bread quality, occurred at higher CO_2 concentrations.

Since soybean is a known source of allergens that can also be hidden in other food commodities, a MALDI-TOF/TOF study was developed [108] to identify potential markers of its presence. Soybean protein extracts were digested with trypsin, separated using an LC system and eventually subjected to MS/MS analysis. Peptides arising from G1 glycinin and β-conglycinin were identified and proposed as potential markers for the detection of soybean protein traces in processed foods. Furthermore, these peptides were found to be stable when subjected to simulated food processing reactions, such as denaturation, Maillard reaction, and oxidation.

Rapid, simple, and reliable MALDI-TOF methods has been also proposed for cultivar characterization of hazelnut kernels through proteomic analysis [109], for geographical origin discrimination (coupled to multivariate analysis) of fermented salted vegetables comparing the relevant mass fingerprints [110] and to understand the effect of organic and inorganic crop nourishments on the nutritional quality of yardlong bean through two-dimensional proteome profiling [111]. MALDI-TOF/TOF was used by Pinto at al. [112] for the untargeted analysis of dark chocolate from cocoa beans of different geographical areas to characterize the relevant metabolites (including sucrose, choline, sphingolipids, phospholipids, peptide material, and polyphenols) based on molecular weights and fragmentation patterns.

2.5. Other Matrices

Product quality, authentication, and identification of adulterants are ongoing challenges facing several other food commodities. Hence, other MALDI-based methods have been developed for the detection of different frauds in various matrices. MALDI coupled to capillary zone electrophoresis was successfully exploited [113] to rapidly detect cheap sweeteners (sucrose from cane or beet, starch hydrolysates) in orange juices by monitoring the variation of concentration ratio of low molecular mass saccharides, such as glucose, fructose, and sucrose, and/or the presence of compounds that are absent in the sugar profiles of citrus fruits.

Two other proteomic studies were devoted to the assessment of authenticity and traceability of wine. A fast approach proposed [114] the direct tryptic digestion and subsequent MALDI analysis of extracts of white wine produced from different *V. vinifera* cultivars. The detailed and peculiar peptide profiles present in the relevant mass spectra were converted into simulated images that represented unique "mass codes" able to display differences between samples, suggesting their potential use for food quality assessment. The comparison of anthocyanin concentration, main organic acids and sugars, and proteomic profiles of berry with different geographical origin, grape variety, and ripening stage was performed by Fraige at al. [115]. In particular, 2-DE combined with MALDI was employed to obtain information about the proteome changes subjecting 128 gel spots to MS analysis, 80 of which were identified as proteins mainly expressed as a response to defense/stress or related to carbohydrate metabolism. A multivariate analysis of protein abundance was able to divide the analyzed samples in different groups according to the considered variables, clearly showing the potential of the approach to provide useful information for wine characterization.

Two strategies for the analysis of honey targeting different compounds were successfully developed to differentiate various types of honey and to detect adulterated samples, respectively. Won and colleagues [116] exploited the molecular weight differences (56 and 59 kDa, respectively) of major proteins of honey produced from different bee species to discriminate between two honey types by 2-DE coupled to MALDI analysis, while Qu at al. [117] targeted oligosaccharide and polysaccharide profiles to reveal adulterations by MALDI MS or MS/MS.

Edible insects are available on the European market in ground form, in which any visual identification is impossible, thus causing concrete risks of adulteration that imposes quality and safety controls of such products. A reliable MALDI method [118] was then developed for the detection of proteins in whole insect powder extracts. Species-specific mass spectra were obtained permitting the easy differentiation of the different species under investigation (buffalo worms, mealworms, crickets, and grasshoppers).

A further common fraud is represented by truffle counterfeiting, since different varieties are almost impossible to recognize visually, and less expensive species absorb the aroma of higher prized ones when closely stored together. MALDI analysis and clustering permitted [119] to distinguish seven *Tuber* species confirming misidentifications in 26% of commercial specimens. To the same purpose, MALDI-TOF was applied [120] for the generation of specific mass spectra from 73 truffle tubers belonging to eight different species. Both studies demonstrated the potential of MALDI-TOF to provide useful information for quality assurance and fraud control in the truffle market in a fast, reliable, and inexpensive way.

The entire food industry, as well as the field of dietary and sport nutritional supplements, suffers the occurrence of several frauds. *Echinacea* species (*E. purpurea* and *E. angustifolia*) are widely used in dietary supplements and their adulteration is a great concern. Greene and colleagues [121] developed a rapid automated sample-preparation system combined with MALDI analysis for the differentiation of *Echinacea* species by the obtained mass fingerprints, which was even able to confirm the published composition of a commercial product and identify signatures suggestive of additional product components. An approach involving in-solution digestion followed by MALDI-TOF was proposed for the fast and simple screening of food supplements declaring high value protein components [122]. In this case, contrary to the products label statements, proteins were absent or present at very low concentrations in most products, clearly indicating the potential of the method for quality control purposes. The results obtained by in solution digestion were validated by a comparison with SDS–PAGE followed by tryptic in-gel digestion.

3. Conclusions

The range of analytes typically targeted in MALDI-TOF MS analyses has expanded over the years from macromolecules, such as peptides and proteins, to smaller molecules, such as lipids, hydrocarbons, sugars, phenols, and many others, widening the possible applications of the technique, which has been projected towards fields not initially contemplated, including food quality evaluation and adulteration detection. Consequently, several MALDI-based applications for the fight against food frauds have been reported in the literature over the years. The higher number of papers have been focused on milk, oils, fish and seafood, meat, fruits, and vegetables. In the case of milk, the target analytes have been mainly proteins and peptides and, to a minor extent, lipids and phospholipids. Oils were mostly characterized by means of triacylglycerols profiles, even if some studies focused on proteins/peptides and phospholipids were also reported. Again, proteomic studies were the most performed in matrices such as fish and meat as well as in fruit and vegetables samples, where smaller polyphenolic and carotenoid compounds were also determined. The MALDI approach has proved to be an excellent system for the determination of food fraud, allowing in many cases the development of reliable, fast, and cost-effective analytical methods, often requiring very simple sample pre-treatment protocols, and sometimes assisted by chemometrics. On these bases, also considering the noticeable number of recent papers available in the literature, a further increase of these applications is likely to be observed in the next future.

Funding: This research received no external funding.

Institutional Review Board Statement: Not applicable.

Informed Consent Statement: Not applicable.

Data Availability Statement: Not applicable.

Conflicts of Interest: The authors declare no conflict of interest.

References

1. Robson, K.; Dean, M.; Haughey, S.; Elliott, C. A comprehensive review of food fraud terminologies and food fraud mitigation guides. *Food Control.* **2021**, *120*, 107516. [CrossRef]
2. European Comission. Food Fraud: What Does It Mean? Available online: https://ec.europa.eu/food/safety/food-fraud/what-does-it-mean_en (accessed on 5 April 2021).
3. Hong, E.; Lee, S.Y.; Jeong, J.Y.; Park, J.M.; Kim, B.H.; Kwon, K.; Chun, H.S. Modern analytical methods for the detection of food fraud and adulteration by food category. *J. Sci. Food Agric.* **2017**, *97*, 3877–3896. [CrossRef]
4. Morin, J.F.; Lees, M. *Food Integrity Handbook, A Guide to Food Authenticity Issues and Analytical Solutions*; Eurofins Analytics France: Nantes, France, 2018.
5. McGrath, T.F.; Haughey, S.A.; Patterson, J.; Fauhl-Hassek, C.; Donarski, J.; Alewijn, M.; van Ruth, S.; Elliott, C.T. What are the scientific challenges in moving from targeted to non-targeted methods for food fraud testing and how can they be addressed?—Spectroscopy case study. *Trends Food Sci. Technol.* **2018**, *76*, 38–55. [CrossRef]
6. Cavanna, D.; Righetti, L.; Elliott, C.; Suman, M. The scientific challenges in moving from targeted to non-targeted mass spectrometric methods for food fraud analysis: A proposed validation workflow to bring about a harmonized approach. *Trends Food Sci. Technol.* **2018**, *80*, 223–241. [CrossRef]
7. Gao, B.; Holroyd, S.E.; Moore, J.C.; Laurvick, K.; Gendel, S.M.; Xie, Z. Opportunities and challenges using non-targeted methods for food fraud detection. *J. Agric. Food Chem.* **2019**, *67*, 8425–8430. [CrossRef] [PubMed]
8. Rodionova, O.; Pomerantsev, A. Chemometric tools for food fraud detection: The role of target class in non-targeted analysis. *Food Chem.* **2020**, *317*, 126448. [CrossRef]
9. Rodriguez-Saona, L.; Allendorf, M. Use of FTIR for rapid authentication and detection of adulteration of food. *Annu. Rev. Food Sci. Technol.* **2011**, *2*, 467–483. [CrossRef] [PubMed]
10. Cozzolino, D. The role of vibrational spectroscopy as a tool to assess economically motivated fraud and counterfeit issues in agricultural products and foods. *Anal. Methods* **2015**, *7*, 9390–9400. [CrossRef]
11. Lohumi, S.; Lee, S.; Lee, H.; Cho, B.-K. A review of vibrational spectroscopic techniques for the detection of food authenticity and adulteration. *Trends Food Sci. Technol.* **2015**, *46*, 85–98. [CrossRef]
12. Sørensen, K.M.; Khakimov, B.; Engelsen, S.B. The use of rapid spectroscopic screening methods to detect adulteration of food raw materials and ingredients. *Curr. Opin. Food Sci.* **2016**, *10*, 45–51. [CrossRef]
13. Sobolev, A.P.; Thomas, F.; Donarski, J.; Ingallina, C.; Circi, S.; Marincola, F.C.; Capitani, D.; Mannina, L. Use of NMR applications to tackle future food fraud issues. *Trends Food Sci. Technol.* **2019**, *91*, 347–353. [CrossRef]
14. Power, A.C.; Néill, C.N.; Geoghegan, S.; Currivan, S.; Deasy, M.; Cozzolino, D. A brief history of whiskey adulteration and the role of spectroscopy combined with chemometrics in the detection of modern whiskey fraud. *Beverages* **2020**, *6*, 49. [CrossRef]
15. Hassoun, A.; Måge, I.; Schmidt, W.F.; Temiz, H.T.; Li, L.; Kim, H.-Y.; Nilsen, H.; Biancolillo, A.; Aït-Kaddour, A.; Sikorski, M.; at al. Fraud in animal origin food products: Advances in emerging spectroscopic detection methods over the past five years. *Foods* **2020**, *9*, 1069. [CrossRef]
16. Valand, R.; Tanna, S.; Lawson, G.; Bengtström, L. A review of Fourier Transform Infrared (FTIR) spectroscopy used in food adulteration and authenticity investigations. *Food Addit. Contam. Part A* **2019**, *37*, 19–38. [CrossRef]

17. Cozzolino, R.; De Giulio, B. Application of ESI and MALDI-TOF MS for triacylglycerols analysis in edible oils. *Eur. J. Lipid Sci. Technol.* **2010**, *113*, 160–167. [CrossRef]
18. Kaufmann, A. The current role of high-resolution mass spectrometry in food analysis. *Anal. Bioanal. Chem.* **2012**, *403*, 1233–1249. [CrossRef] [PubMed]
19. Castro-Puyana, M.; Herrero, M. Metabolomics approaches based on mass spectrometry for food safety, quality and traceability. *Trends Anal. Chem.* **2013**, *52*, 74–87. [CrossRef]
20. Black, C.; Chevallier, O.P.; Elliott, C.T. The current and potential applications of Ambient Mass Spectrometry in detecting food fraud. *Trends Anal. Chem.* **2016**, *82*, 268–278. [CrossRef]
21. Morisasa, M.; Sato, T.; Kimura, K.; Mori, T.; Goto-Inoue, N. Application of matrix-assisted laser desorption/ionization mass spectrometry imaging for food analysis. *Foods* **2019**, *8*, 633. [CrossRef]
22. Creydt, M.; Fischer, M. Food phenotyping: Recording and processing of non-targeted liquid chromatography mass spectrometry data for verifying food authenticity. *Molecules* **2020**, *25*, 3972. [CrossRef]
23. Stachniuk, A.; Sumara, A.; Montowska, M.; Fornal, E. Liquid chromatography–mass spectrometry bottom-up proteomic methods in animal species analysis of processed meat for food authentication and the detection of adulterations. *Mass Spectrom. Rev.* **2021**, *40*, 3–30. [CrossRef]
24. FAO. *Milk Facts*; FAO: Rome, Italy, 2018; Available online: http://www.fao.org/3/I9966EN/i9966en.pdf (accessed on 15 March 2021).
25. FAO. *Food Outlook—Biannual Report on Global Food Markets*; FAO: Rome, Italy, 2020. [CrossRef]
26. Thompson, L.A.; Darwish, W.S. Environmental chemical contaminants in food: Review of a global problem. *J. Toxicol.* **2019**, *2019*, 2345283. [CrossRef]
27. Ulberth, F. Tools to combat food fraud—A gap analysis. *Food Chem.* **2020**, *330*, 127044. [CrossRef]
28. Montgomery, H.; Haughey, S.A.; Elliott, C.T. Recent food safety and fraud issues within the dairy supply chain (2015–2019). *Glob. Food Secur.* **2020**, *26*, 100447. [CrossRef]
29. Cozzolino, R.; Passalacqua, S.; Salemi, S.; Malvagna, P.; Spina, E.; Garozzo, D. Identification of adulteration in milk by matrix-assisted laser desorption/ionization time-of-flight mass spectrometry. *J. Mass Spectrom.* **2001**, *36*, 1031–1037. [CrossRef]
30. Cozzolino, R.; Passalacqua, S.; Salemi, S.; Garozzo, D. Identification of adulteration in water buffalo mozzarella and in ewe cheese by using whey proteins as biomarkers and matrix-assisted laser desorption/ionization mass spectrometry. *J. Mass Spectrom.* **2002**, *37*, 985–991. [CrossRef]
31. Sassi, M.; Arena, S.; Scaloni, A. MALDI-TOF-MS platform for integrated proteomic and peptidomic profiling of milk samples allows rapid detection of food adulterations. *J. Agric. Food Chem.* **2015**, *63*, 6157–6171. [CrossRef]
32. Russo, R.; Rega, C.; Chambery, A. Rapid detection of water buffalo ricotta adulteration or contamination by matrix-assisted laser desorption/ionization time-of-flight mass spectrometry. *Rapid Commun. Mass Spectrom.* **2016**, *30*, 497–503. [CrossRef]
33. Caira, S.; Nicolai, M.A.; Lilla, S.; Calabrese, M.G.; Pinto, G.; Scaloni, A.; Chianese, L.; Addeo, F. Eventual limits of the current EU official method for evaluating milk adulteration of water buffalo dairy products and potential proteomic solutions. *Food Chem.* **2017**, *230*, 482–490. [CrossRef]
34. Liland, K.H.; Mevik, B.-H.; Rukke, E.-O.; Almøy, T.; Skaugen, M.; Isaksson, T. Quantitative whole spectrum analysis with MALDI-TOF MS, Part I: Measurement optimisation. *Chemom. Intell. Lab. Syst.* **2009**, *96*, 210–218. [CrossRef]
35. Liland, K.H.; Mevik, B.-H.; Rukke, E.-O.; Almøy, T.; Isaksson, T. Quantitative whole spectrum analysis with MALDI-TOF MS, Part II: Determining the concentration of milk in mixtures. *Chemom. Intell. Lab. Syst.* **2009**, *99*, 39–48. [CrossRef]
36. Nicolaou, N.; Xu, Y.; Goodacre, R. MALDI-MS and multivariate analysis for the detection and quantification of different milk species. *Anal. Bioanal. Chem.* **2011**, *399*, 3491–3502. [CrossRef]
37. Calvano, C.D.; De Ceglie, C.; Monopoli, A.; Zambonin, C.G. Detection of sheep and goat milk adulterations by direct MALDI-TOF MS analysis of milk tryptic digests. *J. Mass Spectrom.* **2012**, *47*, 1141–1149. [CrossRef]
38. Calvano, C.D.; De Ceglie, C.; Aresta, A.; Facchini, L.A.; Zambonin, C.G. MALDI-TOF mass spectrometric determination of intact phospholipids as markers of illegal bovine milk adulteration of high-quality milk. *Anal. Bioanal. Chem.* **2012**, *405*, 1641–1649. [CrossRef]
39. Piras, C.; Ceniti, C.; Hartmane, E.; Costanzo, N.; Morittu, V.M.; Roncada, P.; Britti, D.; Cramer, R. Rapid liquid AP-MALDI MS profiling of lipids and proteins from goat and sheep milk for speciation and colostrum analysis. *Proteomes* **2020**, *8*, 20. [CrossRef]
40. Cunsolo, V.; Muccilli, V.; Saletti, R.; Foti, S. MALDI-TOF mass spectrometry for the monitoring of she-donkey's milk contamination or adulteration. *J. Mass Spectrom.* **2013**, *48*, 148–153. [CrossRef]
41. Di Girolamo, F.; Masotti, A.; Salvatori, G.; Scapaticci, M.; Muraca, M.; Putignani, L. A Sensitive and effective proteomic approach to identify she-donkey's and goat's milk adulterations by MALDI-TOF MS fingerprinting. *Int. J. Mol. Sci.* **2014**, *15*, 13697–13719. [CrossRef]
42. Calvano, C.D.; Monopoli, A.; Loizzo, P.; Faccia, M.; Zambonin, C. Proteomic approach based on MALDI-TOF MS to detect powdered milk in fresh cow's milk. *J. Agric. Food Chem.* **2013**, *61*, 1609–1617. [CrossRef] [PubMed]
43. Garcia, J.S.; Sanvido, G.B.; Saraiva, S.A.; Zacca, J.J.; Cosso, R.G.; Eberlin, M.N. Bovine milk powder adulteration with vegetable oils or fats revealed by MALDI-QTOF MS. *Food Chem.* **2012**, *131*, 722–726. [CrossRef]
44. Hinz, K.; O'Connor, P.M.; Huppertz, T.; Ross, R.P.; Kelly, A.L. Comparison of the principal proteins in bovine, caprine, buffalo, equine and camel milk. *J. Dairy Res.* **2012**, *79*, 185–191. [CrossRef] [PubMed]

45. Calvano, C.D.; De Ceglie, C.; Zambonin, C.G. Proteomic analysis of complex protein samples by MALDI–TOF mass spectrometry. *Methods Mol. Biol.* **2014**, *1129*, 365–380.
46. England, P.; Tang, W.; Kostrzewa, M.; Shahrezaei, V.; Larrouy-Maumus, G. Discrimination of bovine milk from non-dairy milk by lipids fingerprinting using routine matrix-assisted laser desorption ionization mass spectrometry. *Sci. Rep.* **2020**, *10*, 5160. [CrossRef]
47. Magenis, R.B.; Prudêncio, E.S.; Molognoni, L.; Daguer, H. A control method to inspect the compositional authenticity of minas frescal cheese by gel electrophoresis. *J. Agric. Food Chem.* **2014**, *62*, 8333–8339. [CrossRef]
48. Damário, N.; Delafiori, J.; Ferreira, M.S.; De Oliveira, D.N.; Catharino, R.R. Cheese lipid profile using direct imprinting in glass surface mass spectrometry. *Anal. Methods* **2015**, *7*, 2877–2880. [CrossRef]
49. Rau, J.; Korte, N.; Dyk, M.; Wenninger, O.; Schreiter, P.; Hiller, E. Rapid animal species identification of feta and mozzarella cheese using MALDI-TOF mass-spectrometry. *Food Control.* **2020**, *117*, 107349. [CrossRef]
50. Calvano, C.D.; Jensen, O.N.; Zambonin, C.G. Selective extraction of phospholipids from dairy products by micro-solid phase extraction based on titanium dioxide microcolumns followed by MALDI-TOF-MS analysis. *Anal. Bioanal. Chem.* **2009**, *394*, 1453–1461. [CrossRef]
51. Gasilova, N.; Gassner, A.-L.; Girault, H.H. Analysis of major milk whey proteins by immunoaffinity capillary electrophoresis coupled with MALDI-MS. *Electrophoresis* **2012**, *33*, 2390–2398. [CrossRef]
52. Calvano, C.D.; De Ceglie, C.; Zambonin, C.G. Development of a novel direct in-matrix extraction (DIME) protocol for MAL-DI-ToF-MS detection of glycated phospholipids in heat treated food samples. *J. Mass Spectrom.* **2014**, *49*, 831–839. [CrossRef]
53. Ebner, J.; Baum, F.; Pischetsrieder, M. Identification of sixteen peptides reflecting heat and/or storage induced processes by profiling of commercial milk samples. *J. Proteom.* **2016**, *147*, 66–75. [CrossRef] [PubMed]
54. Arena, S.; Salzano, A.M.; Scaloni, A. Identification of protein markers for the occurrence of defrosted material in milk through a MALDI-TOF-MS profiling approach. *J. Proteom.* **2016**, *147*, 56–65. [CrossRef]
55. Di Francesco, L.; Di Girolamo, F.; Mennini, M.; Masotti, A.; Salvatori, G.; Rigon, G.; Signore, F.; Pietrantoni, E.; Scapaticci, M.; Lante, I.; at al. A MALDI-TOF MS approach for mammalian, human, and formula milks' profiling. *Nutrients* **2018**, *10*, 1238. [CrossRef]
56. Martinez-Gonzalez, M.A.; Martin-Calvo, N. Mediterranean diet and life expectancy; beyond olive oil, fruits, and vegetables. *Curr. Opin. Clin. Nutr. Metab. Care* **2016**, *19*, 401–407. [CrossRef] [PubMed]
57. International Olive Council. Trade Standard Applying to Olive Oils and Olive-Pomace Oils, (2016) COI/T.15/NC No. 3/Rev. 11. Available online: https://www.internationaloliveoil.org/wp-content/uploads/2019/11/COI-T.15-NC.-No-3-Rev.-13-2019-Eng.pdf (accessed on 15 March 2021).
58. Frankel, E.; Bakhouche, A.; Lozano-Sánchez, J.; Segura-Carretero, A.; Fernández-Gutiérrez, A. Literature review on pro-duction process to obtain extra virgin olive oil enriched in bioactive compounds. Potential use of byproducts as alternative sources of polyphenols. *J. Agric. Food Chem.* **2013**, *61*, 5179–5188. [CrossRef]
59. Casadei, E.; Valli, E.; Panni, F.; Donarski, J.; Farrús Gubern, J.; Lucci, P.; Conte, L.; Lacoste, F.; Maquet, A.; Brereton, P.; at al. Emerging trends in olive oil fraud and possible countermeasures. *Food Control* **2021**, *124*, 107902. [CrossRef]
60. Meenu, M.; Cai, Q.; Xu, B. A critical review on analytical techniques to detect adulteration of extra virgin olive oil. *Trends Food Sci. Technol.* **2019**, *91*, 391–408. [CrossRef]
61. Calvano, C.D.; Palmisano, F.; Zambonin, C.G. Laser desorption/ionization time-of-flight mass spectrometry of triacylglycerols in oils. *Rapid Commun. Mass Spectrom.* **2005**, *19*, 1315–1320. [CrossRef] [PubMed]
62. Calvano, C.D.; Aresta, A.; Zambonin, C.G. Detection of hazelnut oil in extra-virgin olive oil by analysis of polar components by micro-solid phase extraction based on hydrophilic liquid chromatography and MALDI-ToF mass spectrometry. *J. Mass Spectrom.* **2010**, *45*, 981–988. [CrossRef]
63. Calvano, C.D.; De Ceglie, C.; D'Accolti, L.; Zambonin, C.G. MALDI-TOF mass spectrometry detection of extra-virgin olive oil adulteration with hazelnut oil by analysis of phospholipids using an ionic liquid as matrix and extraction solvent. *Food Chem.* **2012**, *134*, 1192–1198. [CrossRef]
64. De Ceglie, C.; Calvano, C.D.; Zambonin, C.G. Determination of hidden hazelnut oil proteins in extra virgin olive oil by cold acetone precipitation followed by in-solution tryptic digestion and MALDI-TOF-MS analysis. *J. Agric. Food Chem.* **2014**, *62*, 9401–9409. [CrossRef]
65. Arlorio, M.; Coisson, J.D.; Bordiga, M.; Travaglia, F.; Garino, C.; Zuidmeer, L.; Van Ree, R.; Giuffrida, M.G.; Conti, A.; Martelli, A. Olive oil adulterated with hazelnut oils: Simulation to identify possible risks to allergic consumers. *Food Addit. Contam. Part A* **2010**, *27*, 11–18. [CrossRef]
66. Di Girolamo, F.; Masotti, A.; Lante, I.; Scapaticci, M.; Calvano, C.D.; Zambonin, C.; Muraca, M.; Putignani, L. A Simple and effective mass spectrometric approach to identify the adulteration of the Mediterranean diet component extra-virgin olive oil with corn oil. *Int. J. Mol. Sci.* **2015**, *16*, 20896–20912. [CrossRef]
67. Jergovic, A.M.; Persuric, Z.; Saftic, L.; Kraljevic, S. Pavelic, evaluation of MALDI-TOF/MS technology in olive oil adulteration. *J. Am. Oil Chem. Soc.* **2017**, *94*, 749–757. [CrossRef]
68. Kuo, T.-H.; Kuei, M.-S.; Hsiao, Y.; Chung, H.-H.; Hsu, C.-C.; Chen, H.-J. Matrix-assisted laser desorption/ionization mass spectrometry typings of edible oils through spectral networking of triacylglycerol fingerprints. *ACS Omega* **2019**, *4*, 15734–15741. [CrossRef]

69. Krist, S.; Stuebiger, G.; Bail, S.; Unterweger, H. Detection of adulteration of poppy seed oil with sunflower oil based on volatiles and triacylglycerol composition. *J. Agric. Food Chem.* **2006**, *54*, 6385–6389. [CrossRef]

70. Zambonin, C.G.; Calvano, C.D.; D'Accolti, L.; Palmisano, F. Laser desorption/ionization time-of-flight mass spectrometry of squalene in oil samples. *Rapid Commun. Mass Spectrom.* **2005**, *20*, 325–327. [CrossRef]

71. Calvano, C.D.; Aresta, A.; Palmisano, F.; Zambonin, C.G.; Laser, A. Desorption ionization time-of-flight mass spectrometry in-vestigation on triacylglycerols degradation during thermal stressing of edible oils. *Anal. Bioanal. Chem.* **2007**, *389*, 2075–2084. [CrossRef]

72. Picariello, G.; Paduano, A.; Sacchi, R.; Addeo, F. MALDI-TOF mass spectrometry profiling of polar and nonpolar fractions in heated vegetable oils. *J. Agric. Food Chem.* **2009**, *57*, 5391–5400. [CrossRef]

73. Chapagain, B.P.; Wiesman, Z. MALDI-TOF/MS Fingerprinting of triacylglycerols (TAGs) in olive oils produced in the Israeli Negev Desert. *J. Agric. Food Chem.* **2009**, *57*, 1135–1142. [CrossRef]

74. Vichi, S.; Lazzez, A.; Grati-Kamoun, N.; Caixach, J. Modifications in virgin olive oil glycerolipid fingerprint during olive ripening by MALDI-TOF MS analysis. *LWT Food Sci. Technol.* **2012**, *48*, 24–29. [CrossRef]

75. Shen, Q.; Dong, W.; Yang, M.; Baibado, J.T.; Wang, Y.; Alqouqa, I.; Cheung, H.-Y. Lipidomic study of olive fruit and oil using TiO_2 nanoparticle based matrix solid-phase dispersion and MALDI-TOF/MS. *Food Res. Int.* **2013**, *54*, 2054–2061. [CrossRef]

76. Alves, E.; Melo, T.; Rey, F.; Moreira, A.S.; Domingues, P.; Domingues, M.R. Polar lipid profiling of olive oils as a useful tool in helping to decipher their unique fingerprint. *LWT Food Sci. Technol.* **2016**, *74*, 371–377. [CrossRef]

77. Persuric, Z.; Osuga, J.; Grbac, T.G.; Katalinic, J.P.; Kraljevic Pavelic, S. MALDI-SpiralTOF technology for assessment of triacylglyc-erols in Croatian olive oils. *Eur. J. Lipid Sci. Technol.* **2017**, *119*, 1500375. [CrossRef]

78. Peršurić, Ž.; Saftić, L.; Mašek, T.; Pavelić, S.K. Comparison of triacylglycerol analysis by MALDI-TOF/MS, fatty acid analysis by GC-MS and non-selective analysis by NIRS in combination with chemometrics for determination of extra virgin olive oil geographical origin. A case study. *LWT Food Sci. Technol.* **2018**, *95*, 326–332. [CrossRef]

79. Kaufman, M.; Wiesman, Z. Pomegranate Oil Analysis with Emphasis on MALDI-TOF/MS Triacylglycerol Fingerprinting. *J. Agric. Food Chem.* **2007**, *55*, 10405–10413. [CrossRef]

80. Bail, S.; Stuebiger, G.; Unterweger, H.; Buchbauer, G.; Krist, S. Characterization of volatile compounds and triacylglycerol profiles of nut oils using SPME-GC-MS and MALDI-TOF-MS. *Eur. J. Lipid Sci. Technol.* **2009**, *111*, 170–182. [CrossRef]

81. Saraiva, S.A.; Cabral Marcos, E.C.; Eberlin, N.; Catharino, R.R. Amazonian vegetable oils and fats: Fast typification and quality control via triacylglycerol (TAG) profiles from dry matrix-assisted laser desorption/ionization time-of-flight (MALDI−TOF) mass spectrometry fingerprinting. *J. Agric. Food Chem.* **2009**, *57*, 4030–4034. [CrossRef]

82. Badu, M.; Awudza, A.M.J. Determination of the triacylglycerol content for the identification and assessment of purity of shea butter fat, peanut oil, and palm kernel oil using maldi-tof/tof mass spectroscopic technique. *Int. J. Food Prop.* **2016**, *20*, 271–280. [CrossRef]

83. Mazzeo, M.F.; De Giulio, B.; Guerriero, G.; Ciarcia, G.; Malorni, A.; Russo, G.L.; Siciliano, R.A. Fish authentication by MALDI-TOF mass spectrometry. *J. Agric. Food Chem.* **2008**, *56*, 11071–11076. [CrossRef]

84. Volta, P.; Riccardi, N.; Lauceri, R.; Tonolla, M. Discrimination of freshwater fish species by matrix-assisted laser desorption/ionization time of flight mass spectrometry (MALDI-TOF MS): A pilot study. *J. Limnol.* **2012**, *71*, 164–169. [CrossRef]

85. Barik, S.K.; Banerjee, S.; Bhattacharjee, S.; Das Gupta, S.K.; Mohanty, S.; Mohanty, B.P. Proteomic analysis of sarcoplasmic peptides of two related fish species for food authentication. *Appl. Biochem. Biotechnol.* **2013**, *171*, 1011–1021. [CrossRef]

86. Siciliano, R.A.; D'Esposito, D.; Mazzeo, M.F. Food authentication by MALDI MS: MALDI-TOF MS analysis of fish species. In *Advances in MALDI and Laser-Induced Soft Ionization Mass Spectrometry*; Cramer, R., Ed.; Springer International Publishing: Cham, Switzerland, 2016; pp. 263–277.

87. Spielmann, G.; Huber, I.; Maggipinto, M.; Haszprunar, G.; Busch, U.; Pavlovic, M. Comparison of five preparatory protocols for fish species identification using MALDI-TOF MS. *Eur. Food Res. Technol.* **2017**, *244*, 685–694. [CrossRef]

88. Stahl, A.; Schröder, U. Development of a MALDI–TOF MS-based protein fingerprint database of common food fish allowing fast and reliable identification of fraud and substitution. *J. Agric. Food Chem.* **2017**, *65*, 7519–7527. [CrossRef]

89. Shao, M.; Bi, H. Direct identification of fish species by surface molecular transferring. *Analyst* **2020**, *145*, 4148–4155. [CrossRef]

90. Wang, C.; Bi, H. Super-fast seafood authenticity analysis by one-step pretreatment and comparison of mass spectral patterns. *Food Control* **2021**, *123*, 107751. [CrossRef]

91. Zhao, J.; Li, J.; Wang, J.; Lv, W. Applying different methods to evaluate the freshness of large yellow croaker (*Pseudosciaena crocea*) fillets during chilled storage. *J. Agric. Food Chem.* **2012**, *60*, 11387–11394. [CrossRef]

92. Deng, X.; Lei, Y.; Yu, Y.; Lu, S.; Zhang, J. The discovery of proteins associated with freshness of Coregonus Peled muscle during refrigerated storage. *J. Food Sci.* **2019**, *84*, 1266–1272. [CrossRef] [PubMed]

93. Ulrich, S.; Beindorf, P.; Biermaier, B.; Schwaiger, K.; Gareis, M.; Gottschalk, C. A novel approach for the determination of freshness and identity of trouts by MALDI-TOF mass spectrometry. *Food Control* **2017**, *80*, 281–289. [CrossRef]

94. Campos, A.; Puerto, M.; Prieto, A.; Cameán, A.; Almeida, A.M.; Coelho, A.V.; Vasconcelos, V. Protein extraction and two-dimensional gel electrophoresis of proteins in the marine mussel *Mytilus galloprovincialis*: An important tool for protein expression studies, food quality and safety assessment. *J. Sci. Food Agric.* **2012**, *93*, 1779–1787. [CrossRef]

95. Flaudrops, C.; Armstrong, N.; Raoult, D.; Chabrie, E. Determination of the animal origin of meat and gelatin by MAL-DI-TOF-MS. *J. Food Compos. Anal.* **2015**, *41*, 104–112. [CrossRef]

96. Kiran, M.; Naveena, B.M.; Reddy, K.S.; Shahikumar, M.; Reddy, V.R.; Kulkarni, V.V.; Rapole, S.; More, T.H. Understanding tenderness variability and ageing changes in buffalo meat: Biochemical, ultrastructural and proteome characterization. *Animal* **2016**, *10*, 1007–1015. [CrossRef]

97. Subramaniyan, S.; Kang, D.R.; Jung, Y.-C.; Choi, Y.-I.; Lee, M.J.; Choe, H.S.; Shim, K.S. Comparative studies of meat quality traits and the proteome profile between low pH and high pH muscles in longissimus dorsi of Berkshire. *Can. J. Anim. Sci.* **2017**, *97*, 640–649. [CrossRef]

98. Zheng, N.; Zhu, Z.; Xin, Q.; Zhang, Z.; Miao, Z.; Li, L.; Zhang, L.; Wang, Z.; Huang, Y. Differential protein profiles in duck meat during the early postmortem storage period. *Anim. Sci. J.* **2019**, *90*, 757–768. [CrossRef]

99. Picariello, G.; Ferranti, P.; Chianese, L.; Addeo, F. Differentiation of Vitis vinifera L. and hybrid red grapes by matrix-assisted laser desorption/ionization mass spectrometry analysis of berry skin anthocyanins. *J. Agric. Food Chem.* **2012**, *60*, 4559–4566. [CrossRef] [PubMed]

100. Esquivel-Alvarado, D.; Alfaro-Viquez, E.; Krueger, C.G.; Vestling, M.M.; Reed, J.D. Classification of proanthocyanidin profiles using matrix-assisted laser desorption/ionization time-of-flight mass spectrometry (MALDI-TOF MS) spectra data combined with multivariate analysis. *Food Chem.* **2021**, *336*, 127667. [CrossRef]

101. Pittenauer, E.; Rados, E.; Koulakiotis, N.; Tsarbopoulos, A.; Allmaier, G. Processed stigmas of *Crocus sativus* L. imaged by MALDI-based MS. *Proteomics* **2016**, *16*, 1726–1730. [CrossRef]

102. Aiello, D.; Siciliano, C.; Mazzotti, F.; Di Donna, L.; Athanassopoulos, C.M.; Napoli, A. Molecular species fingerprinting and quantitative analysis of saffron (*Crocus sativus* L.) for quality control by MALDI mass spectrometry. *RSC Adv.* **2018**, *8*, 36104–36113. [CrossRef]

103. Aiello, D.; Siciliano, C.; Mazzotti, F.; Di Donna, L.; Athanassopoulos, C.M.; Napoli, A. A rapid MALDI MS/MS based method for assessing saffron (*Crocus sativus* L.) adulteration. *Food Chem.* **2020**, *307*, 125527. [CrossRef]

104. Pattemore, J.A.; Rice, N.; Marshall, D.F.; Waugh, R.; Henry, R.J. Cereal variety identification using MALDI-TOF mass spec-trometry SNP genotyping. *J. Cereal Sci.* **2010**, *52*, 356–361. [CrossRef]

105. Jin, Z.; Mu, Y.-W.; Sun, J.-Y.; Li, X.-M.; Gao, X.-L.; Lu, J. Proteome analysis of metabolic proteins (pI 4–7) in barley (*Hordeum vulgare*) malts and initial application in malt quality discrimination. *J. Agric. Food Chem.* **2012**, *61*, 402–409. [CrossRef]

106. Jin, Z.; Cai, G.-L.; Li, X.-M.; Gao, F.; Yang, J.-J.; Lu, J.; Dong, J.-J. Comparative proteomic analysis of green malts between barley (*Hordeum vulgare*) cultivars. *Food Chem.* **2014**, *151*, 266–270. [CrossRef] [PubMed]

107. Fernando, N.; Panozzo, J.; Tausz, M.; Norton, R.; Fitzgerald, G.; Khan, A.; Seneweera, S. Rising CO_2 concentration altered wheat grain proteome and flour rheological characteristics. *Food Chem.* **2015**, *170*, 448–454. [CrossRef]

108. Cucu, T.; De Meulenaer, B.; Devreese, B. MALDI based identification of soybean protein markers—Possible analytical targets for allergen detection in processed foods. *Peptides* **2012**, *33*, 187–196. [CrossRef] [PubMed]

109. Ciarmiello, L.F.; Mazzeo, M.F.; Minasi, P.; Peluso, A.; De Luca, A.; Piccirillo, P.; Siciliano, R.A.; Carbone, V. Analysis of different European hazelnut (*Corylus avellana* L.) cultivars: Authentication, phenotypic features, and phenolic profiles. *J. Agric. Food Chem.* **2014**, *62*, 6236–6246. [CrossRef] [PubMed]

110. Yoon, S.R.; Kim, S.H.; Lee, H.W.; Ha, J.H. A novel method to rapidly distinguish the geographical origin of traditional fer-mented-salted vegetables by mass fingerprinting. *PLoS ONE* **2017**, *12*, e0188217. [CrossRef] [PubMed]

111. Kumar, D.K.; Mathew, D.; Nazeem, P.A.; Abida, P.S.; Thomas, C.G. A comparative proteome assay on the quality of yardlong bean pods as influenced by the organic and inorganic nourishment systems. *Acta Physiol. Plant.* **2017**, *39*, 265. [CrossRef]

112. Pinto, G.; Aurilia, M.; Illiano, A.; Fontanarosa, C.; Sannia, G.; Trifuoggi, M.; Lettera, V.; Sperandeo, R.; Pucci, A.P. From un-targeted metabolomics to the multiple reaction monitoring-based quantification of polyphenols in chocolates from different geographical areas. *J. Mass Spectrom.* **2021**, *56*, e4651. [CrossRef] [PubMed]

113. Žídková, J.; Chmelík, J. Determination of saccharides in fruit juices by capillary electrophoresis and matrix-assisted laser desorption/ionization time-of-flight mass spectrometry. *J. Mass Spectrom.* **2001**, *36*, 417–421. [CrossRef]

114. Chambery, A.; del Monaco, G.; Di Maro, A.; Parente, A. Peptide fingerprint of high quality Campania white wines by MALDI-TOF mass spectrometry. *Food Chem.* **2009**, *113*, 1283–1289. [CrossRef]

115. Fraige, K.; González-Fernández, R.; Carrilho, E.; Jorrín-Novo, J.V. Metabolite and proteome changes during the ripening of Syrah and Cabernet Sauvignon grape varieties cultured in a nontraditional wine region in Brazil. *J. Proteom.* **2015**, *113*, 206–225. [CrossRef]

116. Won, S.-R.; Lee, D.-C.; Ko, S.H.; Kim, J.-W.; Rhee, H.-I. Honey major protein characterization and its application to adulteration detection. *Food Res. Int.* **2008**, *41*, 952–956. [CrossRef]

117. Qu, L.; Jiang, Y.; Huang, X.; Cui, M.; Ning, F.; Liu, T.; Gao, Y.; Wu, D.; Nie, Z.; Luo, L. High-throughput monitoring of multiclass syrup adulterants in honey based on the oligosaccharide and polysaccharide profiles by MALDI mass spectrometry. *J. Agric. Food Chem.* **2019**, *67*, 11256–11261. [CrossRef] [PubMed]

118. Ulrich, S.; Kühn, U.; Biermaier, B.; Piacenza, N.; Schwaiger, K.; Gottschalk, C.; Gareis, M. Direct identification of edible insects by MALDI-TOF mass spectrometry. *Food Control.* **2017**, *76*, 96–101. [CrossRef]

119. El Karkouri, K.; Couderc, C.; Decloquement, P.; Abeille, A.; Raoult, D. Rapid MALDI-TOF MS identification of commercial truffles. *Sci. Rep.* **2019**, *9*, 17686. [CrossRef]

120. Klein, M.; Kielhauser, R.; Dilger, T. Optimized rapid and reliable identification of *Tuber* spp. by matrix-assisted laser de-sorption ionization time-of-flight mass spectrometry. *J. Mass Spectrom.* **2020**, *55*, e4655.

121. Greene, L.A.; Isaac, I.; Gray, D.E.; Schwartz, S.A. Streamlining plant sample preparation: The use of high-throughput robotics to process echinacea samples for biomarker profiling by MALDI-TOF mass spectrometry. *J. Biomol. Tech. JBT* **2007**, *18*, 238–244. [PubMed]
122. De Ceglie, C.; Calvano, C.D.; Zambonin, C.G. MALDI-TOF MS for quality control of high protein content sport supplements. *Food Chem.* **2015**, *176*, 396–402. [CrossRef] [PubMed]

applied sciences

Article

The Effect of Edible Coatings on Selected Physicochemical Properties of Cassava Chips

Diofanor Acevedo-Correa [1,*], José Jaimes-Morales [2] and Piedad M. Montero-Castillo [1]

[1] Grupo de Investigación en Innovación y Desarrollo Agropecuario y Agroindustrial (IDAA), Universidad de Cartagena, Campus Piedra de Bolívar, Av. Consulado, # 48-152, P.O. Box 130015 Cartagena de Indias, Colombia; pmonteroc@unicartagena.edu.co

[2] Grupo de Investigación en Medio Ambiente, Alimentos y Salud (MAAS), Universidad de Cartagena, Campus de Zaragocilla, Cra. 50 #24120, P.O. Box 130015 Cartagena de Indias, Colombia; jjaimesm@unicartagena.edu.co

* Correspondence: dacevedoc1@unicartagena.edu.co

Abstract: The objective of this research was to study the effect of edible coatings on the physicochemical properties of cassava chips. The oil and moisture absorption in fried cassava chips that were not coated and in chips that were coated with pectin and whey protein films were determined using a completely randomized experiment design with a 3^3 factorial arrangement. The multifactorial ANOVA analysis of variance showed that all factors had significant statistical differences for moisture loss and oil absorption ($p < 0.05$). The coating type, the control, and the whey protein-coated chips presented a 321% greater oil content on average at 180 °C and 180 s than the pectin-coated chips. The density, heat capacity, and thermal diffusivity had statistical differences at all temperatures ($p < 0.05$). The sensory analysis showed that the coating type affected all sensory parameters, except crispness, as indicated by significant statistical differences ($p < 0.05$). The temperature only influenced the color of the control chips, with statistical differences ($p < 0.05$) at all temperatures.

Keywords: edible coating; cassava chips; physicochemical properties

check for **updates**

Citation: Acevedo-Correa, D.; Jaimes-Morales, J.; Montero-Castillo, P.M. The Effect of Edible Coatings on Selected Physicochemical Properties of Cassava Chips. *Appl. Sci.* **2021**, *11*, 3265. https://doi.org/10.3390/app11073265

Academic Editor: Theodoros Varzakas

Received: 2 March 2021
Accepted: 29 March 2021
Published: 6 April 2021

Publisher's Note: MDPI stays neutral with regard to jurisdictional claims in published maps and institutional affiliations.

1. Introduction

The nutritional composition of cassava is important because it is the main component of the root, which is consumed in less-developed countries. Factors associated with geographic location and environmental conditions influence the nutritional value [1]. The root contains significant amounts of carbohydrates and fiber. In addition, it has significant percentages of minerals such as calcium, iron, and phosphorus; and vitamins such as thiamine, riboflavin, niacin, and vitamin C (ascorbic acid). It also contains large amounts of amino acids such as arginine, glutamic acid, and aspartic acid [2,3]. Current cassava processing methods include peeling, smoking, drying, slicing, grating, fermenting, pressing, soaking, grinding, roasting, and frying, which are carried out to remove/reduce toxic substances, impart flavor, and increase shelf-life [4–6]. The frying method stands out. Deep-fat frying is one of the conventional and most common operations in the preparation of a variety of fried foods, which is used worldwide to create desirable flavors and textures in foods [7,8]. The advantageous sensory characteristics of most deep-fried foods derive from the formation of a composite structure that provides a crispy, porous, oily outer layer and a moist, cooked interior [9–12]. Frying is commonly used in the food industry to produce a range of food products with high consumer acceptance although high-fat contents contribute to obesity and cardiovascular disease. When food absorbs fat, it can change the composition, texture, size, and shape of the food, resulting in a loss of nutrients, specifically vitamins [13]. There is growing interest in methods that could minimize oil uptake and reduce the fat content of fried foods. Hydrocolloids have been applied indiscriminately as a food coating or incorporated into chip formulation and have shown promising fat reductions and moisture retention during frying [14]. Hydrocolloids are natural compounds, such as polysaccharides and proteins, which have some hydrophilic groups. They have been used as food coatings, film-forming materials, and emulsifiers. Before frying potatoes, the application of a hydrocolloid layer on the surface has been considered an effective method to prevent oil absorption and acrylamide formation [15,16].

Edible coatings are currently used as viable alternatives for frying since these substances adhere to the product and form an external barrier that prevents the absorption of fat during immersion frying processes [17]; however, these thermal processes are currently not fully controlled for temperature and processing time, which usually results in a product with inadequate sensory characteristics and a high oil content. In general, they are artisanal and inefficient, so it is not possible to have an exact control of the process variables [18]. Research by Ajo [19] reported that the application of an edible coating with xanthan gum reduced oil absorption by up to 57% and improved the overall quality of the product, including taste, flavor, and crunchiness. One of the alternatives for processing cassava is cooking it with boiling water and then frying it by immersion in oil and coating it with edible films [20] based on proteins, such as whey, isolated soybean, some carbohydrates, pectins, and hydrocolloids that have oil barrier properties [21].

This results in excessive energy and economic expense on an industrial scale and affects the low acceptability of food by consumers since the final presentation is not traditional [22]. Therefore, it is important to investigate the effects of edible coatings in reducing the absorption of oil by products during frying in order to present consumers with a quality product with adequate processing conditions that reduce fat consumption and develop more acceptable, safe, and healthy cassava products [23]. These coatings were tested on cassava chips because they have been applied on the surface of fresh, frozen, and manufactured foods to improve food quality and increase shelf-life; in addition, they present mechanical properties and a barrier to water vapor and gases [24–26]. Furthermore, coatings were applied to decrease moisture loss values during frying since properties such as crispiness are affected, which is a fundamental parameter for consumers [27,28]. Currently, research is needed to transform this product into alternatives and new presentations that allow and expand its consumption. Therefore, this study aimed to characterize the coatings and determine the effect of coatings based on pectin and whey protein on reductions in oil absorption in fried cassava chips.

2. Materials and Methods

2.1. Obtaining the Raw Material

Cassava roots (*Manihot esculenta Crantz*), variety MCol 2215, were used, which were obtained from the supply center in the city of Cartagena. Local, refined, 100% vegetable palm oil was also used.

2.2. Elaboration of the Edible Films

Pectin and Whey Protein Films

The films were prepared by slowly dispersing 4, 8, and 12 g of pectin in 400 mL of distilled water. The solution was heated to 90 °C for 10 min with constant agitation at 120 rpm. Forty milliliter aliquots were distributed in 11.8 cm diameter plates, and the films were dried at 28 °C for 48 h [22]. The whey protein films were prepared by slowly dissolving 18, 22, and 26 g of whey protein along with 6.0, 6.6, and 7.8 g of glycerol (respectively) in 140 mL of distilled water. The solutions were heated in a water bath at 90 °C for 5 min with slow agitation at 120 rpm. 20 mL aliquots were distributed in 11.8 cm diameter plates, and the films were left to dry at 28 °C for 24 h.

2.3. Characterization of Edible Coatings

2.3.1. Permeability to Water Vapor and Thickness

The water vapor permeability (WVP) was determined according to ASTM E-96 standard method 1995. The thickness of each film was evaluated using a digital micrometer, using the arithmetic mean of 10 measurements taken at random over the entire film surface.

2.3.2. Solubility in Water

The water solubility of the films was determined using the gravimetric method of Gontard et al. [29]. The films were cut in 2 cm diameter disks. The initial weight of the samples was obtained after drying for a period of 24 h at a temperature of 105 °C. After the first weighing, the samples were immersed in a container containing 50 mL of distilled water and kept at 120 rpm for 24 h. After this period, the samples were removed and dried at a temperature of 105 °C for another 24 h to obtain the final dry weight. The tests were carried out in triplicate. The water solubility (WS) was expressed with the ratio: WS = (Initial weight − Final weight)/Initial weight × 100.

2.4. Experiment and Statistical Design

For the frying process, a completely randomized experiment design (DCA) was used with 3^3 factorial arrangements, where the factors with their respective levels were: Temperature (140, 160, and 180 °C), time (60, 120, and 180 s), and% of coating (0%, 1% pectin, and 9% whey protein). Each run was done in triplicate, and the results were expressed as the mean and the respective standard deviation. Statistically significant differences were determined in each parameter with a completely randomized analysis of variance (ANOVA) and multiple comparisons using Tukey's HSD test with a significance level of 5% ($p \leq 0.05$). Table 1 shows the experiment treatments.

Table 1. Factors and levels of cassava chips in the experiment design.

Factor	Level		
	Low	Central	High
Type of coating (%)	0	1	9
Time (s)	60	120	180
Temperature (°C)	140	160	180

The responses of the experiment design were the physical-chemical parameters of moisture loss and oil absorption (Table 1).

The mathematical model of the complete experiment design with three factors (α, β, γ) and the interaction is schematized in Equation (1):

$$Y_{ijk} = \mu + \alpha_i + \beta_j + \gamma_r + (\alpha\beta)_{ij} + (\alpha\gamma)_{ir} + (\beta\gamma)_{jr} + (\alpha\beta\gamma)_{ijk} + \varepsilon_{ijk} \tag{1}$$

2.5. Frying by Immersion Process

The cassava samples were cut in cylindrical shapes (6 cm length × 3 cm diameter) and cooked for 15 min using a Laboratory Water Bath (Memmert, Germany), controlling the temperature at 100 °C ± 2 °C. For each 100 g of cassava sample, 600 mL of potable water was used. The edible coatings (1% pectin and 9% whey protein) were applied to the cassava samples with immersion. There was a control without coatings. After the coating, the samples were dried at 25 °C, weighed, and fried by immersion. For the frying process, 10 pre-cooked cassava pieces, 4 × 2 × 2 cm, were used. An electric MKE fryer (Indianapolis, Georgetown Rd, USA) with a capacity of 5 L was used, equipped with a thermostat (scale 0–300 °C) with an accuracy of ±0.1 °C. The process conditions included frying temperatures of 140, 160, and 180 °C, with times of 60, 120, and 180 s. The chips were fried in palm oil and placed in a stainless-steel basket. They were then placed in desiccators.

2.6. Thermophysical Properties

The specific heat (Cp.), density (ρ), conductivity (k), and thermal diffusivity (α) of the samples were determined using the methodology by Choi and Okos [30]. Systematized in a computer software called DEPROTER (Determination of Thermophysical Properties) [31].

2.7. Physicochemical Parameters

The moisture (g water g^{-1} dry solids) was determined according to official method 925.09b AOAC. 3–5 g of each sample were dried in a LT04/5 convection oven at 105 °C to constant weight. Then, to determine the fat content, the dried samples were weighed, and placed in a Soxhlet extraction equipment with petroleum ether for 8 h at 50 ± 2 °C; after the extraction, a rotary evaporator was used to evaporate the solvent from the oil. Then, the fat was obtained with gravimetry and expressed in g/100 g of dry solid according to conventional method 920.39 AOAC [32].

2.8. Sensory Evaluation

For the sensory evaluation, an untrained panel of 40 panelists was used, with an age range of 20–28 years, who were habitual consumers of cassava. They were given the fried cassava chips on a plate, in an appropriate, ventilated room with controlled temperature conditions. A five-point hedonic scale was used in which the panelists indicated their degree of acceptance for the parameters of color, odor, flavor, crispness, fat and, hardness. The scale categories ranged from 1 = I dislike it very much to 5 = I like it very much. Subsequently, the data were entered into a spreadsheet and transformed into numerical scores for analysis. Uncoated cassava chips (control) and chips coated with pectin and whey protein were tested at 140, 160, and 180 °C for 180 s, for a total of nine experiments.

3. Results

3.1. Water Vapor Permeability, Solubility and Thickness of Edible Coatings

The pectin films showed higher solubility and lower water vapor permeability and thickness than the whey protein films. Analyzing the 2 and 3% pectin films showed that they were 45.6% more permeable to water vapor than films produced with 1% pectin. Additionally, the pectin films were 68.9% less permeable to water vapor than the whey protein films. The 1% pectin films had the lowest water vapor permeability and thickness, as compared to the 2% and 3% films. In addition, the 9% whey protein films had the lowest permeation and thickness, as compared to the 11 and 13% films. The pectin coatings were totally soluble in water since, after 24 h of immersion, the films were completely solubilized (Table 2). Similar results were reported by Freitas et al. [22], who also demonstrated that pectin films were completely solubilized in water. Pectin is a highly hydrophilic polysaccharide, which rapidly disintegrates in water. Batista et al. [33] similarly stated that all coatings from fatty acids and pectin were completely soluble.

Table 2. Water permeability, solubility, and thickness of edible coatings obtained from pectin and whey protein.

Coating	(%)	Water Vapor Permeability (g mm/m^2 day kPa)	Water Solubility (%)	Thickness (mm)
Pectin	1	1.95 ± 0.03 [e]	100 [a]	0.025 ± 0.001 [e]
	2	2.74 ± 0.06 [d]	100 [a]	0.052 ± 0.003 [d]
	3	2.94 ± 0.04 [d]	100 [a]	0.068 ± 0.008 [c]
Whey protein	9	6.98 ± 0.12 [c]	30.66 ± 2.65 [c]	0.103 ± 0.001 [b]
	11	7.67 ± 0.27 [b]	32.54 ± 2.93 [c]	0.118 ± 0.002 [a]
	13	9.88 ± 0.24 [a]	36.32 ± 3.14 [b]	0.113 ± 0.001 [a]

Different letters in the same column indicate significant statistical differences ($p < 0.05$).

The solubility of polysaccharide coatings in water is advantageous in situations where the film is consumed with the product, causing low alterations in the sensory properties of the food. The coatings obtained from pectin showed less permeation than with whey, which was not very resistant to water vapor, mainly because of the material's porosity, that is, coatings obtained from whey protein were more porous than those from pectin; this difference was statistically significant and could be related to the thickness of the coatings.

The lower water vapor permeability of the pectin-based coatings could be explained by the lower thickness of the coatings, as compared to the whey protein coatings [34]. Matching results were obtained by Freitas et al. [22], who used pectin and whey protein coatings. A possible explanation for the increased permeability of edible coatings made from whey in this study could be the denaturation of the proteins under the temperature conditions used for their processing, i.e., when the protein was disrupted, it was denatured; the polar amino acids were possibly exposed, which increased the absorption of water.

3.2. Moisture Loss and Fat Absorption

Table 3 shows the multifactorial ANOVA analysis of variance for cassava chips processed with different coatings, frying times, and temperatures. In general, all factors had significant statistical differences for moisture loss ($p < 0.05$). Analyzing the coating type showed a statistically significant effect ($p < 0.05$) on the cassava chips; that is, the coating type had an influence on this parameter. The same phenomenon was observed at different temperatures ($p < 0.05$). Time, on the other hand, did not present statistical differences at 120 and 180 s; that is, the additional time did not affect the increase in the loss of this parameter. Significant statistical differences ($p < 0.05$) were only observed with the shortest process time (60 s), rather than the other two times (120 and 180 s).

Table 3. Influence of coating type, time, and temperature on moisture loss of cassava chips.

Factor		*p*-Value	Contrast	+/− Limits	Difference
	Control [b]		C–P	2.05573	−18.2856 *
Type of coating	Pectin [a]	0.0000	C–WP	2.05573	4.40333 *
	Whey protein [c]		P–WP	2.05573	22.6889 *
	60 [a]		60–120	2.18411	6.06 *
Time (s)	120 [b]	0.000	60–180	2.18411	7.93444 *
	180 [b]		120–180	2.18411	1.87444
	140 [a]		140–160	2.86906	9.78889 *
Temperature (°C)	160 [b]	0.000	140–180	2.86906	14.1189 *
	180 [c]		160–180	2.86906	4.33 *

* Different letters for the same factor in the same column indicate statistical differences.

During the chip frying, statistical differences were observed at 60 s for the control sample at all temperatures and at 180 s for the control and whey protein-coated chips ($p < 0.05$). In the first 60 s of processing time, no statistical differences were observed at temperatures of 160 °C 180 °C. The same was observed at 120 s in the pectin-coated chips and the control chips; on the other hand, at 180 s, statistical differences were observed in the control sample and the whey protein-coated chips ($p < 0.05$). When comparing the uncoated fried samples, the pectin coating treatment decreased the moisture loss of the fried chips by 44.55% at 180 °C and 60 s (Figure 1). The differences in fat and moisture content between the uncoated (control) and coated fried samples were possibly due to the replacement of water by oil after the evaporation process during frying [35]. Evaporation of water by heat transfer during a frying process creates empty spaces in the food that are filled with oil, increasing the oil content of the fried food [36].

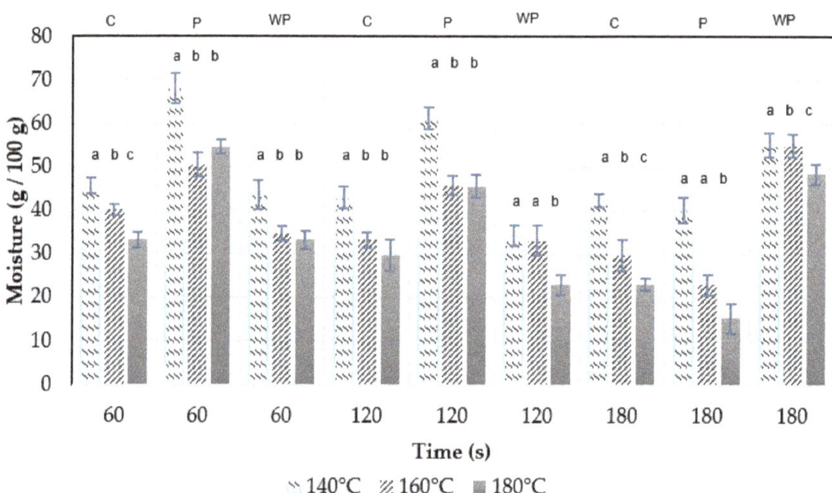

Figure 1. Moisture loss during frying of cassava chips at different temperatures and times: The first column indicates the control sample (C), the second column indicates that pectin (P) was used, and the third column indicates that whey protein (WP) was used. Different letters indicate statistical differences ($p < 0.05$) for the same frying time at different temperatures.

These results can be attributed to the fact that pectins in the outer layer of the samples possibly acted as a protective barrier that closed the pores of the product and prevented water from escaping as steam during the heat treatment, thus preventing the absorption of oil. These results are very important since they can help keep products in good condition in terms of moisture and oil. It is interesting to note that the samples coated with whey protein had a 452% higher oil content than the samples coated with pectin and fried at 160 °C for 60 s (Figure 2), which indicates that this coating was not effective in preventing the product from absorbing oil during and after the immersion frying process.

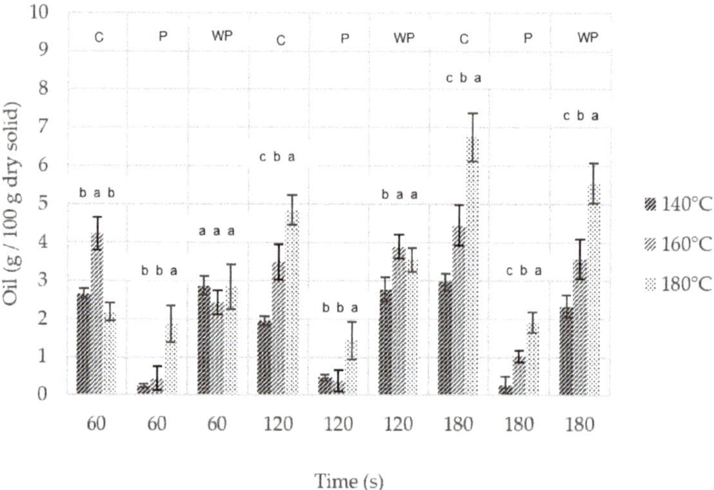

Figure 2. Fat absorption during cassava chip frying at different temperatures and times: The first column indicates the control sample (C), the second column indicates that pectin (P) was used, and the third column indicates that whey protein (WP) was used. Different letters indicate statistical differences ($p < 0.05$) for the same frying time at different temperatures.

In general, the analyses indicated that the cassava chip samples that were treated for a longer time in both the control and the whey protein coatings obtained higher percentages of oil and lower moisture contents (Figures 1 and 2). These results were similar to those reported by Tirado et al. [37] in the immersion frying of tilapia slices and by Alvis et al. [34] during the atmospheric frying of sweet potato slices coated with carboxymethyl cellulose (CMC), where longer processing times meant heat-treated matrices absorbed more oil.

The edible coating containing 1% pectin retained the most moisture in the fried samples. In other words, the use of an edible coating retains moisture in fried foods. The coating can form barriers that prevent moisture loss and reduce fat absorption [38].

The net formed by the pectin edible coating prevented moisture from escaping, thus retaining a higher moisture content in coated samples than in uncoated ones (control) [39].

Table 4 shows the influence of the coating type and frying time and temperature on the oil absorption of the cassava chips. All factors had an influence on the oil absorption of the chips, which was corroborated by the multifactorial ANOVA analysis of variance. For the coating type, significant statistical differences ($p < 0.05$) were observed for oil absorption. The same behavior occurred for frying time and temperature, observing differences between chips coated with pectin and whey protein and the control.

Table 4. Influence of coating type, time, and temperature on oil absorption in cassava chips.

Factor		*p*-Value	Contrast	+/− Limits	Difference
	Control [a]		C–P	0.334673	2.8263 *
Type of coating	Pectin [c]	0.0000	C–WP	0.334673	0.413333
	Whey protein [b]		P–WP	0.334673	−2.41296 *
	60 [c]		60–120	1.6305	−0.335556 *
Time (s)	120 [b]	0.0000	60–180	1.6305	−1.00815 *
	180 [a]		120–180	1.6305	−0.672593 *
	140 [c]		140–160	0.820096	−0.818889
Temperature (°C)	160 [b]	0.0000	140–180	0.820096	−1.60074 *
	180 [a]		160–180	0.820096	−0.782222 *

* Different letters for the same factor in the same column indicate statistical differences.

When analyzing the oil absorption at 60 s, no statistical differences were observed in the whey protein coated chips, but differences were observed in the pectin-coated chips at 180 °C and in the control at 140 °C ($p < 0.05$) (Table 3). At 120 s, the most representative differences were evidenced in the control at all process temperatures; this was also observed at 180 s in the pectin-coated chips, the control, and the whey protein-coated chips at different temperatures. When analyzing coating type, the control and whey protein-coated chips presented 321% more oil content on average at 180 °C and 180 s than the pectin-coated chips (Figure 2).

On the other hand, Freitas et al. [22] investigated the influence of the use of edible coatings from three different hydrocolloids (pectin, whey protein, and soy protein isolate) during the frying of a pre-fried and frozen cassava product. They demonstrated that whey protein showed the best results with respect to fat absorption, with a 27% reduction for the mashed cassava product. The uncoated cassava chips contained the highest oil content, 52.36%. Like the coated cassava chips with a 1% pectin concentration, the oil content of the cassava chips was reduced to 22.31%. The cassava chips treated with a 3% pectin concentration solution had the best or lowest lipid content, 11.75%, where the reduction was up to 40% for the lipid content. This possibly occurred because, during water evaporation, the vapor pressure of water increases the moisture transfer, increasing the porosity of the coating and oil ingress. In addition, high temperatures modify the structure of the material, causing damage to the coating during frying.

Also, Dragich and Krochta [40] stated that chicken strips coated with 10% denatured whey protein isolate (DWPI) resulted in a surprising 30.68% reduction in fat absorption, as compared to a non-whey protein coating. The DWPI solution possibly dehydrated during the frying process to form a film on the outer surface of the coated chicken strips.

Mechanisms that are independent of or in addition to the formation of the barrier film or a possible increase in interfacial tension between chicken with a DWPI solution surface coating and the frying oil may have been responsible for the reduction in fat absorption.

Different results were obtained by Aminlari et al. [41] with the application of whey protein on potato slices, resulting in a 5% reduction in fat absorption. The rate of oil absorption can be affected by changes in interfacial tension through the accumulation of surfactants in the frying oil. The formulation may alter the water holding capacity and consequently affect oil absorption. As a result of the strong hydrogen bonding between water molecules with hydrophilic materials, the displacement of water by oil/fat during the frying process is restricted.

3.3. Thermophysical Properties of the Cassava Chip Control

In general, both the thermal conductivity and diffusivity increased with processing temperature (Table 5), which was due to the fact that the food material conducted heat better, possibly because a temperature increase resulted in dehydration of the product and therefore faster heating. The density, heat capacity, and thermal diffusivity showed statistical differences at all temperatures ($p < 0.05$), indicating that this variable had an effect on these thermal properties. The conductivity increased with increasing temperatures, but no significant statistical differences were observed, which is why temperature had no effect on this parameter [42]. This property is related to the decrease in moisture loss. Sahin et al. [43] stated that the thermal conductivity of food materials depends on porosity, structure, and chemical constituents.

Table 5. Thermophysical properties of the cassava chip control.

Property	Units	Temperatures			
		25 °C	140 °C	160 °C	180 °C
Conductivity	W/m °C	0.47 ± 0.07 [a]	0.48 ± 0.06 [a]	0.49 ± 0.27 [a]	0.53 ± 0.19 [a]
Density	Kg/m^3	1043.11 ± 0.52 [d]	1120.38 ± 0.33 [c]	1124.73 ± 0.9 [b]	1138.44 ± 0.16 [a]
Heat capacity	J/Kg °C	3111.84 ± 0.41 [d]	3118.66 ± 0.98 [c]	3155.75 ± 0.76 [b]	3176.86 ± 0.24 [a]
Thermal Diffusivity	m^2/s	$1.09 \times 10^{-7} \pm 0.30$ [d]	$1.38 \times 10^{-7} \pm 0.10$ [c]	$1.58 \times 10^{-7} \pm 0.67$ [b]	$1.69 \times 10^{-7} \pm 0.66$ [a]

Letters in the same row at different temperatures for each property indicate significant statistical differences ($p < 0.05$).

Several researchers have confirmed that thermophysical properties are important parameters in the description of heat transfer during the heating of solid foods, providing great advantages for information gathering, especially for energy costs and quality assurance in different products [31,44]. It was observed that temperature significantly ($p < 0.05$) affected density, a result that may be related to moisture loss and oil absorption. It was also observed that temperature had a significant influence ($p < 0.05$) on heat capacity and thermal diffusivity. These two thermal properties increased with increasing temperatures. The results of this study were different from those reported by Sosa-Morales et al. [45] during the frying of pork meat, who stated that heat capacity and thermal diffusivity decreased with increasing frying time because of moisture loss.

3.4. Sensory Analysis

Table 6 shows that the coating type affected all sensory parameters, except crispness, which was corroborated by the significant statistical differences ($p < 0.05$). It was also evidenced that temperature only had an influence on the color of the control chips, with statistical differences ($p < 0.05$) at all temperatures.

When analyzing the color of the cassava chips, the highest scores for this parameter were observed at intermediate temperatures (160 °C) with pectin coatings; statistical differences were also observed ($p < 0.05$), this same result was evidenced in the odor (Table 7), which is a parameter that produces a sensation because of volatile substances. At a temperature of 140 °C, no statistical differences were observed in the odor of the cassava chips, and the lowest scores were obtained with the whey protein coated chips. No statistical differences in chip flavor were observed at temperatures of 140 °C and 180 °C;

however, higher scores were observed at 160 °C with the uncoated chips. The greasiness did not show significant differences between the three temperature groups. The crispness of the cassava chips showed significant statistical differences ($p < 0.05$) at 180 °C between the control and the pectin-coated chips, and both coatings had differences($p < 0.05$). The hardness had statistical differences ($p < 0.05$) at 140 °C between the pectin-coated chips and the other treatments. No statistical differences were observed between the control and the whey protein coated chips. The same results were observed at 180 °C. At 160 °C, no significant statistical differences were observed (Table 7).

Table 6. Influence of coating type and temperature on sensory parameters of cassava chips.

Factor	*p*-Value	Color	*p*-Value	Odor	*p*-Value	Flavor	*p*-Value	Greasiness	*p*-Value	Crispness	*p*-Value	Hardness
Type of coating	0.0000	C [a] P [a] WP [a]	0.0512	C [a] P [a] WP [b]	0.0003	C [b] P [a] WP [c]	0.0268	C [b] P [a] WP [b]	0.3574	C [a] P [a] WP [a]	0.0001	C [b] P [a] WP [b]
Temperature (°C)	0.0648	140 [a] 160 [b] 180 [c]	0.4364	140 [a] 160 [a] 180 [a]	0.5617	140 [a] 160 [a] 180 [a]	0.5531	140 [a] 160 [a] 180 [a]	0.6184	140 [a] 160 [a] 180 [a]	0.1961	140 [a] 160 [a] 180 [a]

Different letters in the same column for each factor indicate significant statistical differences. Control chips (C), pectin-coated chips (P) and whey protein chips (WP).

Table 7. Sensory analysis of control cassava chips coated with pectin and whey protein at different frying temperatures.

Experiments	Factors			Sensory Parameters					
	T °C	T (s)	Coating	Color	Odor	Flavor	Greasiness	Crispness	Hardness
1	140	180	Control	3.59 ± 0.58 [c]	3.86 ± 0.78 [a]	3.74 ± 0.38 [a]	3.73 ± 0.65 [a]	3.67 ± 0.52 [a]	3.02 ± 0.34 [b]
2	140	180	Pectin	4.54 ± 0.23 [a]	3.76 ± 0.35 [a]	3.61 ± 0.38 [a]	4.09 ± 0.84 [a]	3.59 ± 0.58 [a]	4.34 ± 0.37 [a]
3	140	180	Whey protein	3.67 ± 0.52 [ab]	3.62 ± 0.51 [a]	3.47 ± 0.45 [a]	3.86 ± 0.22 [a]	4.14 ± 0.43 [a]	3.37 ± 0.28 [b]
4	160	180	Control	3.65 ± 0.43 [b]	4.52 ± 0.37 [b]	4.78 ± 0.19 [a]	3.62 ± 0.36 [a]	3.72 ± 0.88 [a]	3.74 ± 0.44 [a]
5	160	180	Pectin	4.48 ± 0.25 [a]	4.08 ± 0.64 [a]	4.21 ± 0.53 [a]	4.44 ± 0.42 [a]	4.34 ± 0.28 [a]	4.02 ± 0.34 [a]
6	160	180	Whey protein	3.25 ± 0.26 [b]	3.67 ± 0.56 [b]	2.27 ± 0.77 [b]	3.83 ± 0.36 [a]	3.92 ± 0.44 [a]	3.85 ± 0.31 [a]
7	180	180	Control	3.48 ± 0.57 [a]	3.92 ± 0.84 [ab]	4.13 ± 0.55 [a]	3.83 ± 0.36 [a]	3.85 ± 0.11 [b]	3.12 ± 0.73 [b]
8	180	180	Pectin	4.45 ± 0.37 [a]	4.43 ± 0.56 [a]	4.00 ± 0.34 [a]	4.69 ± 0.20 [a]	4.27 ± 0.25 [a]	4.83 ± 0.11 [a]
9	180	180	Whey protein	2.13 ± 0.94 [b]	3.09 ± 0.47 [b]	3.46 ± 0.66 [a]	3.93 ± 0.74 [a]	3.86 ± 0.26 [b]	3.77 ± 0.45 [b]

Different letters in the same column indicate statistical differences between the three coatings tested at each temperature.

In the case of color, higher scores (4.54) were observed at intermediate temperatures (160 °C) (Table 7). The coating favored browning and coloring reactions in the cassava chips. Similar results were obtained by Muhamad and Shaharuddin [46], indicating that chips coated with 2% pectin had the highest score for this parameter; the pectin coating did not change the initial color during frying of the cassava chips, and the characteristic golden color of the chips can be attributed to non-enzymatic browning of the starch. On the other hand, the best rating for odor(4.52) was in the control samples, which was possibly due to the fact that, under these conditions, the chemical components responsible for flavor were enhanced, which was also found by Aguirre et al. [47] in chips of different banana varieties.

Regardless of the type of coating used and the temperature applied, higher greasiness acceptance was observed in those samples coated with pectin, with higher scores at 180 °C; i.e., at higher temperatures, the oil absorption in the cassava chips was reduced. Muhamad and Shaharuddin [47] found that pectin-coated cassava chips samples absorbed less oil than uncoated chips. The substantial reduction in oil absorption can be attributed mainly to the barrier properties because coatings make the surface stronger and more brittle, with fewer small pores, which reduces evaporation and leads to less fat absorption. The crispness did not present statistical differences ($p < 0.05$) in any of the treatments.

4. Conclusions

The pectin films showed higher solubility and lower water vapor permeability and thickness than the whey protein films. Moreover, the pectin films were 68.9% less water vapor permeable than the whey protein films. All factors influenced the moisture loss and oil absorption of the cassava chips. When analyzing the coating type, the control and whey protein-coated chips presented a 321% greater oil content on average at 180 °C and 180 s than the pectin-coated chips. The thermal properties were affected by process temperatures, except for conductivity. The sensory analysis showed that the coating type affected all sensory parameters, except crispness. Temperature only had an effect on the color of the control chips. It would be interesting to conduct future studies using vacuum frying, with different hydrocolloids and oils, to further study the physicochemical properties of cassava chips.

Author Contributions: D.A.-C. methodology, formal analysis, writing—original draft preparation. J.J.-M. validation, formal analysis, writing—original draft preparation. P.M.M.-C. supervision, software, writing—original draft preparation. All authors have read and agreed to the published version of the manuscript.

Funding: This research received no external funding.

Institutional Review Board Statement: Not applicable.

Informed Consent Statement: Not applicable.

Data Availability Statement: The data used to support this study are available from the corresponding author.

Acknowledgments: In this section, you can acknowledge any support given which is not covered by the author contribution or funding sections. This may include administrative and technical support, or donations in kind (e.g., materials used for experiments).

Conflicts of Interest: The authors declare no conflict of interest.

References

1. Bayata, A. Review on Nutritional Value of Cassava for Use as a Staple Food. *Sci. J. Anal. Chem.* **2019**, *7*, 83–91. [CrossRef]
2. Gil, J.L.; Buitrago, A.J.A. La yuca en la alimentación animal. In *La Yuca en el Tercer Milenio: Sistemas Modernos de Producción, Procesamiento, Utilización y Comercialización*, 1st ed.; Ospina, B., Ceballos, H., Alvarez, E., Bellotti, A.C., Lee, C., Arias, B., Cadavid, L.F., Pineda, B., Llano, G.A., Eds.; CIAT: Cali, Colombia, 2002; Volume 1, pp. 527–569.
3. Tewe, O.O.; Lutaladio, N. *The Global Cassava Development Strategy. Cassava for Livestock Feed in Sub-Saharan Africa*, 1st ed.; FAO: Rome, Italy, 2004; pp. 55–57.
4. Nambisan, B.; Sundaresan, S. Effect of processing on the cyanoglucoside content of cassava. *J. Sci. Food Agric.* **1985**, *36*, 1197–1203. [CrossRef]
5. Oyewole, O.B. Cassava processing in Africa. In *Application of Biotechnology in Traditional Fermented Foods. Report of an Ad-hoc Panel of the Board on Science and Technology for International Development*, 1st ed.; Office of International Affairs National Research Council; National Academies Press: Washington, DC, USA, 1992; pp. 89–92.
6. Sukara, E.; Hartati, S.; Ragamustari, S.K. State of the art of Indonesian agriculture and the introduction of innovation for added value of cassava. *Plant Biotechnol. Rep.* **2020**, *14*, 207–212. [CrossRef]
7. Bouchon, P.; Aguilera, J.M.; Pyle, D.L. Structure oil absorption relationships during deep-fat frying. *J. Food Sci.* **2003**, *68*, 2711–2716. [CrossRef]
8. Zamani-Ghalehshahi, A.; Farzaneh, P. Effect of hydrocolloid coatings (Basil seed gum, xanthan, and methyl cellulose) on the mass transfer kinetics and quality of fried potato strips. *J. Food Sci.* **2021**. [CrossRef]
9. Moreira, R. Deep fat frying. In *Heat Transfer in Food Processing: Recent Developments and Applications*, 1st ed.; Yanniotis, S., Sunden, B., Eds.; WIT Press: Boston, MA, USA, 2007; pp. 209–236.
10. Asokapandian, S.; Swamy, G.J.; Hajjul, H. Deep fat frying of foods: A critical review on process and product parameters. *Crit. Rev. Food Sci. Nutr.* **2020**, *60*, 3400–3413. [CrossRef]
11. Liu, Y.; Sun, L.; Bai, H.; Ran, Z. Detection for Frying Times of Various Edible Oils Based on Near-Infrared Spectroscopy. *Appl. Sci.* **2020**, *10*, 7789. [CrossRef]
12. Kita, A.; Nowak, J.; Michalska-Ciechanowska, A. The Effect of the Addition of Fruit Powders on the Quality of Snacks with Jerusalem Artichoke during Storage. *Appl. Sci.* **2020**, *10*, 5603. [CrossRef]
13. Marciniak-Lukasiak, K.; Zbikowska, A.; Marzec, A.; Kozlowska, M. The Effect of Selected Additives on the Oil Uptake and Quality Parameters of Fried Instant Noodles. *Appl. Sci.* **2019**, *9*, 936. [CrossRef]
14. Lumanlan, J.C.; Fernando, W.M.A.D.B.; Jayasena, V. Mechanisms of oil uptake during deep frying and applications of predrying and hydrocolloids in reducing fat content of chips. *Int. J. Food Sci. Technol.* **2020**, *55*, 1661–1670. [CrossRef]
15. Clemens, R.A.; Pressman, P. Food gums: An overview. *Nutr. Today* **2017**, *52*, 41–43. [CrossRef]

16. Al-Asmar, A.; Naviglio, D.; Giosafatto, C.V.L.; Mariniello, L. Hydrocolloid-based coatings are effective at reducing acrylamide and oil content of French fries. *Coatings* **2018**, *8*, 147. [CrossRef]

17. Jiang, Y.; Qin, R.; Jia, C.; Rong, J.; Hu, Y.; Liu, R. Hydrocolloid effects on Nε-carboxymethyllysine and acrylamide of deep-fried fish nuggets. *Food Biosci.* **2021**, *39*, 100797. [CrossRef]

18. Devi, S.; Zhang, M.; Law, C.L. Effect of ultrasound and microwave assisted vacuum frying on mushroom (*Agaricus bisporus*) chips quality. *Food Biosci.* **2018**, *25*, 111–117. [CrossRef]

19. Ajo, R.Y. Application of hydrocolloids as coating films to reduce oil absorption in fried potato chip-based pellets. *Pak. J. Nutr.* **2017**, *16*, 805–812. [CrossRef]

20. Kurek, M.; Ščetar, M.; Galić, K. Edible coatings minimize fat uptake in deep fat fried products: A review. *Food Hydrocoll.* **2017**, *71*, 225–235. [CrossRef]

21. Bouchon, P. Understanding oil absorption during deep-fat frying. In *Advances in Food and Nutrition Research*, 1st ed.; Toldra, F., Ed.; Academic Press: Cambridge, UK, 2009; Volume 57, pp. 209–234.

22. Freitas, D.D.G.C.; Berbari, S.A.G.; Prati, P.; Fakhouri, F.M.; Queiroz, F.P.C.; Vicente, E. Reducing fat uptake in cassava product during deep-fat frying. *J. Food Eng.* **2009**, *94*, 390–394. [CrossRef]

23. Lucas, J.C.; Quintero, V.D.; Leal, J.F.V.; Núñez, L.C. Evaluación de los parámetros de calidad durante la fritura de rebanadas de papa criolla. *Sci. Technol.* **2011**, *2*, 299–304. [CrossRef]

24. Varela, P.; Fiszman, S.M. Hydrocolloids in fried foods. *A review. Food Hydrocoll.* **2011**, *25*, 1801–1812. [CrossRef]

25. Campos, C.A.; Gerschenson, L.N.; Flores, S.K. Development of edible films and coatings with antimicrobial activity. *Food Bioprocess Technol.* **2011**, *4*, 849–875. [CrossRef]

26. Porta, R.; Mariniello, L.; Di Pierro, P.; Sorrentino, A.; Giosafatto, C.V.L. Transglutaminase crosslinked pectin and chitosan-based edible films: A review. *Crit. Rev. Food Sci. Nutr.* **2011**, *51*, 223–238. [CrossRef] [PubMed]

27. Falguera, V.; Quintero, J.P.; Jiménez, A.; Muñoz, J.A.; Ibarz, A. Edible films and coatings: Structures, active functions and trends in their use. *Trends Food Sci. Technol.* **2011**, *22*, 292–303. [CrossRef]

28. Marquez, G.R.; Di Pierro, P.; Esposito, M.; Mariniello, L.; Porta, R. Application of transglutaminase-crosslinked whey protein/pectin films as water barrier coatings in fried and baked foods. *Food Bioprocess Technol.* **2014**, *7*, 447–455. [CrossRef]

29. Gontard, N.; Guilbert, S.; CUQ, J.L. Edible wheat gluten films: Influence of the main process variables on film properties using response surface methodology. *J. Food Sci.* **1992**, *57*, 190–195. [CrossRef]

30. Choi, Y.; Okos, M. Effects of temperature and composition on the thermal properties of foods. *J. Food Process Appl.* **1986**, *1*, 93–101.

31. Alvis, A.; Caicedo, I.; Peña, P. Determinación de Propiedades Termofísicas de Alimentos en Función de la Concentración y la Temperatura empleando un Programa Computacional. *Inf. Technol.* **2012**, *23*, 111–116. [CrossRef]

32. Association International of Analytical Chemists. *Official Methods of Analysis*, 18th ed.; Horwitz, W., Ed.; AOAC International: Gaithersburg, MD, USA, 2005.

33. Batista, J.A.; Tanada-Palmu, P.S.; Grosso, C.R.F. Efeito da adição de ácidos graxos em filmes à base de pectina. *Ciênc. Tecnol. Aliment.* **2005**, *25*, 781–788. [CrossRef]

34. Alvis, A.; González, A.; Arrázola, G. Efecto del Recubrimiento Comestible en las Propiedades de Trozos de Batata (Ipomoea Batatas Lam) Fritos por Inmersión, Parte 2: Propiedades Termofísicas y de Transporte. *Inf. Technol.* **2015**, *26*, 103–116. [CrossRef]

35. Khazaei, N.; Esmaiili, M.; Emam-Djomeh, Z. Effect of active edible coatings made by basil seed gum and thymol on oil uptake and oxidation in shrimp during deep-fat frying. *Carbohydr. Polym.* **2016**, *37*, 249–254. [CrossRef]

36. Kim, D.N.; Lim, J.; Bae, I.Y.; Lee, H.G.; Lee, S. Effect of hydrocolloid coatings on the heat transfer and oil uptake during frying of potato strips. *J. Food Eng.* **2011**, *102*, 317–320. [CrossRef]

37. Tirado, D.; Acevedo, D.; Montero, P. Transferencia de Calor y Materia durante el Freído de Alimentos: Tilapia (*O. niloticus*) y Fruta de Pan (*Artocarpus communis*). *Inf. Technol.* **2015**, *26*, 85–94. [CrossRef]

38. Singthong, J.; Thongkaew, C. Using hydrocolloids to decrease oil absorption in banana chips. *LWT Food Sci. Technol.* **2009**, *42*, 1199–1203. [CrossRef]

39. Ananey-Obiri, D.; Matthews, L.; Tahergorabi, R. Chicken processing by-product: A source of protein for fat uptake reduction in deep-fried chicken. *Food Hydrocoll.* **2020**, *101*, 105500. [CrossRef]

40. Dragich, A.M.; Krochta, J.M. Whey protein solution coating for fat-uptake reduction in deep-fried chicken breast strips. *J. Food Sci.* **2010**, *75*, S43–S47. [CrossRef]

41. Aminlari, M.; Ramezani, R.; Khalili, M.H. Production of protein-coated low-fat potato chips. *Food Sci. Technol. Int.* **2005**, *11*, 177–181. [CrossRef]

42. Ziaiifar, A.M.; Heyd, B.; Courtois, F. Investigation of effective thermal conductivity kinetics of crust and core regions of potato during deep-fat frying using a modified Lees method. *J. Food Eng.* **2009**, *95*, 373–378. [CrossRef]

43. Sahin, S.E.R.P.İ.L.; Sastry, S.K.; Bayindirli, L. Heat transfer during frying of potato slices. *LWT Food Sci. Technol.* **1999**, *32*, 19–24. [CrossRef]

44. Arrazola, G.; Paez, M.; Alvis, A. Composición, Análisis termofísico y análisis sensorial de frutos colombianos: Parte 1: Almendro (*Terminalia Catappa* L.). *Inf. Technol.* **2014**, *25*, 17–22. [CrossRef]

45. Sosa-Morales, M.E.; Orzuna-Espíritu, R.; Vélez-Ruiz, J.F. Mass, thermal and quality aspects of deep-fat frying of pork meat. *J. Food Eng.* **2006**, *77*, 731–738. [CrossRef]

46. Muhamad, I.I.; Shaharuddin, S. Evaluation on Quality Attributes of Pectin Coated-Cassava Chips. *Mater. Today* **2019**, *19*, 1473–1480. [CrossRef]
47. Aguirre, J.C.L.; Castaño, V.D.Q.; Leal, J.F.V.; Artamonov, J.D.M. Evaluación de los parámetros de calidad de chips en relación con diferentes variedades de plátano (*Musa paradisiaca* L.). *Rev. Lasallista* **2012**, *9*, 65–74.

Article

Application of High-Frequency Defrosting, Superheated Steam, and Quick-Freezing Treatments to Improve the Quality of Seafood Home Meal Replacement Products Consisting of the Adductor Muscle of Pen Shells and Common Squid Meat

Bertoka Fajar Surya Perwira Negara [1,2], Seung Rok Kim [1], Jae Hak Sohn [1,2], Jin-Soo Kim [3,*] and Jae-Suk Choi [1,2,*]

1 Seafood Research Center, IACF, Silla University, 606, Advanced Seafood Processing Complex, Wonyang-ro, Amnam-dong, Seo-gu, Busan 49277, Korea; ftrnd12@silla.ac.kr (B.F.S.P.N.); ftrnd4@silla.ac.kr (S.R.K.); jhsohn@silla.ac.kr (J.H.S.)
2 Department of Food Biotechnology, College of Medical and Life Sciences, Silla University, 140, Baegyang-daero 700beon-gil, Sasang-gu, Busan 46958, Korea
3 Department of Seafood and Aquaculture Science, Gyeongsang National University, 38 Cheondaegukchi-gil, Tongyeong-si, Gyeongsangnam-do 53064, Korea
* Correspondence: jinsukim@gnu.ac.kr (J.-S.K.); jsc1008@silla.ac.kr (J.-S.C.); Tel.: +82-557-729-146 (J.-S.K.); +82-512-487-789 (J.-S.C.)

Citation: Negara, B.F.S.P.; Kim, S.R.; Sohn, J.H.; Kim, J.-S.; Choi, J.-S. Application of High-Frequency Defrosting, Superheated Steam, and Quick-Freezing Treatments to Improve the Quality of Seafood Home Meal Replacement Products Consisting of the Adductor Muscle of Pen Shells and Common Squid Meat. *Appl. Sci.* **2021**, *11*, 2926. https://doi.org/10.3390/app11072926

Academic Editor: Theodoros Varzakas

Received: 26 February 2021
Accepted: 22 March 2021
Published: 25 March 2021

Publisher's Note: MDPI stays neutral with regard to jurisdictional claims in published maps and institutional affiliations.

Abstract: We developed a new seafood home meal replacement (HMR) product containing the adductor muscle of the pen shell (AMPS) and common squid meat (CSM) via high-frequency defrosting (HFD), superheated steam, and quick freezing. Test HMR products were produced by mixing defrosted and roasted AMPS, CSM, and sauce in ratios of 27.5, 27.5, and 45.0% (w/w), respectively, followed by quick freezing at $-35\ °C$ in a polypropylene plastic bowl covered with a plastic film. The chemical characteristics, nutritional quality, microbial and sensory properties, and shelf life of the product were examined. The response surface methodology identified the optimal temperature and heating time of the superheated steam for AMPS (220 °C, 1 min) and CSM (300 °C, 1.5 min). Chemical characteristics showed low levels of volatile basic nitrogen (9.45 mg%) and thiobarbituric acid-reactive substances (1.13 mg Malondialdehyde [MDA]/kg). No significant changes ($p < 0.05$) were observed in microbial, color, flavor, taste, texture, and overall acceptance at $-23\ °C$ for 90 days. After reheating, the sensory scores varied from "like moderately" to "like very much." The shelf life of the HMR product was estimated to be 24 months. In conclusion, HFD, superheated steam, and quick freezing successfully improved product quality, with little loss of nutrition and texture.

Keywords: HMR; pen shell; squid meat; superheated steam; high-frequency defrosting

1. Introduction

In recent decades, people have dramatically changed the way they cook food because of the changes in attitude towards food. In the last few years, with increasing lifestyles, the number of people choosing to embrace these convenient food products at the expense of taste and health has steadily increased. Consequently, the popularity of packaged food and instant meals has revolutionized the packaged food industry, with manufacturers seeking new methods to offer fresh and delicious food without inconvenience. Nowadays, home meal replacement (HMR) products that are simple meals to prepare and consume have become increasingly popular in the food and beverage industry. HMR products have the advantage of long shelf life while maintaining the nutrient content of the food. Globally, the demand for HMR products has increased due to increasing single-person households, women's social activities, and the population of elderly people.

Increases in HMR products have encouraged the development of new high-quality seafood HMR products with a long shelf life; these contain the seasoned adductor muscle of the pen shell (AMPS) and common squid meat (CSM). Pen shell (*Atrina pectinata*), a bivalve and common squid (*Todarodes pacificus*), a cephalopod, are two popular and commercially important species. Globally, these organisms are distributed from southeast Africa to Malaysia and New Zealand, the Indo-western Pacific region, Japan, and Korea [1,2]. However, South Korea and Japan are the largest markets for raw and processed products of these two species [3]. *Khi-Jo-Gae* and *O-Jing-Eo* are popular names for *A. pectinate* and *T. pacificus* in Korea. Due to their high nutritional value, these species are often found on the seafood menu. For instance, the AMPS and CSM, the protein-rich, highly nutritional products of pen shell and common squid, have high consumption demand.

As the quality of seafood HMR products continues to improve, it is believed that applying advanced technologies to handle raw material, cook, and freeze during the manufacturing process is likely to result in excellent quality products [4]. To this end, several technologies, including high-frequency defrosting (HFD), superheated steam cooking, and quick freezing techniques, are believed to enhance product quality. HFD, in which the amount of heat generated inside the product and the defrosting time are accurately controlled, can reduce thawing time, inhibit microbial activities, reduce drip loss, and maintain the quality of raw materials [5,6]. Furthermore, oxidation is minimized in superheated steaming due to the lack of oxygen during heating and roasting [7], rendering superheated steam cooking an effective method in the food industry. In addition, superheated steam roasting reduces energy consumption during the roasting process [8] and has been proven to reduce lipid oxidation and preserve food nutrient substances, color, and texture better than traditional cooking methods [7,9–11]. Quick freezing is a key technology for maintaining the quality and prolonging the shelf life of frozen food. This technology successfully inhibits changes in flavor, color, and texture due to oxidative, enzymatic, and microbial changes [12–14].

Though these techniques have been explored in several food products, their applications in seafood-based HMR products are limited. Therefore, we hypothesized that the development of new high-quality fresh product-equivalent seafood HMR products, especially those containing the seasoned AMPS and CSM, could be achieved by employing advanced techniques. Moreover, the shelf life of packaged seafood products is an important factor determining the consumer acceptability and saleability of the product. Therefore, it is essential to decipher the ideal conditions that could prolong the shelf life of HMR products while improving their quality.

In this study, we aimed to elucidate the best method to produce good quality, highly nutritious HMR products with an improved shelf life prepared by mixing roasted AMPS and roasted CSM. Here, we produced the seasoning-mixed AMPS and CSM as test HMR products using a high-frequency defroster, superheated steam, and quick freezing techniques and evaluated its quality characteristics of test HMR products. The findings could unravel the potential of improved technologies, which could be used in the seafood industry to produce high-quality HMR products with little loss of nutrition and texture.

2. Materials and Methods

2.1. Materials

All chemicals used in this study, including potassium carbonate, sulfuric acid, boric acid, sodium hydroxide, trichloroacetic acid, phosphoric acid, N-tert-butyldimethylsilyl-N-methyltrifluoroacetamide with 1% Butyldimethylchlorosilane (MTBSTFA with 1% t-BDMCS), methyl red solution, methylene blue solution, acetonitrile, 2-thiobarbituric acid, ethanol, and sodium bicarbonate, were purchased from Sigma-Aldrich, Inc. (St. Louis, MO, USA). The Difco plate count agar and EC medium used for microbiological analysis were purchased from BD Co. (Franklin Lakes, NJ, USA), whereas Sanita-Kun plates were obtained from JNC Corp. (Tokyo, Japan).

2.2. Experimental Sample

Frozen AMPS and CSM were obtained from EBADA Fishery Co. Ltd. (Busan, Korea). Sample weights were 46–60 g (average 53.7 g) for AMPS and 188–266 g (average 221.8 g) for CSM.

2.3. Drip Loss Analysis

Frozen AMPS and CSM were thawed under three different conditions: room temperature, running water, and HFD. For running water and room temperature thawing, samples were placed in plastic bags. For HFD thawing, the controller of the HFD machine (CHRFT-100, Chamco Co. Ltd., Busan, Korea) was set at 27 MHz for 10 min (with an input power of 11 kW). Drip loss was analyzed using the filter-paper wetness (FPW) method following Kauffman et al. [15]. Quantitative filter paper No. 2 (55 mm; Advantech, Tokyo, Japan) was weighed (y), placed on the samples during thawing, and then filter paper with absorbed fluid was weighed again (x). The weight difference of filter paper was expressed as the weight of the absorbed exudate. Drip loss was quantified as a percentage following this formula:

$$Drip\ loss\ (\%) = \frac{x - y}{x} \times 100$$

2.4. Roasting Treatment

Thawed AMPS and CSM were cut into approximately 3.5 × 0.5 cm and 1 × 5 cm pieces, respectively. The inedible part of the CSM was removed, and the samples were then washed with tap water. Next, all samples were roasted using a superheated steam in an Aero Steam Oven (DFC-560A-2R/L, Naomoto Co., Osaka, Japan) at different temperatures and roasting times. For AMPS, the temperatures used were 192, 200, 220, 240, and 248 °C for 0.53, 0.67, 1.00, 1.33, and 1.47 min, respectively, while for CSM, the temperatures used were 272, 280, 300, 320, and 328 °C for 0.79, 1.00, 1.50, 2.00, and 2.21 min, respectively.

2.5. Sample HMR Product Preparation

First, the stock of sauce was prepared by mixing the ingredients shown in Table 1. A test seafood HMR product was then produced by mixing roasted AMPS, roasted CSM, and sauce at ratios (w/w) of 27.5, 27.5, and 45.0%, respectively. Afterward, 180 g of product was packaged in a polypropylene plastic bowl (New Ecopack Co. Ltd., Jeonju, Korea) and sealed with a plastic film using a tray sealing machine (TPS-TS3T, TPS Co. Ltd., Kyungki-do, Korea) at 180 °C for 5 s. Next, packaged products were frozen using a quick freezer (QF-700, Alpha Tech Co. Ltd., Incheon, Korea) at −35 °C for 10 min, and then stored at −13, −18, and −23 °C in a deep freezer (DF35035, IlShin BioBase Co. Ltd., Dongducheon, Korea) for shelf life estimation.

Table 1. Ingredients used to prepare the sauce (stock), a key ingredient of the new home meal replacement (HMR) product.

Ingredients	(%)
Mixed soy sauce	24.69
Food additive (D-sorbitol liquid)	49.19
Fructose	9.27
Caramel food coloring	Trace
Beverage base (lemon powder)	Trace
Fruit whine	Trace
Starch	Trace
Sodium L-glutamate	Trace
Chilli powder	Trace
Food additive (ethyl p-hydroxybenzoate)	Trace
Purified water	Adequate amount (~16%)

2.6. pH Measurement

Two grams of product were added to 3rd-distilled water at a ratio of 1:9 (w/v), and then homogenized using a homogenizer (SHG-15D, SciLab Co. Ltd., Seoul, Korea). Next, the pH of each sample was measured using a pH meter (ST 3100, Ohaus Co., Parsippany, NJ, USA).

2.7. Measurement of Volatile Basic Nitrogen (VBN)

VBN was performed using the Conway micro diffusion method. Five grams of product were diluted with 25 mL of 3rd-distilled water and homogenized using a vortex for 5 min. Filter paper No. 2 (55 mm; Advantech, Tokyo, Japan) was used to filter the mixture sample. Following filtration, potassium carbonate was added to the filtrate solution at a ratio of 1:1 (v/v) in the outer chamber, while 1 mL of 0.01 M H_2SO_4 was added to the inner chamber of the Conway unit. Conway cells were incubated at 37 °C for 90 min. Brunswick reagent (2–3 drops) was added to the inner chamber and titrated with 0.01 N NaOH.

2.8. Measurement of Thiobarbituric Acid-Reactive Substances (TBARS)

TBARS were measured following the method by Peiretti et al. [16] with modifications. Briefly, 5 g of each sample was homogenized (SHG-15D, SciLab Co. Ltd., Seoul, Korea) in 12.5 mL of TCA solution containing 20% trichloroacetic acid in 2 M phosphoric acid and adjusted to 25 mL with 3rd-distilled water. Next, homogenate samples were centrifuged at 1500 rpm for 10 min, and the upper layer was collected and filtered using filter paper No. 2 (55 mm; Advantech, Tokyo, Japan). The supernatant was then mixed with a 0.005 M thiobarbituric acid solution at a ratio of 1:1 (v/v) and incubated at 95 °C for 30 min in a water bath (JSWB-22TL, JS Research Inc., Gongju City, Korea). The samples were then left to cool to room temperature. Samples (200 µL) and the blank group (3rd-distilled water) were placed on a 96-well plate and measured at 530 nm using a SPECTROstar Nano Microplate Reader (S/N601-0618, BMG Labtech Ltd., Ortenberg, Germany). Malondialdehyde (MDA) bis (dimethyl acetal) was used as the standard.

2.9. Proximate Analysis

Proximate analysis was performed using AOAC standard methods. Moisture was measured following the methods AOAC 952.08 [17] using an oven at 105 °C for 24 h. Ash was determined following the AOAC 938.08 method [18] in a furnace at 550 °C. Sodium was measured following AOAC 971.27 method [19]. Crude protein was determined following AOAC 960.48 method [20]. N content from the test product was then multiplied with 6.26 to obtain the value of crude protein. Calories, carbohydrates, sugars, dietary fiber, crude fat content, cholesterol, vitamin D, potassium iron, and calcium were measured following AOAC 971.10 [21], AOAC 998.18 [22], AOAC 985.29 [23], AOAC 948.15 [24], AOAC 994.10 [25], AOAC 936.14 [26], AOAC 2011.11 [27], AOAC 990.05 [28], and AOAC 984.27 [29], respectively.

2.10. Amino Acid Analysis

Prior to hydrolysis, the protein was extracted following the Kjeldahl method, and the amino acid analysis was performed using the AOAC 994.12b method [30]. Briefly, a 50 µL aliquot of a solution containing a mixture of 91 µg/mL L-amino acids in 0.1 N HCl was dried. Subsequently, 100 µL of neat MTBSTFA, followed by 100 µL of acetonitrile, were added. The mixture was then heated to 100 °C for 4 h. Next, the sample was neutralized with sodium bicarbonate and subjected to gas chromatography-mass spectrometry (GC-MS) analysis using the GCMS-QP2020 system (Shimadzu Corp., Kyoto, Japan). An injection volume of 0.5 µL/min was used for the separations performed on an SLB™-5ms Capillary GC Column (20 m × 0.18 mm I.D., 0.18 µm; Sigma-Aldrich, Inc., St. Louis, MO, USA) using helium as the carrier gas. During the separation, the oven temperature was programmed as follows—initially, it was set at 100 °C for 1 min, then increased to 290 °C at 35 °C/min and held for 3 min, and finally, it was raised to 360 °C at a rate of 40 °C/min and held for

2 min. The temperature for the inlet was set at 250 °C, while the temperature for the mass storage device (MSD) interface was set at 325 °C.

2.11. Fatty Acid Composition Analysis

The fatty acid composition analysis was performed following a hydrolytic method described by Sutikno et al. [11]. Briefly, ether was used to extract fatty acids and methylate them into fatty acid methyl esters (FAMEs). FAMEs were then analyzed using gas chromatography (GCMS-QP2020) fitted with a DB-wax capillary column (30 m × 0.25 mm i.d., 0.25 mm film thickness, Agilent). Helium at a constant linear velocity of 30 cm/s was used as the carrier gas. The split ratio was set at 1/10. During the separation, the column oven was programmed as follows: initial column oven temperature was set at 100 °C; held for 1 min, and increased to at 25 °C/min to 100 °C and after 1 min, and finally, it was raised to 240 °C at 5 °C/min; held for 2 min. The temperature for the inlet was set at 250 °C, while the temperature for Flame Ionization Detector (FID) was set at 270 °C. FAME standard mixture (EN 14078, Paragon Scientific Ltd., Wirral, UK) was used to identify the peak and calculate the response factor. The results were expressed as g/100 g of dry matter.

2.12. Total Bacterial Count (TBC) and Total Coliform Count

TBC was determined according to the instructions of Chen et al. [31]. The sample was mixed with sterile saline at a ratio of 1:9 (*w/v*) in sterilized bags, and homogenized using a Stomacher 400 Circulator (Seward Ltd., West Sussex, UK) for 3 min. Three serial dilutions of an aliquot of the homogenate were plated onto a specific medium. For TBC, Difco plate count agar (BD Co., Franklin Lakes, NJ, USA) was used and incubated at 37 ± 1 °C for 48 h. For the coliform count, EC medium (BD Co., Franklin Lakes, NJ, USA) was used and incubated at 37 ± 1 °C for 24 h; if no gas was observed, results were recorded as negative (-). For *Salmonella* sp. and *Staphylococcus* sp. counts, Sanita-Kun plates (JNC Corp., Tokyo, Japan) were used and incubated at 35 ± 1 °C for 48 h.

2.13. Texture Analysis

A texture analyzer (CT3 4500, Brookfield Engineering Laboratories Inc., Middleboro, MA, USA) operated by TexturePRO CT software (Middleboro, MA, USA) was used to measure hardness. The texture analysis was performed by compressing the sample to 50% of its height using a cylinder glass probe (12.7 mm in diameter and 35 mm in length). The test speed was 0.5 mm/s. The textural analysis was performed at room temperature with triplicate measurements of each sample.

2.14. Sensory Evaluation

Sensory analysis, including color, flavor, aroma, texture, and overall acceptability, were evaluated. Test seafood HMR was reheated using a microwave (RE-M50, Samsung Electronics Co. Ltd., Seoul, Korea) for 1 min 30 s (700 W). Twenty-one trained and certified sensory evaluation panelists (aged 25–40 years) were employed in this test—all of them were researchers at the Industry Academic Cooperation Foundation, Silla University (Busan, Korea). The sensory test was approved by the Silla University Institutional Review Board (Approval No. 1041449-202006-HR-007; 4 October 2020). Each panelist was asked to evaluate and numerically rate all samples. A hedonic scale of 1 to 9 points was used—1 indicating 'remarkably dislike' and 9 indicating 'extreme like.' Five was considered the threshold value; any sample less than a score of 5 was considered unacceptable [32].

2.15. Shelf Life Estimation

The expiration date was established following the Ministry of Food and Drug Safety guidelines, the Republic of Korea. The accelerated experiment to define the expiration date and evaluate the shelf life qualities of the test-HMR products was conducted for 90 days. Test HMRs were stored at −18 °C (distribution temperature) as a control, −13 °C, and −23 °C. Samples were collected seven times over 90 days, and microbiological and physical tests were performed. The number of common bacteria, such as *E. coli, Staphylococcus aureus,* and *Salmonella* spp., and the sensory evaluation scores were used as indicators of quality shelf life. Program simulation (https://www.foodsafetykorea.go.kr, accessed on 17 December 2020) was used to estimate the product's shelf life.

2.16. Statistical Analysis

All experiments were performed in triplicate (n = 3); the data are displayed as mean $\pm$ standard deviation (SD). Drip loss, TBC, and sensory properties were analyzed via a one-way analysis of variance (ANOVA) at a 95% level of probability ($p < 0.05$) using IBM SPSS v 23.0 (IBM, Corp., Armonk, NY, USA) software. Response surface methodology (RSM) was analyzed using Minitab v 14.0 (Minitab Inc., Birmingham, UK). Temperature and heating time were set as independent variables, while overall acceptance and hardness were set as dependent variables.

3. Results and Discussion

3.1. Drip Loss

Drip loss was analyzed via the exudate absorbed into the quantitative filter paper during the thawing process under three different conditions: room temperature, running water, and HFD. Table 2 shows the drip loss results of frozen AMPS and CSM. HFD resulted in the lowest drip loss value of both AMPS and CSM. The drip loss values of AMPS from the high-frequency defroster were significantly different from those thawed using the conventional methods ($p < 0.05$). The highest drip loss value was observed by thawing at room temperature, followed by running water.

Table 2. Drip loss analysis of frozen adductor muscle of the pen shell (AMPS) and common squid meat (CSM) under different conditions.

Raw Materials	Drip Loss (%)		
	Room Temperature	Running Water	High-Frequency Defroster
AMPS	8.15 $\pm$ 0.07 [a]	7.32 $\pm$ 0.14 [ab]	5.74 $\pm$ 0.39 [c]
CSM	4.65 $\pm$ 0.70 [a]	2.80 $\pm$ 0.60 [ab]	2.20 $\pm$ 0.80 [b]

Data are presented as mean $\pm$ SD. Different superscript letters (a–c) show significantly different values according to Duncan's test ($p < 0.05$).

The thawing process for frozen fish and fishery products should be performed as quickly as possible. Water changing from its original position during the thawing process leads to drip loss, resulting in a dry, stringy, and less tasty fish. Nutrient losses, such as proteins, vitamins, and minerals, can occur with drip loss [33], with high drip loss often linked to protein denaturation. In addition, high drip loss also decreases attractiveness, nutritional value, texture, and appearance [34].

Applying technology during the thawing process may improve the quality of fishery products, also likely decreasing thawing time [35]. Furthermore, rapid thawing can maintain fish quality [36] and reduce mechanical damage to cell membranes by decreasing recrystallization [37]. Recrystallization, which can be problematic during both thawing and freezing, leads to cellular damage, increasing drip production [38].

Therefore, HFD is considered an effective option to decrease drip loss and maintain the quality of raw materials in the test HMR product. The results of the present study confirm the effectiveness of HDF compared to conventional methods.

3.2. Optimum Conditions for Roasting

The optimum conditions for roasting were analyzed using RSM. Temperature (X_1) and heating time (X_2) were set as independent variables, while overall acceptance (Y_1) and hardness (Y_2) were set as dependent variables. The model in this study was build using the actual value of independent variables. The model obtained for overall acceptance and hardness of AMPS was $Y_1 = 7.8097 + 0.9016X_2 - 0.9525X_1^2 - 1.4762X_2^2$ with 92.1% of R^2, and $Y_2 = 471.5 + 63.588X_1 + 28.777X_2 + 79.759X_1^2 + 46.134X_2^2$ with 95% of R^2, respectively (Table 3). Three-dimensional response surface plots of AMPS showed an increase in the score of overall acceptance and decrease in hardness value with an increase in time and temperature until an optimum condition was attained (Figure 1a,b). The overall acceptance score started to decline beyond 217.8 °C and 1.08 min (Figure 1a), whereas the hardness value tended to increase by increasing the temperature and time beyond 217.8 °C and 1.08 min (Figure 1b). Therefore, the optimum temperature and time for the superheated steam roasting of AMPS were set at 217.8 °C for 1.08 min (Table 4).

Table 3. Response surface model equations of AMPS and CSM.

Raw Materials	Responses	Quadratic Polynomial Model Equations	R^2
AMPS	Y_1: Overall acceptance (score)	$7.8097 + 0.9016X_2 - 0.9525X_1^2 - 1.4762X_2^2$	0.921
	Y_2: Hardness (g)	$471.5 + 63.588X_1 + 28.777X_2 + 79.759X_1^2 + 46.134X_2^2$	0.950
CSM	Y_1: Overall acceptance (score)	$8.07933 - 0.54590X_2 - 0.76879X_1^2 - 1.54254X_2^2$	0.916
	Y_2: Hardness (g)	$508.67 + 63.44X_1 + 121.91X_2 - 49.13X_1^2 + 47.90X_2^2$	0.965

Y_1: overall acceptance, and Y_2: hardness are the dependent variables; X_1: Temperature (°C); X_2: heating time (min) are the independent variables.

Table 4. Optimal cooking conditions of AMPS and CSM using response surface methodology (RSM)-analyzed superheated steam roasting.

	Raw Material	X_1 Temp (°C)	X_2 Time (min)
AMPS	Coded value	−0.1102	0.2435
	Actual value	217.80	1.08
	Predicted values	Y_1: 7.9086, Y_2: 421.1653	
	Experimental values	Y_1: 7.83 ± 0.55, Y_2: 393.33 ± 87.32	
CSM	Coded value	−0.1964	0.0514
	Actual value	296.07	1.53
	Predicted values	Y_1: 8.00, Y_2: 500.00	
	Experimental values	Y_1: 8.07 ± 0.15, Y_2: 465.33 ± 49.47	

The CSM response surface model is displayed in Table 3. The model obtained for overall acceptance and hardness was $Y_1 = 8.07933 - 0.54590X_2 - 0.76879X_1^2 - 1.54254X_2^2$ with 91.6% of R^2, and $Y_2 = 508.67 + 63.44X_1 + 121.91X_2 - 49.13X_1^2 + 47.90X_2^2$ with 96.5% R^2, respectively. Three-dimensional response surface plots of CSM for overall score and hardness (Figure 1c,d) showed the same pattern as those for AMPS. The overall acceptance score increased with the increasing time and temperature until the optimum condition was attained, while a reverse trend was observed for hardness. Increasing the temperature and time beyond 296.07 °C and 1.53 min decreased the overall acceptance scores (Figure 1c), while a temperature and time beyond 296.07 °C and 1.53 min increased the hardness (Figure 1d). The optimum temperature and time for CSM superheated steam roasting was 296.07 °C for 1.53 min (Table 4).

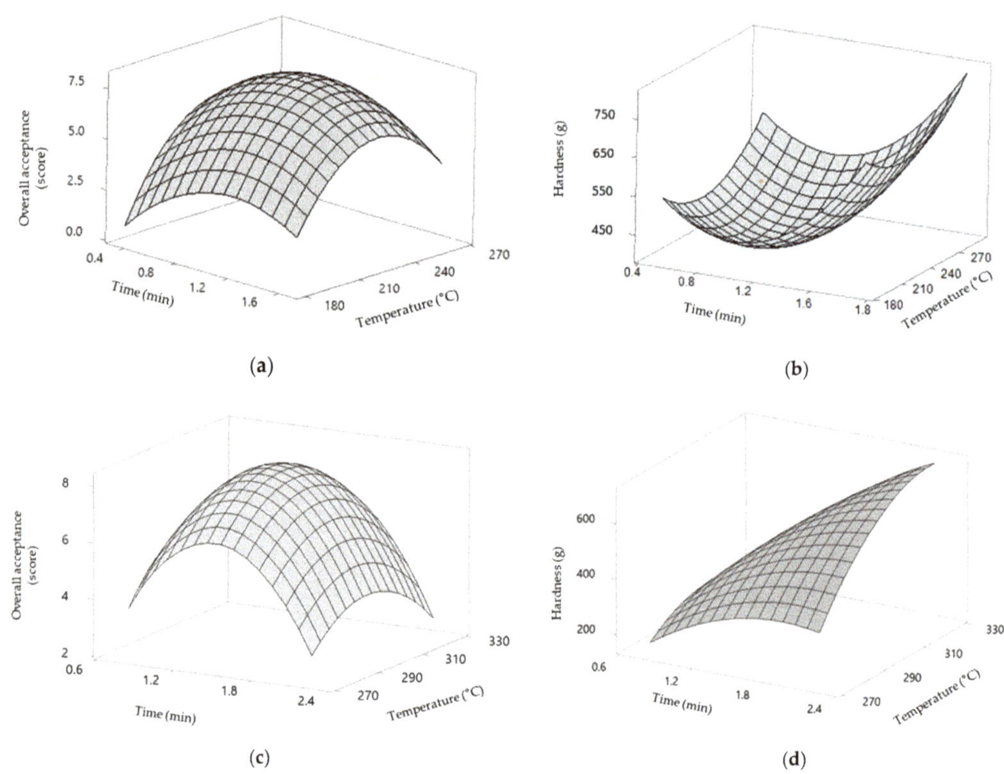

Figure 1. Three-dimensional response surface plots of overall acceptance (**a**) and hardness (**b**) of AMPS and overall acceptance (**c**) and hardness (**d**) of CSM in a superheated steam roaster using RSM.

The optimum temperature and time of AMPS and CSM were 217.8 °C for 1.08 min and 296.1 °C for 1.53 min, respectively. These values were in the range −0.1964 to 0.2435 (Table 4). Moreover, centeral composite design results (Table 5) showed that the optimum practical temperature and time values for superheated steam roasting was 220 °C for 1 min for AMPS and 300 °C for 1.5 min for CSM. In this study, the roasting time for AMPS and CSM was faster than that reported by Mohibbullah et al. [10] and Sutikno et al. [11]. According to Mohibbullah et al. [10], the superheated steam roasting time of AMPS was 4 min at 270 °C to achieve a good panelist-determined sensory evaluation score, which included color, odor, texture, flavor, and overall enjoyment. Moreover, Sutikno et al. [11] proposed the optimum superheated steam roasting time for squid meat was 10 min at 240 °C, as evaluated by panelists based on sensory characteristics.

Table 5. Symbol, experimental range and values of the independent variables in the central composite design.

Raw Materials	Independent	Symbol	Range Level				
			−1.414	**−1**	**0**	**+1**	**+1.414**
AMPS	Temperature (°C)	X_1	192	200	220	240	248
	Time (min)	X_2	0.53	0.67	1.00	1.33	1.47
CSM	Temperature (°C)	X_1	272	280	300	320	328
	Time (min)	X_2	0.79	1.00	1.50	2.00	2.21

The optimum conditions for roasting both AMPS and CSM indicated the best combination of independent and dependent variables. Furthermore, RSM has been effectively used to improve food products [39,40]. Here, we showed RSM successfully optimized superheated steam conditions for efficient production of AMPS and CSM. Moreover, applying RSM to AMPS and CSM roasting resulted in high sensory evaluation values. Taken together, these results indicate the potential of RSM to optimize or choose operating conditions to achieve a particular set of objectives.

3.3. Chemical Characteristics

A test HMR product was prepared by thawing frozen AMPS and CSM using HFD, followed by superheated steam roasting at a particular value for both AMPS and CSM. The test product consisted of 27.5% AMPS, 27.5% CSM, and 45% sauce. Chemical properties, such as pH, VBN, and TBARS, were used to evaluate the quality of the product (Table 6). The pH, VBN, and TBARS of the test HMR were 6.30 $\pm$ 0.13, 9.45 $\pm$ 1.09 mg%, and 1.13 $\pm$ 0.27 mg MDA/kg, respectively. Consistent with the results of Sutikno et al. [11] and Mohibbullah et al. [10], who reported near-neutral pH in seafood products (squid meat and AMPS) following superheated steam roasting, the pH of the test HMR was close to neutral (7). Moreover, for fresh fish and fishery products, an ideal pH level is almost neutral. However, post-mortem period-associated nitrogenous compounds increase the pH, leading to decreased quality [41].

Table 6. Chemical characteristics of test seafood HMR products consisting of a mixture of AMPS, CSM, and sauce.

Sample	pH	VBN (mg%)	TBARS (mg MDA/kg)
HMR Products	6.30 $\pm$ 0.13	9.45 $\pm$ 1.09	1.13 $\pm$ 0.27

Data are presented as mean $\pm$ SD.

A product's VBN value indicates the amount of nitrogen derived from proteolytic bacteria and endogenous enzymes [29], with the VBN test examining the level of protein breakdown and the presence of non-protein nitrogenous compounds. Furthermore, unpleasant smells generally correlate with high VBN values. Superheated steam roasting has been observed to lead to low VBN values. Similar results were reported by Sutikno et al. [11], who found that the VBN of roasted squid meat was lower when using superheated steam roasting compared to other methods. Furthermore, Mohibbullah et al. [10] also reported a lower VBN for superheated steam roasted AMPS.

The low TBARS value of a product labels it as a perfect product according to Yildiz [41], who categorized seafood products into the following three TBARS value-based groups: perfect product (less than 3 mg MDA/kg), good product (3–5 mg MDA/kg), and consumable limit (7–8 mg MDA/kg). Using superheated steam roasting resulted in a low TBARS in squid meat [11] and AMPS [10]. This could be attributed to the reduced oxygen availability during this process resulting in higher inhibition of lipid oxidation and decreased product peroxide (POV) and TBARS values [42].

Collectively, the results of chemical characteristics of the test HMR indicated the quality of the product. According to VBN and TBARS values, test seafood HMR products can be classified as very good.

3.4. Nutritional Quality

Nutritional quality results are shown in Table 7. Proximate composition results, including moisture, crude protein, and ash, were 57.21, 11.6, and 1.62 g/100 g, respectively. Furthermore, mineral contents, such as sodium, calcium, and potassium, were 1.04, 0.19, and 0.001 g/100 g, respectively. Other parameters of nutritional quality, including calories, salt, carbohydrates, sugars, and crude fat, were 167.2 kcal/100 g, 1.03 g/100 g, 28.4 g/100 g, 14.9 g/100 g, and 0.8 g/100 g, respectively.

Table 7. Nutritional values of test seafood HMR products consisting of a mixture of AMPS, CSM, and sauce.

Chemical Composition	Unit	Contain
Ash	g/100 g	1.62
Calcium	g/100 g	0.19
Calories	kcal/100 g	167.2
Carbohydrate	g/100 g	28.4
Cholesterol	g/100 g	0.63
Crude fat	g/100 g	0.8
Crude protein	g/100 g	11.60
Moisture	g/100 g	57.21
Potassium	g/100 g	0.001
Salt	g/100 g	1.03
Saturated fat	g/100 g	0.3
Sodium	g/100 g	1.04
Sugar	g/100 g	14.9
Trans fat	g/100 g	0.0
Vitamin D	µg/100 g	0.0

The nutritional quality of test HMR products indicated that they contained essential nutrition for humans. Wholesome food consists of macronutrients, such as protein, carbohydrates, and fat, thereby offering calories to fuel the body and energy for specific health-maintaining roles. The human body requires nutrition from food, including proteins, carbohydrates, minerals, and crude fat, to participate in daily activities and maintain good health [43]. This study suggests that the product has the essential nutrients required for healthy human body functioning, also contributing to daily nutrition needs.

3.5. Amino Acid Composition

The amino acid composition of the HMR product, consisting of mixed AMPS, CSM, and sauce, is presented in Table 8. The total amino acid content was 7.93 g/100 g, with glutamine as the predominant amino acid. Glutamine and glutamate are abundant amino acids in fish and fishery products [44,45]. In the human body, glutamine is an essential source of energy for the immune system [46].

Non-essential amino acids were dominant (53.84%) compared with essential amino acids (46.16%). Glutamine, aspartate, and alanine were the most abundant non-essential amino acids found, whereas lysine and arginine were the predominant essential amino acids (both at 8.20%). Both essential and non-essential amino acids benefit human health, such as lowering the risk of cardiovascular disease and enhancing the immune system [47]. Furthermore, following digestion, amino acids provide various benefits to human health, including repairing tissue, supporting growth, breaking down food, and providing energy [48].

This study indicates that test seafood HMR products contain amino acids essential for the human body. In addition, the product's amino acid content contributes to the overall daily amino acid requirement.

Table 8. Amino acid analysis of test seafood HMR products consisting of a mixture of AMPS, CSM, and sauce.

Amino Acid	g/100 g	%
Alanine	0.52	6.56
Aspartate	0.81	10.21
Cysteine	0.11	1.39
Glutamine	1.47	18.54
Glycine	0.45	5.67
Proline	0.33	4.16
Serine	0.38	4.79
Tyrosine	0.20	2.52
Total NE	**4.27**	**53.84**
Arginine	0.65	8.20
Histidine	0.15	1.89
Isoleucine	0.31	3.91
Leucine	0.63	7.94
Lysine	0.65	8.20
Methionine	0.19	2.40
Phenylalanine	0.32	4.04
Threonine	0.36	4.54
Tryptophan	0.07	0.88
Valine	0.33	4.16
Total E	**3.66**	**46.16**
Total amino acid	**7.93**	**100**

NE: Non-essential; E: Essential.

3.6. Fatty Acid Composition

Fatty acid composition results are presented in Table 9. Total saturated fatty acids (SFAs) showed the highest value (0.71 g/100 g), followed by polyunsaturated fatty acids (PUFAs) (0.26 g/100 g), and total monounsaturated fatty acids (MUFAs) (0.18 g/100 g). The three predominant fatty acids found in the sample were palmitic acid, docosahexaenoic acid (DHA), and oleic acid. Due to their nutritional value, fatty acids have important health benefits. Palmitic acid, the most common fatty acid in meat, fish, and fishery products, accounted for approximately 50–60% of total fats [49]. Furthermore, DHA is a known agent in treating primary and secondary heart disease and neurological and neuropsychiatric disorders [50]. The Food and Agriculture Organization (FAO) of the United Nations and the World Health Organization (WHO) have given special consideration to fatty acids in food as essential nutrients, and their impact on early growth, development, and nutrition-related chronic diseases in humans.

In processed seafood products, amino acids contribute to the taste of the product [51]. A sweet taste is caused by glycine, alanine, and trimethyl, whereas a bitter taste results from arginine [52]. The fatty acid content in the test seafood HMR product indicated its good nutritional value, thereby suggesting its potential to contribute to daily fatty acid requirements.

Table 9. Fatty acid contents of test seafood HMR products consisting of a mixture of AMPS, CSM, and sauce.

Fatty Acids		(g/100 g)
Lauric acid	C12: 0	0.03
Myristic acid	C14: 0	0.05
Pentadecanoic acid	C15: 0	0.01
Palmitic acid	C16: 0	0.52
Palmitoleic acid	C16: 1	0.00
Heptadecanoic acid	C17: 0	0.01
Stearic acid	C18: 0	0.09
Tricosanoic acid	C23: 0	0.00
Σ SFA		**0.71**
Oleic acid	C18: 1 n−9c	0.10
cis-11-Eicosenoic acid	C20: 1 n−9	0.03
Erucic acid	C22: 1 n−9	0.01
Nervonic acid	C24: 1 n−9	0.04
ΣMUFA		**0.18**
Linolelaidic acid	C18: 2 n−6t	0.00
Linoleic acid	C18: 2 n−6c	0.08
cis-11,14-Eicosadienoic acid	C20: 2 n−6	0.00
Arachidonic acid	C20: 4 n−6	0.01
Σ n−6		*0.09*
cis-11,14,17-Eicosatrienoic acid	C20: 3 n−3	0.00
Eicosapentaenoic acid	C20: 5 n−3	0.03
Docosahexaenoic acid	C22: 6 n−3	0.14
Σ n−3		0.17
ΣPUFA		**0.26**
n−3/n−6		1.89

SFA: saturated fatty acids; MUFA: monounsaturated fatty acids; PUFA: polyunsaturated fatty acids.

3.7. Effect of Storage Duration on Microbiological Changes

Microbiological changes in food can indicate product safety. Table 10 shows the microbial change during storage duration. The score of TBC increased in conjunction with the storage time. At −13 °C, bacterial growth increased significantly ($p < 0.05$) on day 90 (3.20 ± 0.00), while at −18 °C, bacterial growth increased significantly ($p < 0.05$) after days 75 and 90 compared to day 0. However, at −23 °C, bacterial growth was maintained throughout the 90-day storage period. The optimum temperature in which to store products based on the TBC was less than −23 °C. In this study, the TBC of the product was less than 5 log CFU/g, which is acceptable for consumption (International Commission of Microbiological Specification for Food [ICMSF]) [53]. Overall, the findings suggest that low TBC values are influenced by the storage temperature—low temperatures, especially freezing, can decrease microbial activity.

The results of this study align with those of Alizadeh et al. [37], who reported that microbial and enzymatic activities successfully decrease in salmon fillets stored at freezing conditions, therefore maintaining a better nutritional content than those in chilled storage. Furthermore, Pohoro and Ranghoo-Sanmukhiya [54] reported that the TBC of frozen fish sold in Mauritius was lower during storage. Similar results were reported by Gornik et al. [55], who reported that low temperatures during storage result in low TBC values in lobster (Nephrops norvegicus).

Table 10. TBC and microbiology of test seafood HMR products, consisting of a mixture of AMPS, CSM, and sauce, stored at different temperatures.

Temperature	Day	TBC (Log CFU/g)	*S. aureus*	*Salmonella* spp.	*E. coli*
−13 °C	0	2.94 ± 0.10 [ab]	-	-	-
	15	2.85 ± 0.12 [a]	-	-	-
	30	2.94 ± 0.02 [ab]	-	-	-
	45	2.99 ± 0.21 [ab]	-	-	-
	60	2.98 ± 0.03 [ab]	-	-	-
	75	3.11 ± 0.04 [bc]	-	-	-
	90	3.20 ± 0.00 [c]	-	-	-
−18 °C	0	2.94 ± 0.10 [ab]	-	-	-
	15	2.86 ± 0.05 [a]	-	-	-
	30	2.94 ± 0.04 [ab]	-	-	-
	45	2.98 ± 0.03 [b]	-	-	-
	60	2.93 ± 0.03 [ab]	-	-	-
	75	3.07 ± 0.04 [c]	-	-	-
	90	3.11 ± 0.04 [c]	-	-	-
−23 °C	0	2.94 ± 0.10 [a]	-	-	-
	15	2.90 ± 0.03 [a]	-	-	-
	30	2.95 ± 0.03 [a]	-	-	-
	45	2.94 ± 0.05 [a]	-	-	-
	60	2.93 ± 0.07 [a]	-	-	-
	75	2.96 ± 0.04 [a]	-	-	-
	90	2.97 ± 0.03 [a]	-	-	-

TBC: total bacteria count; Data are presented as mean ± SD. Means in each column at each temperature group with different letters ([a–c]) differed significantly according to Duncan's test ($p < 0.05$).

In this study, pathogenic bacteria, such as *S. aureus*, Salmonella spp., and *E. coli*, were not detected during storage at all temperatures. These results suggest that temperature plays an important role in the growth of pathogenic bacteria in fishery products. At low temperatures, microbial activity is suppressed, and most bacteria cannot grow [56,57]; hence, the growth of many spoilage-causing bacteria is prevented via temperature reduction.

3.8. Sensory Evaluation of the Product

Five sensory parameters, including color, flavor, odor, texture, and overall acceptance, were used to evaluate changes in sensory properties following storage and reheating using a microwave. On day 0, sensory evaluation scores of the test HMR products ranged from 8.24 to 8.38, with the highest value for flavor and odor. However, during storage, the scores were decreased. When stored at −13 °C and −18 °C, the sensory evaluation score started to decrease significantly ($p < 0.05$) on day 45 and day 90, respectively, compared to day 0. In contrast, storage at −23 °C did not significantly alter the sensory evaluation scores throughout the 90-day storage period. Sensory evaluation scores ranged from 7.14 to 8.62 (Table 11). According to the hedonic scale, the test HMR was scored by a panelist from "like moderately" to "like very much." These results indicate that these products maintained favorable sensory properties following storage and reheating, likely due to rapid freezing. The quality deterioration of frozen food may be promoted by three factors: ice crystal formation, dehydration of protein molecules, and an increase in solid concentration during the freezing process [58].

Table 11. Sensory evaluation of test seafood HMR products, consisting of a mixture of AMPS, CSM, and sauce, stored at different temperatures.

Temperature	Day	Color	Flavor	Odor	Texture	Overall Acceptance
	0	8.33 ± 0.47 [a]	8.38 ± 0.49 [a]	8.38 ± 0.49 [a]	8.24 ± 0.43 [a]	8.33 ± 0.47 [a]
	15	7.86 ± 0.71 [b]	7.95 ± 0.72 [b]	8.14 ± 0.56 [ab]	8.05 ± 0.65 [ab]	7.89 ± 0.77 [b]
	30	8.00 ± 0.69 [ab]	7.81 ± 0.66 [b]	7.95 ± 0.65 [b]	7.86 ± 0.71 [b]	7.88 ± 0.69 [b]
−13 °C	45	7.48 ± 0.66 [c]	7.14 ± 0.47 [c]	7.48 ± 0.59 [c]	7.43 ± 0.58 [c]	7.33 ± 0.56 [c]
	60	7.38 ± 0.49 [c]	7.29 ± 0.45 [c]	7.52 ± 0.50 [c]	7.24 ± 0.43 [c]	7.33 ± 0.45 [c]
	75	7.29 ± 0.45 [c]	7.05 ± 0.21 [c]	7.19 ± 0.39 [c]	7.24 ± 0.43 [c]	7.19 ± 0.33 [c]
	90	7.14 ± 0.35 [c]	7.21 ± 0.50 [c]	7.33 ± 0.47 [c]	7.19 ± 0.39 [c]	7.18 ± 0.32 [c]
	0	8.33 ± 0.47 [a]	8.38 ± 0.49 [a]	8.38 ± 0.49 [a]	8.24 ± 0.43 [a]	8.33 ± 0.47 [a]
	15	8.10 ± 0.29 [abc]	8.38 ± 0.49 [a]	8.29 ± 0.45 [a]	7.95 ± 0.21 [ab]	8.19 ± 0.29 [ab]
	30	8.29 ± 0.82 [ab]	8.14 ± 0.77 [ab]	8.38 ± 0.72 [a]	8.24 ± 0.81 [a]	8.26 ± 0.81 [a]
−18 °C	45	7.90 ± 0.68 [bc]	7.76 ± 0.61 [bc]	8.00 ± 0.76 [a]	7.81 ± 0.66 [b]	7.86 ± 0.71 [bc]
	60	8.10 ± 0.29 [abc]	7.90 ± 0.29 [bc]	8.17 ± 0.32 [a]	8.00 ± 0.00 [ab]	8.07 ± 0.14 [abc]
	75	8.26 ± 0.61 [ab]	8.12 ± 0.53 [ab]	8.36 ± 0.64 [a]	8.07 ± 0.49 [ab]	8.19 ± 0.45 [ab]
	90	7.76 ± 0.61 [c]	7.57 ± 0.79 [c]	7.95 ± 0.79 [a]	7.76 ± 0.68 [b]	7.76 ± 0.63 [c]
	0	8.33 ± 0.47 [a]	8.38 ± 0.49 [ab]	8.38 ± 0.49 [a]	8.24 ± 0.43 [a]	8.33 ± 0.47 [a]
	15	8.48 ± 0.50 [a]	8.48 ± 0.50 [a]	8.62 ± 0.49 [a]	8.48 ± 0.50 [a]	8.57 ± 0.47 [a]
	30	8.38 ± 0.84 [a]	8.00 ± 0.87 [b]	8.24 ± 0.97 [a]	8.19 ± 1.05 [a]	8.33 ± 0.99 [a]
−23 °C	45	8.48 ± 0.66 [a]	8.38 ± 0.65 [ab]	8.62 ± 0.65 [a]	8.24 ± 0.53 [a]	8.48 ± 0.61 [a]
	60	8.40 ± 0.48 [a]	8.26 ± 0.43 [ab]	8.45 ± 0.49 [a]	8.38 ± 0.49 [a]	8.43 ± 0.47 [a]
	75	8.24 ± 0.53 [a]	8.19 ± 0.59 [ab]	8.43 ± 0.58 [a]	8.19 ± 0.59 [a]	8.31 ± 0.47 [a]
	90	8.19 ± 0.39 [a]	8.00 ± 0.44 [b]	8.38 ± 0.49 [a]	8.10 ± 0.29 [a]	8.17 ± 0.24 [a]

Data are presented as mean ± SD. Means in each column at each temperature group with different letters ([a–c]) differed significantly according to Duncan's test ($p < 0.05$).

Ice crystal formation during freezing alters the physical properties of muscle tissues, thus distorting the structure of the meat [12,59–61]. Furthermore, ice crystal formation in frozen seafood reportedly decreases the sensory properties of the products [62,63]. Interestingly, the size of ice crystals is dependent on the freezing rate. Quick freezing leads to a greater number of fine ice crystal formations within muscle cells, whereas slow freezing leads to large ice crystal formations outside of muscle cells [12,37,64]. As such, the sensory properties of the products were largely preserved after reheating. Therefore, applying quick freezing to test seafood HMR products successfully maintained their sensory properties during storage or after reheating using microwaves.

3.9. Shelf Life Estimation

Changes in sensory and microbial parameters indicate product quality, potentially affecting the shelf life of fish products [65]. On day 0, the sensory characteristics score was above 8.24 for all parameters. During storage at −13 and −20 °C, the product began to lose its sensory characteristics; however, panelists still scored products with decreased sensory characteristics above 7. The product stored at −23 °C showed no significant ($p < 0.05$) differences throughout the 90-day storage period. At −23 °C, test seafood HMR products maintained their sensory characteristics, obtaining a score above 8 from the panelists.

The physicochemical properties of fish and fishery products correlate with their sensory properties and storage time [66]. Moreover, the microbial activity gives flavors and odors to products during the storage period [67], and the decreases in sensory properties could be caused by changes in microbial parameters (Table 10). The TBC of the HMR product at −18 and −13 °C significantly ($p < 0.05$) increased during storage time; however, at −23 °C, there was no significant change ($p < 0.05$). These results suggest that lower temperatures decrease microbial activity in the product, thereby increasing shelf life.

Moreover, the microbial quality of seafood can be classified into the following three TBC value-based groups: satisfactory (TBC < 5 log CFU/g), acceptable ($\geq$5 TBC < 6 log CFU/g), and unsatisfactory (TBC $\geq$ 6 log CFU/g) [68]. Therefore, based on the TBC results of this study, the test seafood HMR product could be classified as satisfactory.

The test seafood HMR product's expiry date and overall acceptance values were established using TBC and the simulator program. It identified the shelf life of test seafood HMR products was 29.62 months when the sensory evaluation, as the main criteria, had the highest statistical value. However, the final shelf life was set to 24 months by multiplying it with the safety factor (0.82), which considers temperature changes during production, purchase, storage, and consumption by the consumer; hence, the quality of HMR products could be maintained for 24 months. The shelf life of fishery products is influenced by temperature and limited by biochemical and microbiological changes. Low temperatures may reduce microbial activity, enzymatic autolysis, and oxidation in fishery products, thereby prolonging shelf life [69–72]. Furthermore, although frozen storage shelf life has constraints, it may vary from a few weeks to years [73].

4. Conclusions

This study revealed that the technological improvements produced high-quality test HMR products by mixing AMPS and CSM. HFD maintained the nutritional quality of the raw materials with little drip loss. Optimization of superheated steam roasting using RSM successfully roasted AMPS and CSM at the optimum temperature and time, resulting in good overall acceptance and hardness. Moreover, quick freezing prevented nutrition and texture loss during storage and after reheating. Overall, the application of HFD, superheated steam, and quick freezing successfully produced test HMR products with high nutrition, good texture, and long shelf life, suggesting the potential of the methods reported here for use in the seafood industry to produce new HMR products.

Author Contributions: Conceptualization, J.-S.C.; methodology, J.-S.C., J.-S.K.; software, S.R.K.; formal analysis, J.H.S.; writing—original draft preparation, B.F.S.P.N.; writing—review and editing, J.-S.C.; visualization, J.-S.K.; supervision, J.-S.C., J.-S.K. All authors have read and agreed to the published version of the manuscript.

Funding: This study was funded by the Ministry of Oceans and Fisheries, the Republic of Korea, under project no. PJT201277.

Institutional Review Board Statement: Not applicable.

Informed Consent Statement: Not applicable.

Data Availability Statement: Data supporting reported results are available upon request.

Conflicts of Interest: The authors declare no conflict of interest.

References

1. FAO. FAO Species Identification Guide for Fishery Purpose. 2018. Available online: http://www.fao.org/fishery/species/3567/e (accessed on 27 December 2020).
2. Rosewater, J. The family *Pinnidae* in the Indo-Pacific. *Indo-Pac. Mollusca* **1961**, *1*, 175–226.
3. Choi, K.; Lee, C.I.; Hwang, K.; Kim, S.W.; Park, J.H.; Gong, Y. Distribution and migration of Japanese common squid, *Todarodes pacificus*, in the southwestern part of the East (Japan) Sea. *Fish. Res.* **2008**, *91*, 281–290. [CrossRef]
4. Grabowska, S. Improvement of the production process in the industry 4.0 context. *MAPE* **2018**, *1*, 55–62. [CrossRef]
5. Chamchong, M.; Datta, A. Thawing of foods in microwave oven I. Effect of power levels and power cycling. *Int. Microw. Power Inst.* **1999**, *34*, 9–21. [CrossRef]
6. Taher, B.J.; Farid, M.M. Cyclic microwave thawing of frozen meat: Experimental and theoretical investigation. *Chem. Eng. Process.* **2001**, *40*, 379–389. [CrossRef]
7. Yu, L.C.; Zzaman, W.; AkandaM, J.H.; Yang, T.A.; Easa, A.M. Influence of superheated steam cooking on proximate, fatty acid profile, and amino acid composition of catfish (*Clarias batrachus*) fillet. *Turkish J. Fish. Aquat. Sci.* **2017**, *17*, 1387–1395. [CrossRef]
8. Alfy, A.; Kiran, B.V.; Jeevitha, G.C.; Hebbar, H.U. Recent Developments in Superheated Steam Processing of Foods—A Review. *Crit. Rev. Food Sci. Nutr.* **2014**, *56*, 2191–2208. [CrossRef]

9. Abdulhameed, A.A.; Yang, T.A.; Abdulkarim, A.A. Kinetics of Texture and Colour Changes in Chicken Sausage during Superheated Steam Cooking. *Pol. J. Food Nutr. Sci.* **2016**, *66*, 199–209. [CrossRef]
10. Mohibbullah, M.; Won, N.E.; Jeon, J.H.; An, J.H.; Park, Y.; Kim, H.; Bashir, K.M.I.; Park, S.M.; Kim, Y.S.; Yoon, S.J.; et al. Effect of superheated steam roasting with hot smoking treatment on improving physicochemical properties of the adductor muscle of pen shell (*Atrina pectinate*). *Food Sci. Nutr.* **2018**, *6*, 1317–1327. [CrossRef] [PubMed]
11. Sutikno, L.A.; Bashir, K.M.I.; Kim, H.; Park, Y.; Won, N.E.; An, J.H.; Jeon, J.H.; Yoon, S.J.; Park, S.M.; Sohn, J.H.; et al. Improvement in Physicochemical, Microbial, and Sensory Properties of Common Squid (*Todarodes pacificus*) by Superheated Steam Roasting in Combination with Smoking Treatment. *J. Food Qual.* **2019**, 8721725. [CrossRef]
12. Leygonie, C.; Britz, T.J.; Hoffman, L.C. Impact of freezing and thawing on the quality of meat. *Meat Sci.* **2012**, *91*, 93–98. [CrossRef]
13. Li, F.F.; Wang, B.; Liu, Q.; Chen, Q.; Zhang, H.W.; Xia, X.F.; Kong, B.H. Changes in myofibrillar protein gel quality of porcine longissimus muscle induced by its stuctural modification under different thawing methods. *Meat Sci.* **2019**, *147*, 108–115. [CrossRef]
14. Wang, B.; Du, X.; Kong, B.; Liu, Q.; Li, F.; Pan, N.; Xia, X.; Zhang, D. Effect of ultrasound thawing, vacuum thawing, and microwave thawing on gelling properties of protein from porcine *longissimus dorsi*. *Ultrason. Sonochem.* **2019**, *64*, 104860. [CrossRef]
15. Kauffman, R.G.; Eikelenboom, G.; van der Wal, P.G.; Merkus, G.; Zaar, M. The use of filter paper to estimate drip loss of porcine musculature. *Meat Sci.* **1986**, *18*, 191–200. [CrossRef]
16. Peiretti, P.G.; Medana, C.; Visentin, S.; Giancotti, V.; Zunino, V.; Meineri, G. Determination of carnosine, anserine, homocarnosine, pentosidine and thiobarbituric acid reactive substances contents in meat from different animal species. *Food Chem.* **2011**, *126*, 1939–1947. [CrossRef]
17. AOAC 952.08. *Official Methods of Moisture Analysis*, 17th ed.; Horwitz, W., Ed.; AOAC International: Washington, DC, USA, 2000; 2200p.
18. AOAC 938.08. *Official Methods of Moisture Analysis*, 17th ed.; Horwitz, W., Ed.; AOAC International: Washington, DC, USA, 2000; 2200p.
19. AOAC 971.27. *Official Methods of Moisture Analysis*, 17th ed.; Horwitz, W., Ed.; AOAC International: Washington, DC, USA, 2000; 2200p.
20. AOAC 960.48. *Official Methods of Moisture Analysis*, 17th ed.; Horwitz, W., Ed.; AOAC International: Washington, DC, USA, 2000; 2200p.
21. AOAC 971.10. *Official Methods of Moisture Analysis*, 17th ed.; Horwitz, W., Ed.; AOAC International: Washington, DC, USA, 2000; 2200p.
22. AOAC 998.18. *Official Methods of Moisture Analysis*, 17th ed.; Horwitz, W., Ed.; AOAC International: Washington, DC, USA, 2000; 2200p.
23. AOAC 985.29. *Official Methods of Moisture Analysis*, 17th ed.; Horwitz, W., Ed.; AOAC International: Washington, DC, USA, 2000; 2200p.
24. AOAC 948.15. *Official Methods of Moisture Analysis*, 17th ed.; Horwitz, W., Ed.; AOAC International: Washington, DC, USA, 2000; 2200p.
25. AOAC 994.10. *Official Methods of Moisture Analysis*, 17th ed.; Horwitz, W., Ed.; AOAC International: Washington, DC, USA, 2000; 2200p.
26. AOAC 936.14. *Official Methods of Moisture Analysis*, 17th ed.; Horwitz, W., Ed.; AOAC International: Washington, DC, USA, 2000; 2200p.
27. AOAC 2011.11. *Official Methods of Moisture Analysis*, 19th ed.; Horwitz, W., Ed.; AOAC International: Washington, DC, USA, 2012; 1399p.
28. AOAC 990.05. *Official Methods of Moisture Analysis*, 17th ed.; Horwitz, W., Ed.; AOAC International: Washington, DC, USA, 2000; 2200p.
29. AOAC 984.27. *Official Methods of Moisture Analysis*, 17th ed.; Horwitz, W., Ed.; AOAC International: Washington, DC, USA, 2000; 2200p.
30. AOAC 994.12b. *Official Methods of Moisture Analysis*, 17th ed.; Horwitz, W., Ed.; AOAC International: Washington, DC, USA, 2000; 2200p.
31. Chen, Y.; Wu, S.; Pan, S. Effect of water-soluble chitosan in combination with glutathione on the quality of pen shell adductor muscles. *Int. J. Biol. Macromol.* **2015**, *72*, 1250–1253. [CrossRef]
32. Li, M.; Wang, W.; Fang, W.; Li, Y. Inhibitory effects of chitosan coating combined with organic acids on *Listeria monocytogenes* in refrigerated ready-to-eat shrimps. *J. Food Prot.* **2013**, *76*, 1377–1383. [CrossRef]
33. Gökoglu, N.; Topuz, O.K.; Buyukbenli, H.A.; Yerlikaya, P. Inhibition of lipid oxidation in anchovy oil (*Engraulis encrasicholus*) enriched emulsions during refrigerated storage. *J. Food Sci. Technol.* **2012**, *47*, 1398–1403. [CrossRef]
34. Otto, G.; Roehe, R.; Looft, H.; Thoelking, L.; Kalm, E. Comparison of different methods for determination of drip loss and their relationships to meat quality and carcass characteristics in pigs. *Meat Sci.* **2004**, *68*, 401–409. [CrossRef]
35. Cai, L.; Cao, M.; Cao, A.; Regenstein, J.; Li, J.; Guan, R. Ultrasound or microwave vacuum thawing of red seabream (*Pagrus major*) fillets. *Ultrason. Sonochem.* **2018**, *47*, 122–132. [CrossRef]

36. Venugopal, V. *Seafood Processing Adding Value through Quick Freezing, Retortable Packaging, and Cook-Chilling*; Taylor & Francis: New York, NY, USA, 2006; 485p.
37. Alizadeh, E.; Chapleau, N.; De Lamballerie, M.; LeBail, A. Effects of freezing and thawing processes on the quality of Atlantic salmon (*Salmo salar*) fillets. *J. Food Sci.* **2007**, *72*, 279–284. [CrossRef]
38. Ambrosiadis, I.; Theodorakakos, N.; Georgakis, S.; Lekas, S. Influence of thawing methods on the quality of frozen meat and the drip loss. *Fleischwirtschaft.* **1994**, *74*, 284–287.
39. Beggs, K.L.H.; Bowers, J.A.; Brown, D. Sensory and physical characteristics of reduced-fat turkey frankfurters with modified corn starch and water. *J. Food Sci.* **1997**, *62*, 1240–1244. [CrossRef]
40. Pappa, I.C.; Bloukas, J.G.; Arvanitoyannis, I.S. Optimisation of salt, olive oil and pectin for low-fat frankfurters produced by replacing pork backfat with olive oil. *Meat Sci.* **2000**, *56*, 81–88. [CrossRef]
41. Yildiz, P.O. Effect of essential oils and packaging on hot smoked rainbow trout during storage. *J. Food Process. Preserv.* **2015**, *39*, 806–815. [CrossRef]
42. Huang, T.C.; Ho, C.T.; Fu, H.Y. Inhibition of lipid oxidation in pork bundles processing by superheated steam frying. *J. Agric. Food Chem.* **2004**, *52*, 2924–2928. [CrossRef]
43. Murthy, W.S.; Singh, S.N. Nutritional requirements for human adaptation in extreme environments. *Proc. Indian Natn Sci Acad.* **2003**, *69*, 485–506.
44. Li, X.; Rezaei, R.; Li, P.; Wu, G. Composition of amino acids in feed ingredients for animal diets. *Amino Acids* **2011**, *40*, 1159–1168. [CrossRef]
45. Li, P.; Wu, G. Composition of amino acids and related nitrogenous nutrients in feedstuffs for animal diets. *Amino Acids* **2020**, *52*, 523–542. [CrossRef]
46. Calder, P.C.; Yaqoob, P. Glutamine and the immune system. *Amino Acids* **1999**, *17*, 227–241.
47. Takahashi, T.; Eri, T.; Ram, B.S.; Meester, F.D.; Wilczynska, A.; Wilson, D.; Juneja, L.R. Essential and non-essential amino acids in relation to glutamate. *Open Nutraceuticals J.* **2011**, *4*, 205–212. [CrossRef]
48. Akram, M.; Asif, H.M.; Uzair, M.; Akhtar, N.; Madni, A.; Shah, S.M.A.; Hasan, Z.U.; Ullah, A. Amino acids: A review article. *J. Med. Plant. Res.* **2011**, *5*, 3997–4000.
49. Innis, S.M.; Dyer, R. Dietary triacylglycerols with palmitic acid (16:0) in the 2-position increase 16:0 in the 2-position of plasma and chylomicron triacylglycerols, but reduce phospholipid arachidonic and docosahexaenoic acids, and alter cholesteryl ester metabolism in formula-Fed piglets. *J. Nutr.* **1997**, *127*, 1311–1319.
50. Watson, R.R.; De Meester, F.; Zibadi, S.; Crawford, M.A.; Wang, Y.; Lehane, C.; Ghebremeskel, K. Fatty Acid Ratios in Free-Living and Domestic Animals. In *Modern Dietary Fat Intakes in Disease Promotion*; Humana Press: Totowa, NJ, USA, 2010; pp. 95–108.
51. Tirtawijaya, G.; Park, Y.; Won, N.E.; Kim, H.; An, J.H.; Jeon, J.H.; Park, S.M.; Yoon, S.J.; Sohn, J.H.; Kim, J.S.; et al. Effect of steaming and hot smoking treatment combination on the quality characteristics of hagfish (*Myxine glutinosa*). *J. Food Process. Preserv.* **2020**, *44*, 1–12. [CrossRef]
52. Sarower, M.G.; Hasanuzzaman, A.F.M.; Biswas, B.; Abe, H. Taste producing components in fish and fisheries products: A review. *IJFF* **2012**, *2*, 113–121.
53. International Commission of Microbiological Specification for Food. *Microorganisms in Food 2. Sampling for Microbiological Analysis: Principles and Specific Applications*, 2nd ed.; University of Toronto Press: Toronto, ON, Canada, 1986; 131p.
54. Pohoroo, A.; Ranghoo-Sanmukhiya, V.M. Food-borne bacterial load in fresh and frozen fish sold in Mauritius. *Int. Food Res. J.* **2017**, *24*, 2193–2200.
55. Gornik, S.G.; Albalat, A.; Macpherson, H.; Birkbeck, H.; Neil, D.M. The effect of temperature on the bacterial load and microbial composition in Norway lobster (*Nephrops norvegicus*) tail meat during storage. *J. Appl. Microbiol.* **2011**, *111*, 582–592. [CrossRef]
56. Liu, D.S.; Liang, L.; Xia, W.S.; Regenstein, J.M.; Zhou, P. Biochemical and physical changes of grass carp (*Ctenopharyngodon idella*) fillets stored at −3 and 0 degree C. *Food Chem.* **2013**, *140*, 105–114. [CrossRef]
57. Lu, H.; Luo, Y.K.; Zhou, Z.Y.; Bao, Y.L.; Feng, L.G. The quality changes of Songpu mirror carp (*Cyprinus carpio*) during partial freezing and chilled storage. *J. Food Process. Preserv.* **2014**, *38*, 948–954. [CrossRef]
58. Shenouda, S.Y.K. Theories of protein denaturation during frozen storage of fish flesh. *Adv. Food Res.* **1980**, *26*, 275–311.
59. Ertbjerg, P.; Puolanne, E. Muscle structure, sarcomere length and influences on meat quality: A review. *Meat Sci.* **2017**, *132*, 139–152. [CrossRef] [PubMed]
60. Ngapo, T.M.; Babare, I.H.; Reynolds, J.; Mawson, R.F. Freezing rate and frozen storage effects on the ultrastructure of samples of pork. *Meat Sci.* **1999**, *53*, 159–168. [CrossRef]
61. Martino, M.N.; Otero, L.; Sanz, P.D.; Zaritzky, N.E. Size and location of ice crystals in pork frozen by high-pressure-assisted freezing as compared to classical methods. *Meat Sci.* **1998**, *50*, 303–313. [CrossRef]
62. Jiang, Q.; Nakazawa, N.; Hu, Y.; Osako, K.; Okazaki, E. Micro- structural modification and its effect on the quality attributes of frozen-thawed bigeye tuna (*Thunnus obesus*) meat during salting. *LWT* **2019**, *100*, 213–219. [CrossRef]
63. Hanzawa, R.; Fukuda, Y. Improvement of quality by new loin feezing tuna. *Refrigeration* **2006**, *81*, 204–207.
64. Benjakul, S.; Bauer, F. Biochemical and physicochemical changes in catfish (*Silurusglanis linne*) muscle as influenced by different freeze-thaw cycles. *Food Chem.* **2001**, *72*, 207–217. [CrossRef]
65. Abbas, K.A.; Mohamed, A.; Jamilah, B.; Ebrahimian, M. A review on correlations between fish freshness and pH during cold storage. *Am. J. Biochem. Biotechnol.* **2008**, *4*, 416–421. [CrossRef]

66. Šimat, V.; Bogdanović, T.; Krželj, M.; Soldo, A.; Maršić-Lučić, J. Differences in chemical, physical and sensory properties during shelf life assessment of wild and farmed gilthead sea bream (*Sparus aurata*, L.). *J. Appl. Ichthyol.* **2012**, *28*, 95–101. [CrossRef]
67. Kumar, P.; Ganguly, S. Role of vacuum packaging in increasing shelf life in fish processing technology: A Review. *Asian J. Biol. Sci.* **2014**, *9*, 109–112.
68. Miguéis, S.; Santos, C.; Saraiva, C.; Esteves, A. Evaluation of ready to eat sashimi in northern Portugal restaurants. *Food Control* **2015**, *47*, 32–36. [CrossRef]
69. George, R.M. Freezing processes used in the food industry. *Trends Food Sci. Tech.* **1993**, *4*, 134–138. [CrossRef]
70. Piñeiro, C.; Barros-Velázquez, J.; Aubourg, S.P. Effects of newer slurry ice systems on the quality of aquatic food products: A comparative review versus flake-ice chilling methods. *Trends Food Sci. Tech.* **2004**, *15*, 575–582. [CrossRef]
71. Lawrie, R.A.; Ledward, D.A. *Lawrie's Meat Science*, 7th ed.; Woodhead Publishing Limited: Cambridge, UK, 2006; 464p.
72. Noor, R.; Acharjee, M.; Ahmed, T.; Das, K.K.; Paul, L.; Munshi, S.K. Microbiological study of major sea fish available in local markets of Dhaka City Bangladesh. *J. Microbiol. Biotech. Food Sci.* **2013**, *2*, 2420–2430.
73. Bremner, H.A. A critical look at whether "freshness" can be determined. *J. Aquat. Food Prod. Technol.* **2000**, *9*, 5–25. [CrossRef]

applied sciences

Article

Improved Hot Smoke Processing of Chub Mackerel (*Scomber japonicus*) Promotes Sensorial, Physicochemical and Microbiological Characteristics

Md. Masud Rana [1,†], Md. Mohibbullah [1,†], Na Eun Won [2], Md. Abdul Baten [1], Jae Hak Sohn [2], Jin-Soo Kim [3,*] and Jae-Suk Choi [2,4,*]

1   Department of Fishing and Post Harvest Technology, Sher-e-Bangla Agricultural University, Dhaka 1207, Bangladesh; rana.fpht@sau.edu.bd (M.M.R.); mmohib.fpht@sau.edu.bd (M.M.); mabaten.fpht@sau.edu.bd (M.A.B.)
2   Seafood Research Center, Silla University, #605, Advanced Seafood Processing Complex, Wonyang-ro, Amnam-dong, Seo-gu, Busan 49277, Korea; ftrnd2@silla.ac.kr (N.E.W.); jhsohn@silla.ac.kr (J.H.S.)
3   Department of Seafood and Aquaculture Science, Gyeongsang National University, Tongyeong-si 53064, Korea
4   Department of Food Biotechnology, Division of Bioindustry, College of Medical and Life Sciences, Silla University, Busan 46958, Korea
*   Correspondence: jinsukim@gnu.ac.kr (J.-S.K.); jsc1008@silla.ac.kr (J.-S.C.); Tel.: +82-55-772-9146 (J.-S.K.); +82-51-248-7789 (J.-S.C.)
†   M.M.R. and M.M. contributed equally.

check for **updates**

Citation: Rana, M.M.; Mohibbullah, M.; Won, N.E.; Baten, M.A.; Sohn, J.H.; Kim, J.-S.; Choi, J.-S. Improved Hot Smoke Processing of Chub Mackerel (*Scomber japonicus*) Promotes Sensorial, Physicochemical and Microbiological Characteristics. *Appl. Sci.* **2021**, *11*, 2629. https://doi.org/10.3390/app11062629

Academic Editor: Theodoros Varzakas

Received: 14 February 2021
Accepted: 12 March 2021
Published: 16 March 2021

Publisher's Note: MDPI stays neutral with regard to jurisdictional claims in published maps and institutional affiliations.

Abstract: Chub mackerel (CM), *Scomber japonicus*, is a commercially important fish species in pacific countries including South Korea and its rapid quality deterioration by various spoilage mechanisms while marketed has been reported, leading to a dramatic decline of the market price. To overcome this problem, a combination of superheated steam roasting (270 °C for 4 min) and hot smoking (70 °C) on CM fillets was applied to impart extending shelf-life at the market level. Using different sawdust with time-dependent smoking revealed that Oak sawdust at 25 min of optimized smoking time significantly ($p < 0.05$) provided the highest sensory properties (appearance, odor, color, texture and overall preferences), improved physicochemical, microbial, and nutritional properties, and subsequently, promoted shelf life of processed CM during the storage period at 10 °C for up to 34 days. Moreover, the processed CM offered high nutritional value, especially, essential and non-essential amino acids were found to be 13.14 and 15.48 g/100 g of CM fillets, and also reduced the trimethylamine-N-oxide level to an acceptable limit, indicating its quality and safety with high nutritional standards to end-point users upon consumption.

Keywords: chub mackerel; smoking treatment; sensory analysis; physiochemical characteristics; microbiological quality; biochemical analysis

1. Introduction

Worldwide, fish and fish-based products are exceptionally important elements of the human diet in respect to their unique taste and nutritional properties such as plethora of proteins and polyunsaturated fatty acids (PUFAs), which make them more precious than other terrestrial foods [1]. Based on the demand of healthy fish-food items, fish processing industries have been developed in the recent past to offer a variety of consumer products in which quality and safety are being prioritized [2]. The reason is that fish flesh is extremely vulnerable to spoilage [3]. The spoilage mechanisms in fish are attributed to the presence of a high amount of free amino acid, non-protein nitrogenous compounds (NPN) and a high post mortem pH, which eventually support the growth of microorganisms [4]. In addition, endogenous enzymatic activity, oxidation of lipids and proteins are responsible for the quality deterioration of fish and fish products [5,6]. Collectively, the following spoilage mechanisms result in the release of various metabolites that negatively affect the sensorial attributes of the product, and simultaneously, shorten the shelf-life of products, especially in storage conditions.

Chub Mackerel (CM), *Scomber japonicus*, is a migratory coastal-pelagic fish species, distributed from subtropical to temperate zones of the world ocean [7]. CM has been considered as an oily fish and one of the important commercial fish species in western North Pacific [8], and it is estimated that about 10–25% of the total annual marine fish catch are from CM in Korea [9]. CM is generally marketed as a raw, dressed or processed condition like frozen fillet, and salted fish in the retail markets of South Korea [10]. Due to the presence of a rich source of PUFAs, despite applying various processing techniques, CM deteriorates quickly especially in tropics, so many quality and safety issues have arisen on CM products while marketed or during the storage period.

Smoking is one of the traditional and common methods of preserving fish [11], which is the combination of salting, drying and cooking processes [12]. Smoked products have an attractive and unique color and flavor, which is because of phenolic compounds and other volatile components released from plant-derived sawdust materials [13]. Additionally, superheated steam technology was employed to dry CM fillets before being processed for hot smoking. Since it is colorless and transparent hot-dry gas, the result of dried products has been evident in the preparation of foods with good textural and chemical properties [14]. When a press-cake of mackerel was prepared with superheated steam, it showed low moisture and high fatty acids contents in a previous study [15]. In the present study, we hypothesized that the combination of superheated steam and hot smoking on CM could provide the final products of delicacy in taste by adding attractive and favorable color, flavor and textural attributes.

The hot smoke (70 °C) processed products are tastier and have longer shelf lives [16], and such treatment improves physicochemical, sensorial and microbiological qualities in processed fish fillet [17], due to the presence of many beneficial complex compounds released from incomplete combustion of wood materials such as phenols, esters, ethers, alcohols, and ketones. Despite their benefits, hot smoked processed products can lead to excessive shrinkage, buckle or even splitting, which undergo decreased yield and loss of nutrients due to denaturation of dietary proteins of fish [18]. So, the optimization of time and temperature in smoking process technology is extremely important to ensure that the products are of premium quality during storage. The physical (i.e., textural profile), chemical (i.e., pH, VBN, TBARS, and TMA), sensorial (i.e., organoleptic attributes), nutritional (i.e., proximate composition and fatty acids) and microbiological (i.e., TBC) characteristics have been assessed by the precedent research, where new or modified processed technology was applied to improve fish-based products [19,20]. However, a very limited number of studies have been investigated on CM, *Scomber japonicus*, or CM-based products. Considering the delicacy in taste and higher nutritional value, CM was selected for the current study aiming at the development of improved process technology of CM fillets with the combination of superheated steam roasting and hot smoking, for enhancing qualities of the end-product based on sensorial, physicochemical, microbiological, and nutritional aspects.

2. Materials and Methods

2.1. Sample Collection and Processing

The CM fish was bought from the local fish market, South Korea, stored with ice flakes, rinsed in the laboratory, and longitudinally cut to form a fish fillet. The size of the fillet section was 37.4 × 7.8 × 1.5 cm and the average weight of the sample was 355.2 ± 20.91 g.

2.2. Pretreatment with Superheated Steam Roasting

The CM fillets were placed in the AERO Steam Oven chamber (DFC-560A-2R/L, Naomoto Corporation, Osaka, Japan), maintaining the temperature at 270 °C for 4 min as followed by the previous method of Mohibbullah et al. [21]. Afterward, the roasted CM fillets were transferred into the polyamide bags, vacuum-packed using a vacuum sealer machine (Sambo Tech. Corporation, Gyeonggi-do, Korea), and finally stored at 4 °C until otherwise/further use.

2.3. Smoking Treatment with Variant Sawdust

The 24-h dip treatment was conducted for CM fillets with a brine solution of 8% NaCl CM samples were then dried by placing at 30 °C temperature for 30 min and allowed to drain out excess water. After removing water, the fillets were submitted to smoking treatment. The samples were placed inside the smoking kiln (Braai Smoker, Bradley, Canada) and generated by combusting different sawdust (Apple, Chestnut, oak, Cherry and Walnut) materials. During the smoking process, the smoke temperature was recorded using a sensitive thermometer at different time intervals such as 20, 25 and 30 min. The desired temperature was 70 °C and the smoked CM fillets were stored at 4 °C until analysis.

2.4. Sensorial Properties

The parameters such as color, odor, flavor, and overall acceptance of treated and untreated smoked products were observed and all the CM fillet samples were impartially coded preceding the sensory evaluation. These characteristics were judged by ten panelists with ages between 25 and 40, who were previously well trained about the evaluation method. During the sensory evaluation, the changes in color, flavor, odor, and overall acceptances were assessed by a 9-point hedonic scale (1 for extremely dislike and 9 for extremely like). The value 5 was considered as the threshold limit and values less than 5 were considered as rejected [22].

2.5. Physical Properties

2.5.1. Evaluation of Odor Intensity

Following the procedure of Macagnano et al. [23], a total of 5 g of sample was taken in a 50 mL conical flask, the lid closed with parafilm and fitted with an odor concentration meter (XP-329, New Cosmos Electric Co. Ltd., Osaka, Japan). The odor intensity was calculated by observing the signal (peak), which is expressed as an arbitrary unit.

2.5.2. Calculation of Weight Loss

The Goulas and Kontominas [24] method was used to calculate the weight loss of CM fillets. It was done by evaluating the differences between the weight of the samples before and after oven drying.

2.5.3. Texture Analysis

The texture of CM fillets was assessed by Brookfield Texture Analyzer (Massachusetts, USA), equipped with a computer software system (Texture PRO CT, Middleboro, USA). A series of parameters, explicitly hardness, chewiness, cohesiveness and springiness, were estimated by following the experiment of Ganesan and Benjakul [25].

2.5.4. Evaluation of Color

The CM fillet color was assessed by the CM-700d Konica Minolta (Tokyo, Japan) instrument. Using a Hunter system value, the surface color of each sample was calculated as in previous reports [26–28].

2.6. Chemical Properties

2.6.1. Determination of pH

The homogenized CM sample (4 g) was thoroughly mixed with 45 mL of distilled water (DW) for 2 min and well centrifuged (SHG-15D, SciLab, and Seoul, Korea). The pH of the sample was measured using a pH meter (OHAUS STARTER 2100, Seoul, Korea) with a glass electrode.

2.6.2. Determination of Volatile Base Nitrogen (VBN)

The VBN value is measured by a Conway micro diffusion method. A total of 5 g of sample was mixed with 25 mL of distilled water, filtrated, and mixed with potassium carbonate, following the method of Ali et al. [29].

2.6.3. Determination of Thiobarbituric Acid-Reactive Species (TBARS)

A total of 5 g CM fillet sample was taken and thoroughly mixed with 12.5 mL TBARS solution containing 20 trichloroacetic acid and 2 M phosphoric acid, followed by filtration, incubation at 95 °C for 30 min, and reading the absorbance value at 530 nm wavelength [30].

2.6.4. Determination of Trimethylamine (TMA) Analysis by Gas Chromatography-Mass Spectrometry (GC/MS)

The presence of TMA in spoiling marine organisms is due to the bacterial degradation of TMAO. The TMA is used as a marker to assess the seafood quality because it is the primary component to producing fishy off-odor [31]. The GC/MS method was performed to estimate the value of trimethylamine N-oxide (TMAO) in both raw and smoked CM products, as followed by Mohibbullah et al. [21]. In this analysis, 10 g of fish sample was mixed with 10 mL of distilled water, sonicated for 20 min, filtrated, and then volatilized using GC instrument (Agilent 7890B GC, Agilent Technologies Inc., Santa Clara, CA, USA) at an oven temperature of 240 °C which increased from 40 to 210 °C following the flow rate of 10 °C/min. Procedures for quantification of TMAO content in raw and smoked CM fillet were followed by comparing the standard compound (Sigma Aldrich, St. Louis, MO, USA) run side-by-side.

2.7. Microbiological Properties: Total Bacterial Count (TBC) and Total Coliform

The ten-fold serial dilution method was used to determine the bacterial load of the CM fillet sample. A 1 g of sample was taken and transferred to a test tube containing 9 mL of sterile saline solution, spread quickly on the surface of plate count agar (Difco, Franklin Lakes, NJ, USA), as followed by Chen et al. [26]. The total coliform count was estimated following the preferred method of the US Food and Drug Administration (FDA) [27].

2.8. Nutritional Properties
2.8.1. Analysis of Nutritional Composition

The nutritional composition of smoked CM fillets was determined by following the method of Horwitz et al. [32], as performed in the Traditional Microorganism Resources Center, Daegu, South Korea.

2.8.2. Analysis of Amino Acid by HPLC Method

The samples of hot smoked CM fillets were duplicated in measurement for the estimation of amino acid content using a High-Performance Liquid Chromatography (HPLC; LC-20A, Shimadzu Corp., Kyoto, Japan) at Pukyong National University laboratory, Busan, South Korea. This method was followed by the previously published procedure of Gheshlaghi et al. [33].

2.9. Statistical Analysis

The statistical analysis was performed by IBM SPSS software, version 21.0 (IBM, Corp., New York, NY, USA). The data of different biochemical parameters were triplicated and calculated as mean $\pm$ standard deviation (std). Results were considered to be statistically significant at 95%, where the significance level was set at $p < 0.05$.

3. Results
3.1. Effect of Different Sawdust Smoke on Sensory Analysis

The color and flavor of CM were developed through the hot smoke of different sawdust. To identify the best smoke material in the processing of CM fillets, two different smoking duration of 20 and 25 min were maintained in the present study, as followed by the previous reports [20,21]. The results significantly ($p < 0.05$) improved sensorial features when CM fillet was smoked by oak sawdust, followed by apple, cherry, and walnut sawdust at 20 and 25 min, respectively (Figure 1). Therefore, based on screening results with different sawdust materials, oak sawdust was selected as the best smoking material for further study.

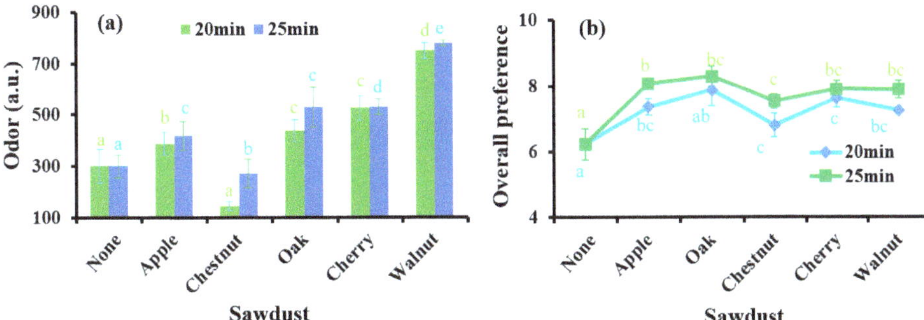

Figure 1. Effect of various sawdust (apple, chestnut, oak, cherry, and walnut) materials on instrumental odor intensity (**a**), and overall preference based on sensory attributes (**b**) in hot smoked chub mackerel (CM) fillet at two smoking times (20 and 25 min). Data were expressed as mean ± std ($n = 10$), where different letters were statistically different ($p < 0.05$).

3.2. Effect of Time-Dependent Smoking on Odor and Sensory Evaluation

The consumer's acceptability and preferences of combined treatment of superheated steam and hot smoked CM were evaluated by an instrumental odor analysis and sensory-based 9-point hedonic scale. The results indicated that the odor intensity significantly ($p < 0.05$) improved with the increased smoke time at 0, 10, 15, 20, 25, and 30 min. However, in the sensory analysis as performed by skilled panelists, 20 and 25 min smoke times were found to have more acceptable sensorial features such as appearance, odor, taste, and overall preferences and were optimized for the next experiment (Figure 2).

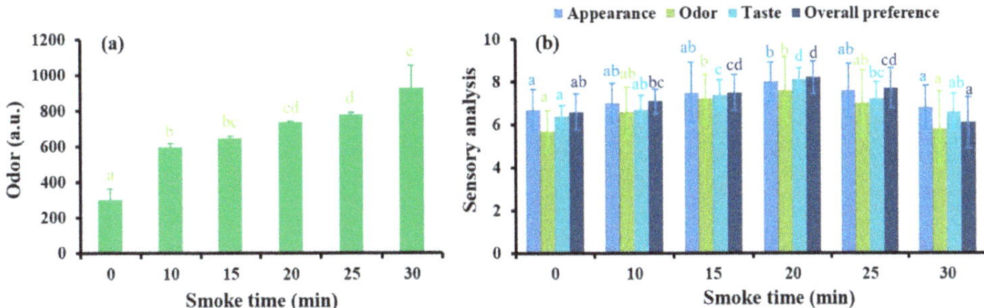

Figure 2. Changes of odor intensity (**a**), and sensory quality (**b**) in CM fillet with different smoke times (0, 15, 20, 25, and 30 min) by oak sawdust. The data were stated as mean ± std ($n = 10$), where different letters were statistically different ($p < 0.05$).

3.3. Effect of Smoke Temperature on Physical Features of CM Fillet

3.3.1. Effect of Smoking Time on Weight Loss

The weight loss of processed CM revealed that there were no significant differences at 20, 25, and 30 min of smoking. In the present study, it was evident that smoking time being applied on processed CM had a non-significant effect on weight loss (Figure 3a).

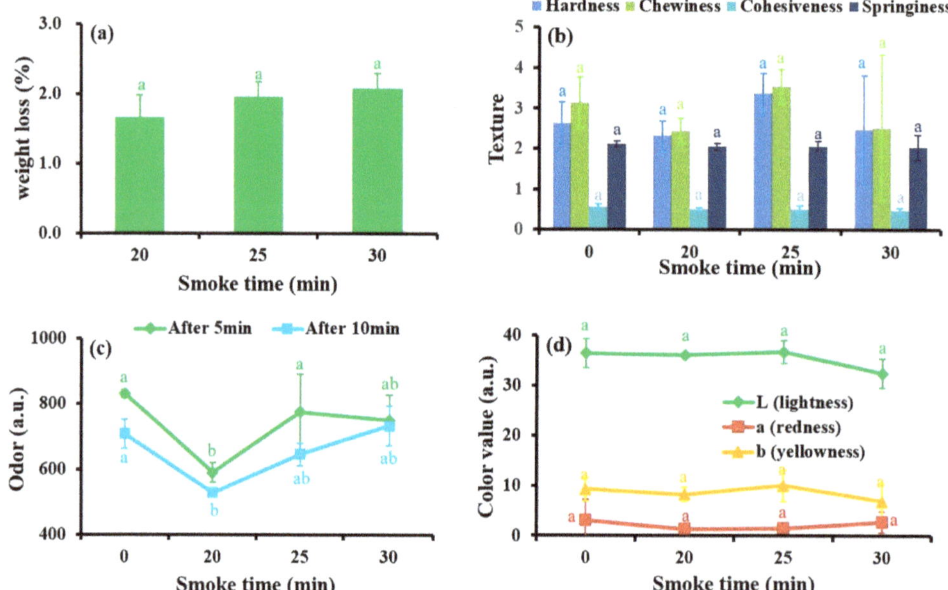

Figure 3. Changes of weight loss (**a**), texture analysis (**b**), odor (**c**), and color value (**d**) in hot smoked CM fillet products with different smoking time. The data were stated as mean ± std (*n* = 10), where different letters were statistically different (*p* < 0.05).

3.3.2. Effect of Smoking Time on Texture

Instrumental texture analysis of processed food products is the most essential and important measurement in the assessment of quality for consumer acceptance. Here, in time-course smoking of 0, 20, 25, and 30 min, there were no significant changes in the textural properties of hardness, chewiness, cohesiveness, and springiness of hot smoked CM, as compared with non-smoking CM product (Figure 3b).

3.3.3. Effect of Smoking Time on Odor

The odor of hot smoked CM was estimated at each 5 min and 10 min of the interval, which significantly (*p* < 0.05) decreased in the smoking time of 20 min. However, after 25 min smoking, odor intensity was similar to that of non-smoked CM, indicating that hot smoking did not influence the odor quality at this condition. (Figure 3c).

3.3.4. Effect of Smoking Time on Color

The color attribute is considered as one of the most important parameters of processed fish products, which directly reflects consumer choice. Instrumental color values are depicted as L for lightness (lightness/darkness: higher/lower), a for red/green (higher/lower), b for yellow/blue (higher/lower). All color parameter values obtained from the color meter showed insignificant differences with smoking time (0 to 30 min), as compared with non-smoked CM. The results suggested that an improved hot smoking process had no obvious effects on color changes of CM fillets, maintaining the original characteristics of the product appearance (Figure 3d).

3.4. Effect of Smoking Treatment on Biochemical Properties of Processed CM

3.4.1. Effect of Smoking Time on pH

pH in processed fish products is used as an indicator for determining the quality of fishery products. Here, the pH level in smoked CM fillet significantly decreased (*p* < 0.05) with smoking time at 20, 25 and 30 min, respectively, and reached its acceptability limit (pH = 6.32–6.22) (Figure 4a).

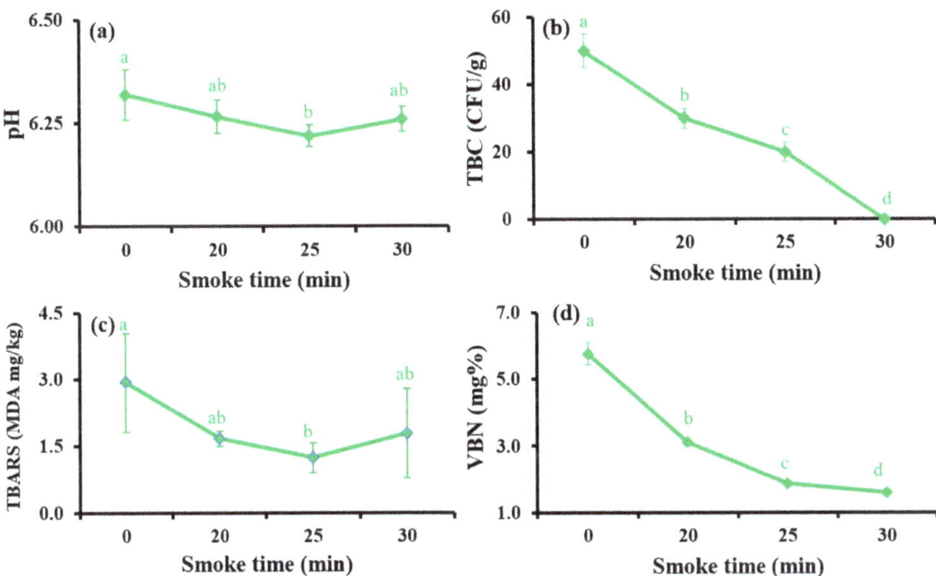

Figure 4. Changes of pH (**a**), total bacterial count (TBC) (**b**), thiobarbituric acid reactive substances (TBARS) (**c**), and volatile base nitrogen (VBN) value (**d**) in hot smoked CM fillet with different smoking times. The data were stated as mean ± std (n = 10), where different letters were statistically different ($p < 0.05$).

3.4.2. Effect of Smoking Time on TBC and Coliform

The microbial quality of processed fish products is an important aspect of food safety to end-point consumers. Therefore, it is very important to assess the TBC and Coliform bacteria in hot smoked CM for ensuring the quality. The present study found a significant ($p < 0.05$) decrease in bacterial biomass that was assessed in smoked CM fillet with the extension of smoke time (Figure 4b).

3.4.3. Effect of Smoking Time on TBARS Changes

Thiobarbituric acid reactive substances (TBARS) are a convenient pathway to determine the level of fat oxidation in processed fish products and are expressed by measuring the malondialdehyde (MDA) content of the sample. The result showed that a significant ($p < 0.05$) decrease in MDA content from 0 to 25 min of smoking time and again a non-significant increment in MDA content at 30 min smoking time were found in processed CM fillet (Figure 4c).

3.4.4. Effect of Smoking Time on VBN Changes

In this experiment, it was found that the VBN of hot smoked CM significantly ($p < 0.05$) decreased with smoking time and the lowest VBN value was observed at 30 min (Figure 4d).

3.4.5. Effect of Smoking Time on Sensory Attributes

As performed by the trained panelists, sensory evaluation on appearance, odor, taste and overall preferences were found to be significantly ($p < 0.05$) increased at 25 min and further declined at 30 min smoking time (Figure 4e). Thus, we optimized 25 min smoke time for further analysis of CM fillets on the effects of storage conditions.

3.5. Effect of Smoking on Storage Conditions of Processed CM

3.5.1. Analysis of Smoked CM Microbiological Quality on Storage Duration

The growth of food-borne microorganisms is greatly influenced by temperature. Herein, the microbiological quality of hot smoked CM fillets was characterized by TBC and coliform count and evaluated in two different storage conditions of 10 and 15 °C temperature. As shown in Table S1, no bacterial growth (CFU/g) was developed during the storage period up to 34 days, indicating that shelf-life of hot smoked CM could be extended up to 34 days without invasion of bacterial biomass at both storage temperatures 10 and 15 °C.

3.5.2. Effect of Storage Duration on VBN Changes

The hot smoke processed CM fillets were kept at two storage conditions, 10 and 15 °C temperature for 0 to 34 days. According to Figure 5a, the VBN value significantly ($p < 0.05$) increased with the extension of storage duration. The VBN value of smoked CM was assessed in between the ranges of 13–20.3 (mg %) and 13–26 (mg %) at 10 and 15 °C, respectively, throughout the storage period of 0–34 days. The study indicated that the 10 °C storage condition had better performance for preserving hot smoked CM fillets.

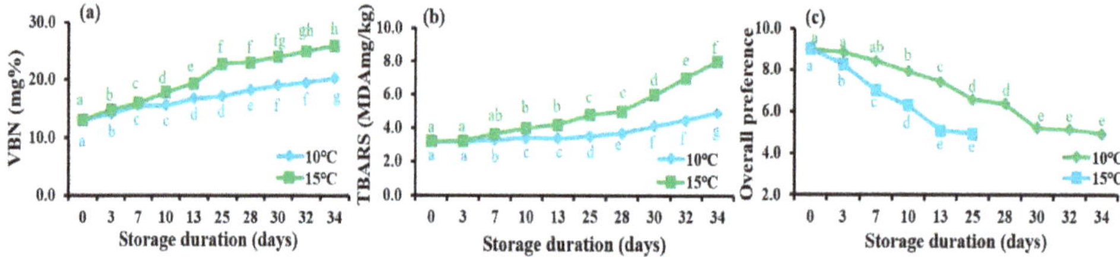

Figure 5. Changes of VBN (**a**), TBARS (**b**) and overall preference (**c**) value in hot smoked CM fillet at two different temperatures (10 and 15 °C) in a storage period 0 to 34 days. The data were stated as mean ± std ($n = 10$), where different letters were expressed as statistically different ($p < 0.05$).

3.5.3. Effect of Storage Duration on TBARS Changes

The TBARS value of hot smoked CM fillet was found to be significantly ($p < 0.05$) increased during the entire storage time at both storage conditions (10 and 15 °C) and varied between 3.2–4.9 and 3.2–8 mg/g, respectively (Figure 5b). This observation supported that CM fillets at 25 min smoking were shown to have extended shelf-life when stored at 10 °C temperature condition.

3.5.4. Effect of Storage Duration on Overall Sensory Preferences

The overall sensory preferences of CM fillets were significantly ($p < 0.05$) decreased with the increase of storage time, from 0 to 34 days, at two different storage conditions 10 and 15 °C (Figure 5c). The overall preference of smoked CM remained at an acceptable condition up to 32 days at 10 °C storage temperature, whereas CM fillets at 15 °C storage temperature remained safe up to 13 days and then rapidly declined in quality.

3.6. Effect of Hot Smoking on TMA Changes

TMA content was found as 2.25 ± 0.29 and 1.25 ± 0.06 µg per 100 g sample of both raw and hot smoked CM, respectively (Figure 6). The study revealed that hot smoking of CM fillets had a promising suppressive effect on TMA content, which might be the cause of inhibiting the growth of specific spoilage bacteria.

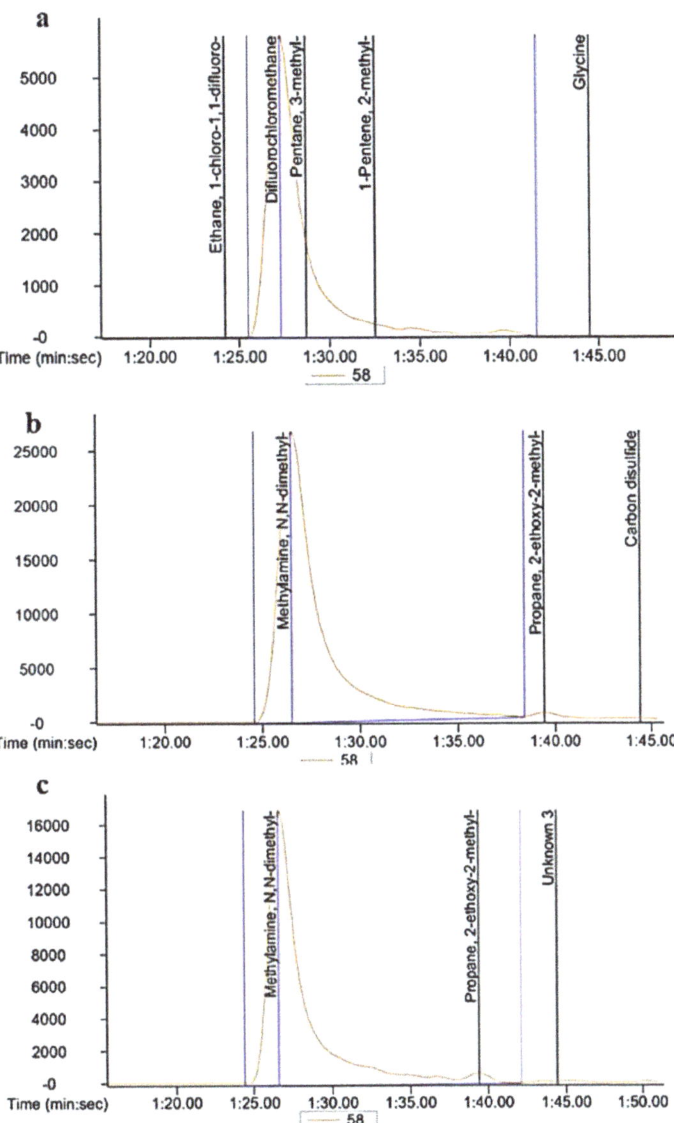

Figure 6. Effect of smoking on TMA content in CM fillets. The results were compared with (**a**) standard, (**b**) raw CM, and (**c**) hot smoked CM.

3.7. Effect of Hot Smoking on Preserving Nutritional Attributes

The nutritional quality or proximate composition of processed fish products varies with processing methods (e.g., drying, salting, and smoking), handling, and storage conditions. The protein, lipid and carbohydrate content of processed CM were 23.880, 30.288, and 0.664 g/100 g, respectively. The minerals viz. potassium, sodium calcium, and iron contents of hot smoked CM were reported as 296.394, 270.803, 67.168, and 1.214 mg/100 g, respectively (Table 1).

Table 1. The nutritional quality of hot smoked CM fillet.

Test Items	Unit	Test Results
Calories	kcal/100 g	280.765
Sodium	mg/100 g	270.803
Carbohydrate	g/100 g	0.664
Sugars	g/100 g	0.0037
Crude fat	g/100 g	30.288
Trans fat	g/100 g	0.0054
Saturated fat	g/100 g	4.830
Cholesterol	mg/100 g	57.236
Crude Protein	g/100 g	23.880
Potassium	mg/100 g	296.394
Calcium	mg/100 g	67.168
Iron	mg/100 g	1.214
Vitamin D	mg/100 g	ND

3.8. Effect of Hot Smoking Treatment on Determination of amino Acid in Smoked CM Fillet

The amino acid analysis is an important parameter for evaluating the protein quality of fish samples. This study identified 17 amino acids including 9 essential and 8 non-essential amino acids in hot smoked CM fillets (Table 2). Among essential amino acids, lysine (2252.2 mg/100 g), leucine (2153.3 mg/100 g) and arginine (1736.3 mg/100 g) were found higher, while that of glutamic acid (3691.9 mg/100 g), proline (2746.0 mg/100 g), and aspartic acid (2213.2 mg/100 g) were higher in the contents of non-essential amino acids found in hot smoked CM. These findings suggested that collective treatment of superheated steam roasting and hot smoking of CM fillets enriched amino acid profiles along with promoting physicochemical properties and antimicrobial performances, which ultimately rendered extension of shelf-life during the storage period.

Table 2. Amino acid content (mg/100 g) in hot smoked CM fillet.

Amino Acid	Hot Smoked CM mg/100 g
Valine	1347.2
Methionine	762.2
Isoleucine	997.0
Leucine	2153.3
Phenylalanine	966.7
Lysine	2252.2
Histidine	1259.4
Arginine	1736.3
Threonine	1665.4
$\sum$**EAA**	**13,139.7**
Aspartic acid	2213.2
Serine	1223.5
Glutamic acid	3691.9
Glycine	1565.1
Alanine	1349.0
Cystine	1769.4
Proline	2746.0
Tyrosine	924.1
$\sum$**Non EAA**	**15,481.9**

4. Discussions

Because of the large number of nutrients, including protein, long-chain omega-3 polyunsaturated fatty acids (*n*-3 PUFAs), and micronutrients including selenium, iodine, potassium, vitamin D and B, fishes are widely considered as a healthy and balanced diet for dietary uptake [34]. Fish is a highly perishable culinary commodity because of the fact that it becomes extremely vulnerable to oxidation resulting in the development of off-odor and -flavor and subsequent spoilage takes place while marketed [35]. Therefore, it is of need to develop or improve preservation and processing techniques more efficiently for keeping quality as well as extending the shelf-life of fish and fishery products. This study was set to the effects of combined treatment of superheated steam roasting and hot smoking (70 °C) of CM in improving sensory, physicochemical, nutritional, and microbiological qualities even at storage conditions, intending to achieve the best quality product for the end-point consumers.

Sensory evaluation is one of the quickest, easiest, and most common methods to evaluate the quality of fish that can be done by human senses such as sight (color), tactility (to test the flesh elasticity), and olfaction (odor) [14]. The present study conducted a sensory assessment technique to know the consumer preferences of hot smoking (70 °C) of Chub mackerel fillet at the different smoking time (0, 20, 25 and 30 min) with different sawdust materials (Apple, Oak, Chestnut, Cherry and Walnut). The study found a supreme sensory quality (appearance, odor, color, texture and overall preferences) product at 25 min smoke time and oak sawdust as the best smoke generating materials. This study complies with the study of hot smoked Wels catfish observed by Küçükgülmez et al. [36]. Similar results were also found in our previous studies that oak sawdust provided good sensorial attributes while smoking of adductor muscle of pen shell *Atrina pectinate* and pacific saury *Cololabis saira* [20,21]. The characteristic flavor and aroma of hot smoked CM are predominantly achieved through the absorption of many volatile components from wood smoke, among which phenolics might be the most influential.

The firmness of the fish products is considered as one of the most important quality parameters to determine consumer preference [37]. The weight loss of the hot smoked CM was found unchanged at the different smoke times (20, 25, and 30 min), which indicates good quality products, and this study was supported by Baten et al. [16,19]. Textural quality significantly affects the overall quality assessment and textural profile, which includes the assessment of hardness, springiness, cohesiveness, chewiness, adhesiveness, and gumminess of a fish product [18]. The texture attributes of hot smoked CM remained constant throughout the smoking time (20, 25, and 30 min), maintaining unique sensorial quality for the consumers, which was more or similar to the study of Mohibbullah et al. [20]. The instrumental odor and color value play an important role in the acceptance or rejection of a fish product by consumers [38]. The time and temperature of smoking treatment had no significant impact on odor and color of processed CM and the findings were similar to the previous study [39]. The pH is an important parameter for assessing the quality of the fish product, as low pH (<6.2) early in the storage indicates good nutritional status and increased shelf-life in the latter stage [18,40]. The present study found significantly lower and acceptable pH at a 25 min smoke and this is in accordance with the study of [30]. Wood smoke has antibacterial properties that retard the growth of bacteria [13,24] and this study showed a similar effect when smoking time significantly decreased bacterial load.

The TBARS is a by-product of lipid peroxidation and its assay measures the MDA level in the fish sample. The lowered level of MDA content was observed in a decreasing trend with increasing smoke time. The previous study reported that smoking treatment decreased TBARS values, where the consumability range is suggested as no more than 7-8 MDA mg/kg [30]. The VBN is considered as one of the prime indicators of raw and processed fish products to measure the degree of freshness [41]. The VBN value of hot smoked CM showed a significant decline in increasing smoke time and remained in an acceptable range as compared with the standard VBN value of fresh fish being 5–20 mg N/100 g muscle [42]. The hot smoking method significantly decreased the VBN value of Japanese Spanish mackerel, *Scomberomorus niphonius* [17]. In storage conditions (10 and 15 °C), the VBN and TBARS values were found to show an increasing trend, which is in compliance with the study of Mohibbullah et al. [21] and Oğuzhan Yildiz [30]. However, the VBN and TBARS values of hot smoked CM of shelf-life retained an acceptable value of

17.7 mg% and 3.7 MDA mg/kg, respectively, after 32 days at 10 °C storage temperature. The results concluded that hot smoked CM provided excellent freshness conditions and did not exceed the maximum permissible limit of VBN and TBARS values up to the storage period of 32 days. The overall sensorial impacts of CM were significantly ($p < 0.05$) decreased with increasing of storage duration either at the 10 or 15 °C storage condition, which is an agreement with the reports suggesting that the sensory scores of smoked bonito, catfish, gwamegi and mackerel decreased with the extension of storage time [43–45]. The suppression of bacterial (TBC and Coliform) growth was observed throughout the storage period at both storage conditions (10 °C and 15 °C). Superheated steam roasting with hot smoking treatment prevented the growth of bacterial proliferation [13,24]. It is expected that the hot smoking process may release phenolic compounds, having antioxidant and anti-microbial actions on the surface of CM fillets.

The TMA is a very efficient index for determining the degree of spoilage in fresh, processed and lightly preserved seafood [46]. The TMAO is converted into TMA in dead fish due to the action of intrinsic enzymes and bacterial action [31,47]. Considering the TMAO content in seafood, it appears to range depending on age, species, environmental factors and harvesting time [48]. The hot smoking with oak sawdust decreased the TMA content to 1.25 µg per 100 g in smoked CM while comparing with raw CM. A similar trend of results was shown by Mohibbullah, et al. [21] and Baten et al. [17,20]. The level of TMA in hot smoked CM fillets was considered safe for human consumption, when the result was compared with the maximum consumable limit of TMA value of 10 mg/100 g [48]. The nutritional content of CM was high and was found to be comparable with Mohibbullah et al. [21]. It is evident that hot smoking treatment increased the crude protein, crude lipid, crude fiber, ash, and other minerals contents by reducing the moisture content, as can be seen in the present CM product and which is comparable to other fish and meat products [49–51]. In addition, the amino acid profile provides essential and non-essential amino acids, and it is determined that the quality of a protein is also beneficial to the human diet [52]. The amino acids and peptides are important components for the development of seafood flavor [53]. The hot smoked Chub mackerel consists of 17 amino acids, of which 9 are essential acids and 8 non-essential amino acids. The results observed in the present study are to some extent similar to Adeyeye et al. [12], who found 18 amino acids in five hot smoked processed fish. Superheated steam roasting with hot smoking treatment significantly increased the lysine and other essential amino acids in CM products, when compared with other processing techniques [54].

5. Conclusions

The improved process technology of CM, namely a combination of superheated steam roasting (270 °C for 4 min) with hot smoking (70 °C), was employed in the present study. Among different sawdust materials to smoke superheated steam roasted CM fillets with time-dependent smoking, oak sawdust showed better sensory characteristics at an optimal smoke time of 25 min. The processed CM fillets with time-dependent smoke provided desirable odor intensity, negligible impact on weight loss, insignificant color changes, and acceptable textural properties, especially, after smoking at 25 min. With this condition, the processed CM also suppressed pH, TBC, VBN, and TBARS values to an acceptable limit at the smoke duration of 25 min. At 25 min smoking time, hot smoked CM was extremely good, and able to preserve its chemical properties and inhibit microbial spoilage successfully at 10 °C up to 32 days in storage condition. Considering the quantity and quality, hot smoked CM posed a significant source of valuable nutrients (protein, lipid, and minerals), especially essential and non-essential amino acids. Moreover, hot smoked CM contained a lower and negligible amount of TMA compared to raw CM and may be considered as a safe item for human consumption. Therefore, this combined process technology could be effectively utilized by processors to increase the storage-ability of CM with safe and healthy conditions at the consumer level.

Supplementary Materials: The following are available online at https://www.mdpi.com/2076-3417/11/6/2629/s1, Table S1: Microbial growth in storage condition of hot smoked CM.

Author Contributions: Conceptualization, J.-S.K., J.H.S. and J.-S.C.; methodology, J.-S.C. and J.H.S.; validation, J.H.S. and J.-S.C.; formal analysis, M.M.R., N.E.W., M.M. and M.A.B.; investigation, N.E.W. and J.-S.C.; resources, M.M.R., N.E.W. and J.-S.C.; data curation, M.M. and J.-S.C.; writing—original draft preparation, M.M.R., M.A.B. and M.M.; writing—review and editing, M.M. and J.-S.C.; visualization, M.M. and J.-S.C.; supervision, J.-S.K., J.H.S. and J.-S.C.; project administration, J.-S.K., J.H.S. and J.-S.C.; funding acquisition, J.-S.K., J.H.S. and J.-S.C. All authors have read and agreed to the published version of the manuscript.

Funding: This study was funded by the Ministry of Oceans and Fisheries, Korea, under the project no. PJT200885 entitled "Development and commercialization of traditional seafood products based on the Korean coastal marine resources".

Institutional Review Board Statement: Not applicable.

Informed Consent Statement: Not applicable.

Data Availability Statement: Not applicable.

Conflicts of Interest: The authors declare no conflict of interest.

References

1. Mei, J.; Ma, X.; Xie, J. Review on natural preservatives for extending fish shelf life. *Foods* **2019**, *8*, 490. [CrossRef]
2. Hicks, D.T. Seafood safety and quality: The consumer's role. *Foods* **2016**, *5*, 71. [CrossRef]
3. Hong, H.; Regenstein, J.M.; Luo, Y. The importance of ATP-related compounds for the freshness and flavor of post-mortem fish and shellfish muscle: A review. *Crit. Rev. Food Sci. Nutr.* **2017**, *57*, 1787–1798. [CrossRef] [PubMed]
4. Chaillou, S.; Chaulot-Talmon, A.; Caekebeke, H.; Cardinal, M.; Christieans, S.; Denis, C.; Desmonts, M.H.; Dousset, X.; Feurer, C.; Hamon, E.; et al. Origin and ecological selection of core and food-specific bacterial communities associated with meat and seafood spoilage. *ISME J.* **2015**, *9*, 1105–1118. [CrossRef] [PubMed]
5. Jiang, D.; Liu, Y.; Jiang, H.; Rao, S.; Fang, W.; Wu, M.; Yuan, L.; Fang, W. A novel screen-printed mast cell-based electrochemical sensor for detecting spoilage bacterial quorum signaling molecules (N-acyl-homoserine-lactones) in freshwater fish. *Biosens. Bioelectron.* **2018**, *102*, 396–402. [CrossRef]
6. Hematyar, N.; Rustad, T.; Sampels, S.; Kastrup Dalsgaard, T. Relationship between lipid and protein oxidation in fish. *Aquac. Res.* **2019**, *50*, 1393–1403. [CrossRef]
7. Lee, K.; Go, S.; Jung, S. Simulation-based yield-per-recruit analysis of chub mackerel *Scomber japonicus* in Korean waters. *Korean J. Fish. Aquat. Sci.* **2018**, *51*, 313–320. [CrossRef]
8. Go, S.; Lee, K.; Jung, S. A Temperature-Dependent Growth Equation for Larval Chub Mackerel (*Scomber japonicus*). *Ocean Sci. J.* **2020**, *55*, 157–164. [CrossRef]
9. Hwang, S.D.; Kim, J.Y.; Lee, T.W. Age, growth, and maturity of chub mackerel of Korea. *N. Am. J. Fish. Manag.* **2008**, *28*, 1414–1425. [CrossRef]
10. Bak, T.J.; Jeon, C.H.; Kim, J.H. Occurrence of anisakid nematode larvae in chub mackerel (*Scomber japonicus*) caught off Korea. *Int. J. Food Microbiol.* **2014**, *191*, 149–156. [CrossRef] [PubMed]
11. Swatawati, F.; Suzuki, T.; Dewi, E.N. The effect of liquid smoke on the quality and omega-3 fatty acids content of tuna fish (*Euthynnus affinis*). *J. Coast. Dev.* **2000**, *3*, 573–579.
12. Adeyeye, S.A.; Fayemi, O.E.; Adebayo-Oyetoro, A.O. Amino acid, vitamin and mineral profiles of smoked fish as affected by smoking methods and fish types. *J. Culin. Sci. Technol.* **2019**, *17*, 195–208. [CrossRef]
13. Bashir, K.M.; Kim, J.S.; An, J.H.; Sohn, J.H.; Choi, J.S. Natural food additives and preservatives for fish-paste products: A review of the past, present, and future states of research. *J. Food Qual.* **2017**, 1–31. [CrossRef]
14. Yu, L.C.; Zzaman, W.; Akanda, M.J.; Yang, T.A.; Easa, A.M. Influence of superheated steam cooking on proximate, fatty acid profile, and amino acid composition of catfish (*Clarias batrachus*) fillets. *Turk. J. Fish. Aquat. Sci.* **2017**, *17*, 935–943. [CrossRef]
15. Bórquez, R.M.; Canales, E.R.; Quezada, H.R. Drying of fish press-cake with superheated steam in a pilot plant impingement system. *Dry. Technol.* **2008**, *26*, 290–298. [CrossRef]
16. Aremu, M.O.; Namo, S.B.; Salau, R.B.; Agbo, C.O.; Ibrahim, H. Smoking methods and their effects on nutritional value of African Catfish (*Clarias gariepinus*). *Open Nutraceuticals J.* **2013**, *6*. [CrossRef]
17. Baten, M.A.; Won, N.E.; Mohibbullah, M.; Yoon, S.J.; Hak Sohn, J.; Kim, J.S.; Choi, J.S. Effect of hot smoking treatment in improving Sensory and Physicochemical Properties of processed Japanese Spanish Mackerel *Scomberomorus niphonius*. *Food Sci. Nutr.* **2020**, *7*, 3957–3968. [CrossRef]
18. Sutikno, L.A.; Bashir, K.M.; Kim, H.; Park, Y.; Won, N.E.; An, J.H.; Jeon, J.H.; Yoon, S.J.; Park, S.M.; Sohn, J.H.; et al. Improvement in physicochemical, microbial, and sensory properties of common Squid (*Todarodes pacificus* Steenstrup) by superheated steam roasting in combination with smoking treatment. *J. Food Qual.* **2019**, *2019*, 8721725. [CrossRef]

19. Huang, X.H.; Qi, L.B.; Fu, B.S.; Chen, Z.H.; Zhang, Y.Y.; Du, M.; Dong, X.P.; Zhu, B.W.; Qin, L. Flavor formation in different production steps during the processing of cold-smoked Spanish mackerel. *Food Chem.* **2019**, *286*, 241–249. [CrossRef]
20. Baten, M.A.; Won, N.E.; Sohn, J.H.; Kim, J.S.; Mohibbullah, M.; Choi, J.S. Improvement of Sensorial, Physicochemical, Microbiological, Nutritional and Fatty Acid Attributes and Shelf Life Extension of Hot Smoked Half-Dried Pacific Saury (*Cololabis saira*). *Foods* **2020**, *9*, 1009. [CrossRef] [PubMed]
21. Mohibbullah, M.; Won, N.E.; Jeon, J.H.; An, J.H.; Park, Y.; Kim, H.; Bashir, K.M.; Park, S.M.; Kim, Y.S.; Yoon, S.J.; et al. Effect of superheated steam roasting with hot smoking treatment on improving physicochemical properties of the adductor muscle of pen shell (*Atrina pectinate*). *Food Sci. Nutr.* **2018**, *5*, 1317–1327. [CrossRef] [PubMed]
22. Li, M.; Wang, W.; Fang, W.; Li, Y. Inhibitory effects of chitosan coating combined with organic acids on *Listeria monocytogenes* in refrigerated ready-to-eat shrimps. *J. Food Prot.* **2013**, *76*, 1377–1383. [CrossRef] [PubMed]
23. Macagnano, A.; Careche, M.; Herrero, A.; Paolesse, R.; Martinelli, E.; Pennazza, G.; Carmona, P.; D'amico, A.; Di Natale, C. A model to predict fish quality from instrumental features. *Sens. Actuators B Chem.* **2005**, *111*, 293–298. [CrossRef]
24. Goulas, A.E.; Kontominas, M.G. Effect of salting and smoking-method on the keeping quality of chub mackerel (*Scomber japonicus*): Biochemical and sensory attributes. *Food Chem.* **2005**, *93*, 511–520. [CrossRef]
25. Ganesan, P.; Benjakul, S. Effect of glucose treatment on texture and colour of pidan white during storage. *J. Food Sci. Technol.* **2014**, *51*, 729–735. [CrossRef]
26. Chen, Y.; Wu, S.; Pan, S. Effect of water-soluble chitosan in combination with glutathione on the quality of pen shell adductor muscles. *Int. J. Biol. Macromol.* **2015**, *72*, 1250–1253. [CrossRef]
27. Feng, P.; Weagant, S.D.; Grant, M.A.; Burkhardt, W.; Shellfish, M.; Water, B. BAM: Enumeration of Escherichia Coli and the Coliform Bacteria. Bacteriological Analytical Manual. 2002. Available online: http://www.fda.gov/Food/FoodScienceResearch/LaboratoryMethods/ucm064948.htm (accessed on 2 January 2021).
28. Bashir, K.M.; Park, Y.J.; An, J.H.; Choi, S.J.; Kim, J.H.; Baek, M.K.; Kim, A.; Sohn, J.H.; Choi, J.S. Antioxidant properties of *Scomber japonicus* hydrolysates prepared by enzymatic hydrolysis. *J. Aquat. Food Prod. Technol.* **2018**, *27*, 107–121. [CrossRef]
29. Ali, M.Y.; Hossain, S.Z.; Rashed, M.A.; Khanom, M.; Sarower, M.G. Protein loss due to post-harvest handling of shrimp (*Penaeus monodon*) in the value chain of Khulna region in Bangladesh. *Int. J. Eng.* **2013**, *3*, 2305–8269.
30. Oğuzhan Yildiz, P. Effect of essential oils and packaging on hot smoked rainbow trout during storage. *J. Food Process. Preserv.* **2015**, *39*, 806–815. [CrossRef]
31. Venugopal, V. (Ed.) *Seafood Processing: Adding Value through Quick Freezing, Retortable Packaging and Cook-Chilling*; CRC Press: Boca Raton, FL, USA, 2005.
32. Horwitz, W.; Chichilo, P.; Reynolds, H. *Official Methods of Analysis of the Association of Official Analytical Chemists*; Association of Official Analytical Chemists: Washington, DC, USA, 1970.
33. Gheshlaghi, R.; Scharer, J.M.; Moo-Young, M.; Douglas, P.L. Application of statistical design for the optimization of amino acid separation by reverse-phase HPLC. *Anal. Biochem.* **2008**, *383*, 93–102. [CrossRef]
34. Weichselbaum, E.; Coe, S.; Buttriss, J.; Stanner, S. Fish in the diet: A review. *Nutr. Bull.* **2013**, *38*, 128–177. [CrossRef]
35. Sampels, S. The effects of processing technologies and preparation on the final quality of fish products. *Trends Food Sci. Technol.* **2015**, *44*, 131–146. [CrossRef]
36. Küçükgülmez, A.; Eslem Kadak, A.; Celik, M. Fatty acid composition and sensory properties of Wels catfish (*Silurus glanis*) hot smoked with different sawdust materials. *Int. J. Food Sci. Technol.* **2010**, *45*, 2645–2649. [CrossRef]
37. Oz, A.T.; Ulukanli, Z.; Bozok, F.; Baktemur, G. The postharvest quality, sensory and shelf life of a *Garicus Bisporus* in active map. *J. Food Process. Preserv.* **2015**, *39*, 100–106. [CrossRef]
38. Nollet, L.M.; Toldrá, F. *Handbook of Seafood and Seafood Products Analysis*; CRC Press: Boca Raton, FL, USA, 2009.
39. Zzaman, W.; Bhat, R.; Yang, T.A.; Easa, A.M. Influences of superheated steam roasting on changes in sugar, amino acid and flavour active components of cocoa bean (*Theobroma cacao*). *J. Sci. Food Agric.* **2017**, *97*, 4429–4437. [CrossRef]
40. Abbas, K.A.; Mohamed, A.; Jamilah, B.; Ebrahimian, M. A review on correlations between fish freshness and pH during cold storage. *Am. J. Biochem. Biotechnol.* **2008**, *4*, 416–421. [CrossRef]
41. Senapati, M.; Sahu, P.P. Onsite fish quality monitoring using ultra-sensitive patch electrode capacitive sensor at room temperature. *Biosens. Bioelectron.* **2020**, *168*, 112570. [CrossRef]
42. Huss, H.H. *Fresh Fish—Quality and Quality Changes: A Training Manual Prepared for the FAO/DANIDA Training Programme on Fish Technology and Quality Control*; Food & Agriculture Organization: Rome, Italy, 1988.
43. Koral, S.; Köse, S.; Tufan, B. The effect of storage temperature on the chemical and sensorial quality of hot smoked Atlantic bonito (*Sarda sarda*, Bloch, 1838) packed in aluminium foil. *Turk. J. Fish. Aquat. Sci.* **2010**, *10*, 439–443. [CrossRef]
44. Yanar, Y. Quality changes of hot smoked catfish (*Clarias gariepinus*) during refrigerated storage. *J. Muscle Foods* **2007**, *18*, 391–400. [CrossRef]
45. Kolodziejska, I.; Niecikowska, C.; Januszewska, E.; Sikorski, Z.E. The microbial and sensory quality of mackerel hot smoked in mild conditions. *LWT Food Sci. Technol.* **2002**, *35*, 87–92. [CrossRef]
46. Iheagwara, M.C. Effect of ginger extract on stability and sensorial quality of smoked mackerel (*Scomber scombrus*) fish. *J. Nutr. Food Sci.* **2013**, *3*, 199. [CrossRef]
47. Rodríguez, C.J.; Besteiro, I.; Pascual, C. Biochemical changes in freshwater rainbow trout (*Oncorhynchus mykiss*) during chilled storage. *J. Sci. Food Agric.* **1999**, *79*, 1473–1480. [CrossRef]

48. Stansby, M.E. *Industrial Fishery Technology; A Survey of Methods for Domestic Harvesting, Preservation, and Processing of Fish Used for Food and for Industrial Products*; Reinhold Pub: New York, NY, USA, 1963; pp. 339–349. [CrossRef]

49. Oparaku, N.F.; Mgbenka, B.O. Effects of electric oven and solar dryer on a proximate and water activity of *Clarias gariepinus* fish. *Eur. J. Sci. Res.* **2012**, *81*, 139–144.

50. Ahmed, A.; Dodo, A.; Bouba, A.M.; Clement, S.; Dzudie, T. Influence of traditional drying and smoke-drying on the quality of three fish species (*Tilapia nilotica*, *Silurus glanis* and *Arius parkii*) from Lagdo Lake, Cameroon. *J. Anim. Vet. Adv.* **2011**, *10*, 301–306.

51. Foline, O.F.; Rachael, A.M.; Iyabo, B.E.; Fidelis, A.E. Proximate composition of catfish (*Clarias gariepinus*) smoked in Nigerian stored products research institute (NSPRI): Developed kiln. *Int. J. Fish. Aquac.* **2011**, *3*, 96–98.

52. Pratama, R.I.; Rostini, I.; Rochima, E. Amino Acid Profile and Volatile Flavour Compounds of Raw and Steamed Patin Catfish (*Pangasius hypophthalmus*) and Narrow-barred Spanish Mackerel (*Scomberomorus commerson*). In *IOP Conference Series: Earth and Environmental Science*; IOP Publishing: Bristol, UK, 2018; Volume 116, No. 1; p. 012056. [CrossRef]

53. Deng, Y.; Luo, Y.; Wang, Y.; Zhao, Y. Effect of different drying methods on the myosin structure, amino acid composition, protein digestibility and volatile profile of squid fillets. *Food Chem.* **2015**, *171*, 168–176. [CrossRef] [PubMed]

54. Huda, N.; Dewi, R.S.; Ahmad, R. Proximate, color and amino acid profile of Indonesian traditional smoked catfish. *J. Fish. Aquat. Sci.* **2010**, *5*, 106–112. [CrossRef]

applied sciences

Article

Effect of Chitosan Nanoemulsion on Enhancing the Phytochemical Contents, Health-Promoting Components, and Shelf Life of Raspberry (*Rubus sanctus* Schreber)

Shirin Rahmanzadeh Ishkeh [1], Habib Shirzad [1,*], Mohammad Reza Asghari [1], Abolfazl Alirezalu [1], Mirian Pateiro [2] and José M. Lorenzo [2,3,*]

[1] Department of Horticultural Sciences, Faculty of Agriculture, Urmia University, Urmia P.O. Box 165-5715944931, Iran; shirinrahmanzade@yahoo.com (S.R.I.); m.asghari@urmia.ac.ir (M.R.A.); a.alirezalu@urmia.ac.ir (A.A.)
[2] Centro Tecnológico de la Carne de Galicia, Parque Tecnológico de Galicia, rúa Galicia n° 4, San Cibrao das Viñas, 32900 Ourense, Spain; mirianpateiro@ceteca.net
[3] Área de Tecnología de los Alimentos, Facultad de Ciencias de Ourense, Universidad de Vigo, 32004 Ourense, Spain
* Correspondence: h.shirzad@urmia.ac.ir (H.S.); jmlorenzo@ceteca.net (J.M.L.)

Citation: Ishkeh, S.R.; Shirzad, H.; Asghari, M.R.; Alirezalu, A.; Pateiro, M.; Lorenzo, J.M. Effect of Chitosan Nanoemulsion on Enhancing the Phytochemical Contents, Health-Promoting Components, and Shelf Life of Raspberry (*Rubus sanctus* Schreber). *Appl. Sci.* **2021**, *11*, 2224. https://doi.org/10.3390/app11052224

Academic Editor: Theodoros Varzakas

Received: 11 February 2021
Accepted: 1 March 2021
Published: 3 March 2021

Publisher's Note: MDPI stays neutral with regard to jurisdictional claims in published maps and institutional affiliations.

Abstract: Due to high water content and perishability, the raspberry fruit is sensitive to postharvest fungal contamination and postharvest losses. In this study, chitosan was used as an edible coating to increase the storage of raspberries, and nanotechnology was used to increase chitosan efficiency. The fruit was treated with an emulsion containing nanoparticles of chitosan (ECNPC) at 0, 2.5, and 5 g L^{-1}, and stored for 9 d. Decay extension rate, fruit phytochemical contents, including total phenolics, flavonoids, and anthocyanin content, phenylalanine ammonia-lyase (PAL), and guaiacol-peroxidase enzymes and antioxidant activity, and other qualitative properties were evaluated during and at the end of storage. After 9 d of storage, the highest amounts of phenolics compounds, PAL enzyme activity, and antioxidant activity were observed in fruit treated with ECNPC at 5 g L^{-1}. The highest levels of total phenol, PAL enzyme activity, and antioxidant activity were 57.53 g L^{-1}, 118.88 μmol/min *trans*-cinnamic acid, and 85.16%, respectively. ECNPC can be considered as an effective, safe, and environmentally friendly method for enhancing fruit phytochemical contents, postharvest life, and health-promoting capacity.

Keywords: edible coating; nanoemulsion; guaiacol peroxidase; shelf life; anthocyanins; phenylalanine ammonia-lyase

1. Introduction

Fruit and vegetables play an important role in improving human health. Among different food crops, colorful fruit and berries, including raspberries, are rich in different antioxidant and anti-stress compounds with powerful anti-cancer and anti-inflammatory attributes [1–3]. *Rubus sanctus* Schreber is a member of the family Rosaceae and its subfamily is Rosoideae. Berry fruits contain abundant phenolic compounds such as phenolic acids, flavonoids, and anthocyanins, and they are an excellent source of antioxidants such as vitamins (A, and ascorbic acid). In addition, they are useful for the treatment of various diseases, particularly diabetic patients [4].

However, this fruit is very perishable and undergoes substantial changes in antioxidant and phytochemical contents during postharvest storage and handling stages. Water loss, softening, decay extension, and metabolic activities are the main causes for a decrease in fruit quality, phytochemical contents, and marketability [5]. The content of bioactive compounds in *Rubus* depends chiefly on the sowing method, variety, harvest season, and postharvest handling. Several investigations indicate that organic farming systems followed by postharvest conditions have a considerable influence on the quality of berry fruit production [6]. The decrease in nutritional properties and marketability of fruit and vegetables during postharvest handling and storage causes significant economic damages to the producers [7]. On the other hand, with the increase in public awareness regarding the adverse effects of the use of chemical residues on human health and environmental safety in recent years, there has been a lot of growth in the market for the demand for organic and chemical-free horticultural products [8]. However, products produced for fresh consumption are highly susceptible to major mechanisms of losses, including, enzymatic bleaching, water loss, and microbial contamination [9].

To increase the postharvest shelf life of harvested fruit and vegetables, there are some useful treatments, including modified atmosphere storage, changed atmosphere packaging, edible coatings, and the use of different natural compounds [10]. Among the treatments mentioned, the use of edible coatings is one of the promising methods for preventing water loss, decreasing metabolic and enzymatic activities and maintaining the aroma and flavor of the crops; because of a relative permeability to different respiratory gases, these coatings prevent the adverse effects of common modified atmosphere storages and packages, such as creating an unpleasant odor and smells [7].

Edible coatings are renewable compounds, including lipids, polysaccharides, and proteins, which are responsible for decreasing the exchange of water vapor, gases, etc., and many compounds used in postharvest technologies such as antimicrobial agents, antioxidants, dyes, and authorized food additives can also be added to them [11]. In addition, a new approach for enhancing fruit phytochemicals, quality attributes, and postharvest life is the use of natural compounds, such as plant growth regulators (PGR) and phytohormones as alternatives to chemical treatments during the production and postharvest stages of food crops [12,13].

Chitosan is a polymer of (4,1) β-N-acetyl-d-glucosamine derived from the chitin of crustaceans, insects, and fungi, which can play an important role in the physiology of plants and harvested crops [14]. It can be used as an edible coating for many harvested horticultural products [15]. In addition to having semi-permeability behavior against respiratory gases, resulting in decreased respiration and metabolic activities, chitosan has direct antimicrobial, bactericidal, and antivirus properties that reduce the need for the use of chemical compounds and increase the safety of the food products [15,16]. Due to its semi-permeability, chitosan has a relative permeability to water vapor and is a good inhibitor of oxygen exchange, thereby altering the internal atmosphere of the product and increasing the shelf life of the product [16,17]. As a PGR, chitosan has been shown to enhance the natural resistance mechanisms of the plants and harvested crops against different pathogens and stress conditions [18].

Moreover, the use of nanotechnology in edible coatings and packaging can increase their efficiency compared to conventional edible coatings and improve the quality of edible coatings by reducing the particle size and the pores of the coatings [19]. The small size of nanoemulsion particles has two important effects—(1) it increases the stability and physicochemical properties of the coating and (2) it can increase the biological activity of lipophilic materials by increasing the surface area to the mass unit rate [20]. Nanoemulsions enhance the bioavailability of bioactive materials [21] and the bactericidal properties of antimicrobials because they can pass as active elements across the bio-membranes [22]. Nanoemulsions are now widely used to cover, protect, and transfer lipophilic materials on foods, fruit, and vegetables [23,24].

Raspberry are rich sources of different powerful antioxidants including anthocyanins, ellagic acid, gallic acid, catechins, camphor, and salicylic acid, with high capacity in the prevention of unwanted damages of free radicals and reactive oxygen species (ROS) to cell membranes and other structures in the human body. Previously mentioned antioxidant compounds may reduce a person's risk of heart disease by preventing platelet buildup and lowering blood pressure by anti-inflammatory agents. *Rubus* antioxidant compounds and anti-inflammatory agents are correlated with cancer protection by reducing the reproduction of cancer tumors. Due to the high levels of antioxidant compounds such as vitamin C and polyphenols, raspberries have an oxygen-radical absorbance capacity of about 30 mM to 100 mM per 100 gr [25]. Ascorbic acid or vitamin C provides extracellular and intracellular antioxidant activity mainly by scavenging reactive oxygen species [26].

Raspberries are also a great source of riboflavin, folate, niacin, magnesium, potassium, and copper. The presence of these nutrients and their synergistic role in improving human health have led to the fact that this fruit is a good source to answer daily needs for various micronutrients (anthocyanins, polyphenols, ascorbic acid, fiber, proteins, vitamins, and minerals) [27]. This fruit is one of the most popular berry fruit, but due to very high sensitivity to different postharvest losses, such as rapid water loss, mechanical damage, and pathogen sensitivity, they have a very short shelf life, leading to rapid phytochemical depletion and quality deterioration [28].

The purpose of this research was to increase the shelf life and maintain phytochemical contents and quality of this valuable fruit with the use of chitosan nanoparticle coatings as a natural safe compound. We studied the effects of this coating at different concentrations on decay extension rate, weight loss, different quality parameters of the fruit, and different resistance mechanisms, including total antioxidant activity and different antioxidant fractions. Moreover, the effects of this coating on fruit taste and flavor as important factors determining consumer acceptance were studied.

2. Materials and Methods

2.1. Preparation of the Fruit for the Treatment

Raspberry fruit (*Rubus sanctus* Schreber) were collected in August at commercial maturity from Khan Valley, Urmia, with a longitude of 45°07′09″, the latitude of 37°19′16″ and an elevation above sea level of 1392 m. Raspberry shrubs were wild-growing and organic. Sampling was carried out early in the morning. The fruit was selected for the similarity of size, maturity, and color, and the damaged and unshaped fruit samples were removed. The samples were then transferred in a fridge of 4 ± 1 °C to the Department of Horticulture Science of Faculty of Agriculture of Urmia University (Urmia, Iran). The *Rubus sanctus* Schreber species identification was conducted by botanist Dr. Shahram Bahadori. The voucher herbarium samples of the collected *Rubus sanctus* Schreber species have been deposited at the herbarium of Urmia University of Medical Sciences (HUPS-359). For each experimental unit, 5 g of fruit was used. Preliminary measurements of the indices of the fruit were carried out on harvest day (day 0) and the fruit was treated with different emulsions containing nanoparticles of chitosan (ECNPC) on the same day of harvest.

2.2. Preparation of Chitosan Nanoemulsion

Chitosan (degree of acetylation (85%), molecular weight (50,000–80,000 Da), and size of nano chitosan particles: <50 nm) nanoemulsion was purchased from Nano Novin Polymer Co. (Sari, Iran). Before applying the treatment, to ensure the quality of the edible coating, the coating was sent to Shahid Beheshti University (Tehran, Iran) to determine the particle size. The dynamic light scattering (DLS) device (NanoPhox 90–246 V, Sympatec GmbH, Clausthal-Zellerfeld, Germany) was used to determine the size of nanoemulsion particles. The type of device lamp was He-Ne Laser 623 nm, and the range of measurement of this device was 1–10,000 nm.

2.3. Application of the Nanoemulsion in Raspberry

For the treatment of the fruit with chitosan nanoemulsion, concentrations of 0, 2.5, and 5 g L^{-1} were prepared. The fruit (60 g of fruit for each repetition) were immersed in containers containing chitosan nanoemulsion for three minutes, and after drying for 8–10 min placed in the pre-sterilized disposable plastic containers (polyethylene terephthalate; 23 cm × 17 cm × 8.5 cm) with four vent holes in the sides (2 mm in diameter). The lid of the containers (clamshell package) was sealed with parafilm to prevent the exchange of air and the packages were stored at cold storage of 4 ± 1 °C with relative humidity (RH) of 85–95% for 9 d (Figure 1).

Figure 1. Effect of chitosan nanoemulsion on the shelf life of raspberry.

2.4. Quality Evaluation

2.4.1. Fruit Titratable Acidity (TA), Total Soluble Solids (TSS), and pH

Total acidity was calculated by the titration method using sodium hydroxide (0.1 N) in terms of citric acid. In other words, it was titrated with soda solution (0.1 N) to reach pH 8.3. After applying, the value of the used soda was introduced in the following formula, and the acidity was calculated based on g of citric acid in 2 L of fruit extract and then, converted to percent [29]:

$$TA = \frac{100 \times M \times N \times V}{S \times n} \tag{1}$$

where TA = acidity value based on g of citric acid in 2 L sample extract, M = molecular weight of the dominant acid, n = dominant acid capacity, V = volume of the used soda, S = amount of the extract used, and N = normality of the used soda.

TSS of the sample extract was measured by a manual refractometer (ATAGO, Tokyo, Japan) at laboratory temperature. The distilled water was used to be calibrated the refractometer. A pH meter (pH-Meter CG 824, SCHOTT, Hofheim, Germany) was used to measure the pH of the fruit juice.

2.4.2. Weight Loss of the Fruit

The digital scale (CANDGL300) was used to measure the value of the weight loss of the fruit. For this purpose, the difference in the weight of the fruit was calculated after

3 d, 6 d, and 9 d of storage, and the value of the weight loss compared to the first day was obtained [30].

$$\text{Weight Loss (\%)} = \frac{\text{Primary weight} - \sec \text{ondary weight}}{\text{Primary weight}} \times 100 \qquad (2)$$

2.4.3. Fruit Firmness

A TA-XT plus texture analyzer (Stable Micro Systems Ltd., Godalming, UK) was used to determine the firmness of raspberries. The probe was set at a speed of 2 mm/s for test. The value of pressure (N) that was introduced into the tip of the firmness tester due to the resistance of the fruit tissue was read on the device [31].

2.4.4. Decay Extension Rate and Fruit Sensory Evaluation

The effect of chitosan nanoemulsion at different concentrations on decay extension rate and consumer acceptance was evaluated by 72 evaluators (24 male and 48 females) with prior experience about the sensory attributes of this type of product. Measurement of the decay extension rate was performed by observation, and their average opinions were discussed and recorded. In this regard, score 1 was given to the samples with the lowest decay extension rate and score 10 was given to the highest rate.

The fruit panel test was also conducted for sensory evaluation. A randomized (complete) block design was conducted. The panelists tested the fruit taste and flavor and their average opinions were analyzed. Before each session, panelists were informed about the objectives of the study and the instructions to complete the test. The samples were individually labeled with aleatory numbers and randomly served to panelists situated in individual cabins during the sessions. For increasing the accuracy of sensory analysis between each testing crackers (unsalted) and water were utilized. The scores were ranged from 1 to 10, i.e., score 1 was given to samples with better taste and score 10 was given to those with the lowest sensory indices. The measurements were performed for the third, sixth, and ninth day separately [30].

2.5. Measurement of Total Phenolics Content (TPC)

TPC was determined using Folin–Ciocalteu reagent according to the method of Alirezalu et al. [32]. A total of 30 μL concentrated extract was poured into the test tube and 90 μL distilled water was added. Then, 600 μL of 10% Folin was added, and after 10 min, 480 μL of 7.5% sodium carbonate was added, and after placing it for 30 min in the dark at room temperature, a spectrophotometer (UV-1800, Shimatzu Corporation, Kyoto, Japan) was used to read the absorbance at 760 nm. Gallic acid was used as a standard. The total phenol content of extracts was expressed as g of gallic acid equivalent (GAE) per L of the fruit extract.

2.6. Measurement of Total Flavonoid Content (TFC)

To measure the total flavonoid content, 50 μL of the concentrated extract was poured into the test tube and 150 μL of 5% sodium nitrite was added. After 5 min, 300 μL of 10% aluminum chloride was added, and again, after 5–10 min, 1000 μL of 1% NaOH were mixed to the resulting solution and was brought to a volume of 5 mL with the deionized water and the absorbance of the resulting mixture at 380 nm was read, compared to the control. Quercetin was used to be drawn the standard curve. The total flavonoid content of the total extracts was expressed as g of quercetin equivalent per L of the fruit extract [33].

2.7. Total Anthocyanin Content (TAC) Evaluation

The pH-difference method was used to measure the total anthocyanin content. Firstly, two buffers were prepared with pH 1 and 4. Then, 2.5 mL of buffer 1 was poured into the test tube. Afterward, 100 μL of the extract was added to the solution poured into the test tube and the absorbance at two wavelengths of 700 nm and 530 nm was read. Then, 2.5 mL of buffer 2 (pH 4.5) was poured into another test tube and 100 μL of the extract was added, and the absorbance at two wavelengths of 700 nm and 530 nm was read. Finally, the following formula was used to be calculated the total absorbance of each of the extracts [34]:

$$A = (A530 - A700)\, pH = 1 - (A530 - A700)\, pH = 4.5 \tag{3}$$

Total anthocyanin content was calculated by g of cyanidin 3-O-glucoside equivalent per kg fresh weight and according to the following formula:

$$TAC = \frac{A \times MW \times V \times DF \times 100}{\varepsilon \times 100} \tag{4}$$

where A = Absorbance, MW = Molecular weight, DF = Dilution factor, and ε = Molar absorbance.

2.8. Measurement of Total Antioxidant Activity

2,2-Diphenyl-1-Picrylhydrazyl-Hydrate (DPPH) Method

The 2,2-diphenyl-1-picrylhydrazyl-hydrate (DPPH) method was used to be evaluated the antioxidant activity. For this purpose, 2000 µL of the DPPH (pre-prepared) solution was poured into the sterilized test tubes. Then, a specific amount of the fruit extract of each of the samples was added, and the resulting solution was shaken at room temperature and in the dark for 30 min. The value of the absorbance of the resulting solution was read by a spectrophotometer at a wavelength of 517 nm. The above method was also used to be prepared the control, but instead of the extract, 80% methanol was used and calculated according to the following formula [35]:

$$RSA = \frac{(Abs\ control)t = 30\ min - (Abs\ sample)t = 30\ min}{(Abs\ control)t = 30\ min} \times 100 \tag{5}$$

Ferric Reducing Antioxidant Power (FRAP) Method

In measuring the total antioxidant activity by the Ferric reducing antioxidant power (FRAP) method, 50 µL of the concentrated extract of raspberry fruit and 3 mL of fresh FRAP reagent (300 mM sodium acetate buffer with the acidity of 3.6, ferric-tris pyridyl-s-triazine 3, and ferric chloride) were mixed, and the resulting mixture was placed in a warm water bath (37 °C) for 30 min, and the value of the absorbance was read by a spectrophotometer at 593 nm, compared to the control. Iron sulfate was used to be drawn the standard curve and the results of the data were expressed as mol Fe^{+2}/L extract [36].

2.9. Evaluation of Enzymes Activity

Phenylalanine Ammonia-Lyase (PAL) Enzyme

The method of Karthikeyan et al. [37] was used under slight modifications to measure the activity of the phenylalanine ammonia-lyase (PAL) enzyme. For this purpose, 0.5 g of fresh fruit texture was squeezed by using 1.5 mL extraction buffer (0.1 M borate buffer, 0.1% polyvinylpyrrolidone, and 1.4 mM mercaptoethanol) at pH 7. It was then centrifuged at 12 °C for 15 min at 12,000× *g*. After completion of the centrifuge, the supernatant was used to measure by the enzyme. To measure the enzyme, sample contents contained 30 µL of the enzymatic extract, 1 mL assay buffer (0.1 M borate buffer, 0.1% polyvinylpyrrolidone, and 1.4 mM mercaptoethanol) at pH 8.8, and 1 mL L-phenylalanine (12 mM), were placed in a warm water bath (Benmari method, 30 °C) for 30 min, and the absorbance was read by a spectrophotometer at 290 nm. The measurement of the PAL enzyme activity was performed by the Beer–Lambert law with an extinction coefficient of 9630 l.cm-µ in µmol/min *trans*-cinnamic acid.

Guaiacol Peroxidase (G-POD) Enzyme Activity

To measure the activity of the guaiacol peroxidase (G-POD) enzyme; 0.1 g of the fresh fruit texture was weighed and placed in a Chinese mortar on ice. Then, 1.5 mL of extraction buffer (125 mM potassium phosphate buffer) at pH 7.8 was added, and after grinding, the supernatant was poured into a micro-tube and centrifuged at 4 °C for 10 min at 15,000× *g*. To measure the activity of the enzyme, 200 µL of the enzymatic extract was poured into the test tube, and 200 µL of 22 mM guaiacol was added. Afterward, 2 mL of 1250 mM potassium phosphate buffer was added. Then, for reading, the above solution was poured into a cell, and to prevent the rapid reaction of hydrogen peroxide with the present solution, hydrogen peroxide was added to a spectrophotometer and the absorbance was read at 0 min and 1 min after the addition of hydrogen peroxide at 470 nm, and its unit was expressed as µmol $H_2O_2.min^{-1}.mL^{-1}$ extract [38].

2.10. Statistical Analysis

The experiment was conducted as a completely randomized design with 3 (ECNPC levels) × 3 (storage times) × 4 replications. The data analysis was performed using the SAS software (SAS Institute Inc., Cary, NC, USA). Duncan's multiple range test was used for the comparison of the mean data. The Friedman test based on a completely randomized block design was also used for the analysis of the decay and taste data.

3. Results

3.1. The Coating Structure of Chitosan Nanoemulsion

In Figure 2, the interval of the particle size of nanoemulsion chitosan is shown by the DLS device. The results of this analysis indicated that most chitosan nanoemulsion particles had a size ranging from 15 nm to 150 nm. This particle size range is approximately close to the size range of chitosan nanoemulsion particles reported in previous studies [19,39].

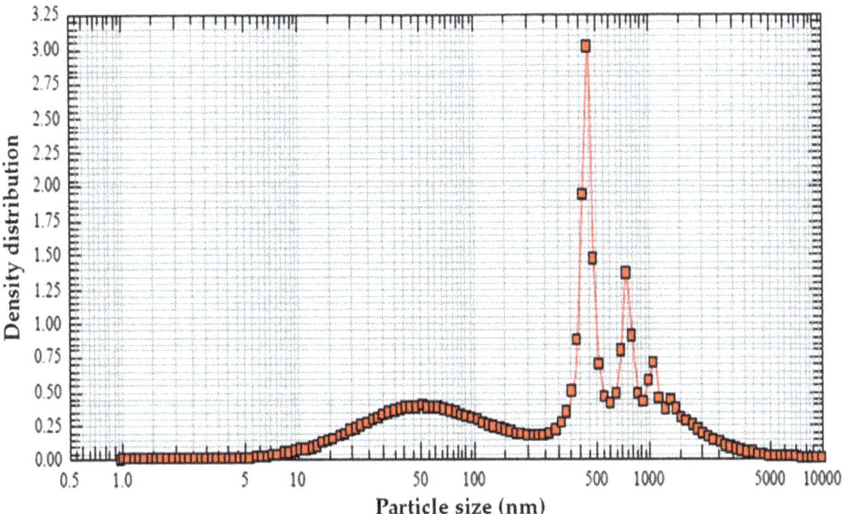

Figure 2. Particle size obtained by dynamic light scattering (DLS) analysis based on size and intensity of emulsion containing nanoparticles of chitosan (ECNPC).

3.2. TA, TSS, and pH

Fruit TA decreased during storage. TSS content showed an increasing trend during storage, and ECNPC significantly decreased the rate of increase in fruit TSS during storage (Table 1). The highest TA (1.25%) was recorded after 6 d in fruit treated with ECNPC at 5 g L^{-1}, and the lowest (0.90%) value was recorded in control fruit after 9 d of cold storage. In both storage periods, the highest TA was seen in fruit treated with 5 g L^{-1} ECNPC ($p < 0.01$). Moreover, the rate of soluble solids varied between 17.33% to 12.18%, which the highest amount was observed in control on the 9 d after storage, and the lowest was observed after 3 d in samples treated with a concentration of 2.5 g L^{-1} ECNPC ($p < 0.01$).

Table 1. Effect of different chitosan nano-emulsion treatment on the quality attributes of raspberry at harvest and in storage time at $4 \pm 1\,^{\circ}C$.

Days	ECNPC (g L^{-1})	TA (%)	TSS (%)	pH
0	Harvest day	1.31 ± 0.01	12.06 ± 0.02	3.23 ± 0.03
3	0	1.16 ± 0.03 [ab]	13.03 ± 0.37 [e]	3.96 ± 0.18 [ab]
6	0	1.04 ± 0.11 [abc]	16.10 ± 0.40 [de]	3.91 ± 0.06 [ab]
9	0	0.90 ± 0.06 [c]	17.33 ± 0.30 [a]	4.05 ± 0.02 [a]
3	2.5	1.19 ± 0.09 [ab]	12.66 ± 0.25 [de]	3.84 ± 0.07 [ab]
6	2.5	1.08 ± 0.05 [abc]	14.96 ± 0.15 [b]	3.88 ± 0.03 [ab]
9	2.5	0.98 ± 0.03 [bc]	13.58 ± 0.40 [d]	3.96 ± 0.06 [ab]
3	5	1.24 ± 0.04 [a]	12.18 ± 0.29 [e]	3.02 ± 0.09 [c]
6	5	1.25 ± 0.06 [a]	13.01 ± 0.04 [c]	3.84 ± 0.05 [ab]
9	5	1.02 ± 0.04 [bc]	12.96 ± 0.02 [de]	3.77 ± 0.04 [b]
		**	**	**

[a–e] Mean values with different letters indicate significant differences among samples. Data obtained from four replications, mean $\pm$ standard error, ** is significant at the 1% (Duncan's multiple range test).

ECNPC at both concentrations reduced the fruit pH compared to the control fruit so that the highest pH was observed in control fruit in all storage times, and the lowest was observed in treated fruit. With an increase in ECNPC concentration, the effects on decreasing the pH were increased ($p < 0.01$) (Table 1).

3.3. Weight Loss

ECNPC significantly affected fruit weight loss during cold storage (Table 2) ($p < 0.01$). The percentage of weight loss was enhanced in all treatments in storage time, and it was lower in samples treated with ECNPC. With an increase in ECNPC concentration, its effects on retaining fruit weight were increased.

Table 2. Effect of different chitosan nano-emulsion treatments on the quality attributes of raspberry at harvest and in storage time at $4 \pm 1\,^{\circ}C$.

Days	ECNPC (g L^{-1})	Loss Weight (%)	Firmness (N)	Decay	Panel Test
0	Harvest day	0.00 ± 0.00	0.17 ± 0.00	1.00 ± 0.00	1.00 ± 0.00
3	0	0.14 ± 0.00 [d]	0.13 ± 0.01 [bc]	4.90 ± 0.04 [e]	4.06 ± 0.08 [e]
6	0	0.21 ± 0.01 [bc]	0.13 ± 0.01 [bc]	6.36 ± 0.06 [d]	6.85 ± 0.06 [c]
9	0	0.26 ± 0.01 [a]	0.12 ± 0.00 [bc]	9.56 ± 0.02 [a]	9.00 ± 0.00 [a]
3	2.5	0.13 ± 0.00 [de]	0.23 ± 0.01 [a]	2.46 ± 0.02 [g]	3.88 ± 0.03 [ef]
6	2.5	0.19 ± 0.00 [c]	0.23 ± 0.01 [a]	4.83 ± 0.02 [e]	7.55 ± 0.53 [bc]
9	2.5	0.26 ± 0.01 [a]	0.16 ± 0.01 [b]	8.38 ± 0.01 [b]	8.58 ± 0.03 [a]
3	5	0.12 ± 0.00 [de]	0.14 ± 0.01 [b]	1.33 ± 0.02 [h]	3.20 ± 0.21 [f]
6	5	0.11 ± 0.00 [e]	0.14 ± 0.02 [b]	3.56 ± 0.02 [f]	4.96 ± 0.02 [e]
9	5	0.23 ± 0.01 [b]	0.09 ± 0.01 [c]	7.86 ± 0.02 [c]	8.16 ± 0.01 [ab]
		**	**	**	

[a–f] Mean values with different letters indicate significant differences among samples. Data obtained from four replications, mean $\pm$ standard error, ** is significant at the 1% (Duncan's multiple range test). $Chi^2_{Decay} = 31.93$ and $Chi^2_{Panel\ test} = 31.20$.

3.4. Firmness

There was a significant difference between treatments and control fruit ($p < 0.01$). The highest firmness value was observed after 3 d of storage in fruit treated with 2.5 g L^{-1} ECNPC (Table 2).

3.5. Decay Extension and Fruit Taste

A statistically significant difference was recorded between the fruit treated with different levels of ECNPC and the control fruit at all evaluation times (Table 2) ($p < 0.01$). The results show that the rate of decay extension in all treatments was high during storage, and the chitosan nanoemulsion treatments reduced the rate of this increasing trend. With the increase in ECNPC concentration the effects on decreasing decay extension rate were increased ($p < 0.01$) (Table 2).

The coating, in a concentration-dependent manner, significantly retained fruit taste and flavor quality ($p < 0.01$) (Table 2). The taste value of the samples varied between 3.2 and 9, which the highest value was related to the control sample after 9 d of storage indicating the lowest taste and flavor rate. On the contrary, the highest taste and flavor rate was observed in 5 g L^{-1} chitosan nanoemulsion coating after 3 d of storage (Table 2).

3.6. Total Phenol Content

Fruit total phenol content was significantly affected by the treatment 5 g L^{-1} chitosan nanoemulsion, and the treated fruit showed the highest phenolics at all evaluation times (Figure 3a) ($p < 0.01$). After 9 d of storage, the highest content of total phenol (57.53 g L^{-1}) was related to the samples treated with 5 g L^{-1} chitosan nanoemulsion, and the lowest rate (26.78 g L^{-1}) was related to the samples treated with 2.5 g L^{-1} chitosan nanoemulsion after 3 d of storage (Figure 3a).

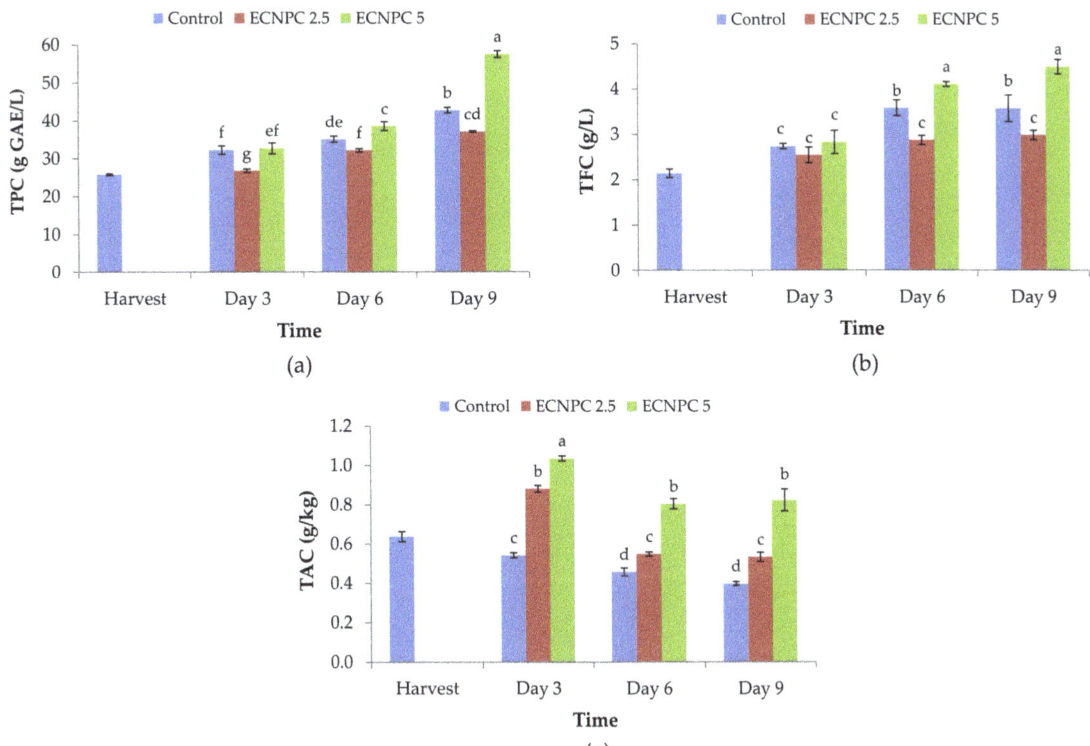

Figure 3. Effect of chitosan nano-emulsion coatings treatment (at 0 g L $^{-1}$, 2.5 g L^{-1}, and 5 g L^{-1}) on (**a**) total phenolics content, (**b**) total flavonoid content, and (**c**) total anthocyanin content of raspberries stored for 9 d (at 4 ± 1 °C with 90–95% RH). Control refers to untreated raspberries. The data shown are the mean $\pm$ standard error of four replicates. Different letters indicate statistical significance ($p < 0.01$).

3.7. Total Flavonoid Content

The total flavonoid content of all fruit in all treated and control samples showed an increasing trend during storage. At all evaluation times, the highest total flavonoid content was recorded in fruit treated with ECNPC at 5 g L^{-1}, but the increase in flavonoid content of fruit treated with 2.5 g L^{-1} was not significant (Figure 3b).

3.8. Total Anthocyanin Content

At all evaluation times, the samples treated with chitosan nanoemulsion had more anthocyanin content than the control samples, and among the treatments, 5 g L^{-1} chitosan nanoemulsion was more effective than 2.5 g L^{-1} in enhancing the anthocyanin content of the fruit. The amount of this parameter varied from 1.0355 g/kg in the samples treated with 5 g L^{-1} chitosan nanoemulsion on the third day after storage to 0.3985 g/kg on the ninth day in the control samples ($p < 0.01$) (Figure 3c).

3.9. Antioxidant Activity

3.9.1. DPPH Method

Changes in antioxidant capacity in the raspberry fruit treated with chitosan nanoemulsion are shown in Figure 4a. According to the results, this index showed an increasing trend in all treatments during storage so that the highest content (85.16%) was related to the fruit treated with 5 g L^{-1} chitosan nanoemulsion after 9 d of storage, while the lowest (56.41%) was observed in fruit treated with the same concentration of chitosan nanoemulsion after 3 d of storage. Meanwhile, there was a statistically significant difference between the treated and control samples ($p < 0.01$).

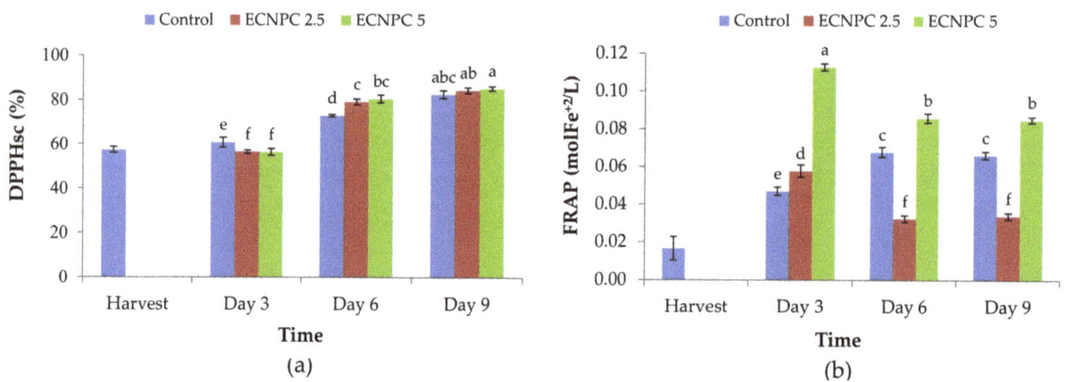

Figure 4. Effect of chitosan nano-emulsion coatings (at 0 g L^{-1}, 2.5 g L^{-1}, and 5 g L^{-1}) on antioxidant activity (**a**) 2,2-diphenyl-1-picrylhydrazyl-hydrate (DPPH) and (**b**) Ferric reducing antioxidant power (FRAP) of raspberries stored for 9 d (at 4 ± 1 °C with 90–95% RH). Control refers to untreated raspberries. The data shown are the mean ± standard error of four replicates. Different letters indicate statistical significance ($p < 0.01$).

3.9.2. FRAP Method

The results of the analysis of variance of the antioxidant activity data using the FRAP method showed that there was a statistically significant difference between the samples treated with chitosan nanoemulsion and the control samples at a 1% level. In other words, there was a significant difference between chitosan nanoemulsion levels and the control samples ($p < 0.01$). However, there was no statistically significant difference between the sixth and the ninth day in any of the treatments (chitosan nanoemulsion of 2.5 and 5 g L^{-1} and the control). Furthermore, the antioxidant activity level varied from 0.11275 mol/L to 0.03238 mol/L (Figure 4b).

3.10. Anti-Stress and Antioxidant Enzymes Activity

3.10.1. PAL Enzyme Activity

PAL enzyme activity in all treatments, except for the 2.5 g L^{-1} treatment with chitosan nanoemulsion on the third day, showed an increasing trend in different during storage ($p < 0.01$). Moreover, the highest enzyme activity (118.88 μmol/min *trans*-cinnamic acid) among the samples was related to the treatment of 5 g L^{-1} chitosan nanoemulsion in the ninth day after storage, and the lowest (49.52 μmol/min *trans*-cinnamic acid) was observed for the treatment of 2.5 g L^{-1} chitosan nanoemulsion on the third day after storage (Figure 5a).

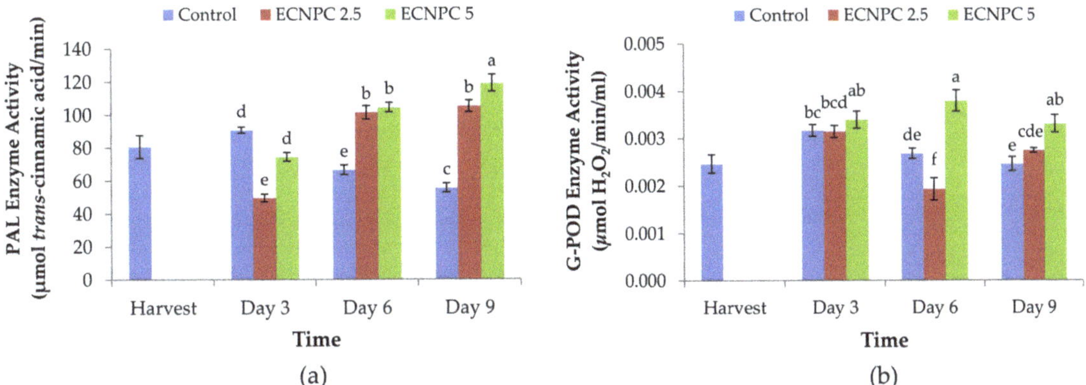

Figure 5. Effect of chitosan nano-emulsion coatings (at 0 g L^{-1}, 2.5 g L^{-1}, and 5 g L^{-1}) on antioxidant activity: (**a**) phenylalanine ammonia-lyase (PAL) enzyme activity and (**b**) guaiacol peroxidase (G-POD) enzyme activity of raspberries stored for 9 d (at 4 ± 1 °C with 90–95% RH). Control refers to untreated raspberries. The data shown are the mean $\pm$ standard error of four replicates. Different letters indicate statistical significance ($p < 0.01$).

3.10.2. G-POD Enzyme Activity

G-POD enzyme activity varied from 0.001 μmol H$_2$O$_2$.min^{-1}.mL^{-1} extract to 0.003 μmol H$_2$O$_2$.min^{-1}.mL^{-1} extract ($p < 0.01$) in different treatments and during storage. According to Figure 4b, treatment of fruit with 5 g L^{-1} chitosan nanoemulsion was more effective in enhancing the G-POD enzyme activity during storage ($p < 0.01$) (Figure 5b).

4. Discussion

Acidity is an important parameter in maintaining fruit quality, which is directly related to the concentration of dominant organic acids in samples [40]. Organic acids are used by the respiration reactions to provide the necessary energy for the normal activities of the cells in storage time [41]. Therefore, the acid content and the pH of the fruit reflect the status of the fruit from the senescence point of view. According to the results of this study, the total acidity content showed a decreasing trend and the pH showed an increasing trend over time, which can be due to the consumption of the organic acid in the respiration process and their conversion to sugars [31]. ECNPC, as any other coating, reduces the respiration and ethylene production rates by restricting the gas exchange resulting in elevated CO$_2$ and decreased O$_2$, thereby reducing the consumption of organic acids and preventing an increase in fruit pH [42].

Change in soluble solids depends on metabolic activities and the activity of different cell wall degrading enzymes. An increase in metabolic activities leads to enhanced ethylene production and a subsequent increase in the activity of degrading enzymes resulting in a dramatic increase in the TSS content of the fruit [43]. Therefore, any decrease in metabolic activities will decrease the ethylene biosynthesis and action resulting in the prevention of a dramatic increase in TSS content. By restricting the gas exchange and making a modified atmosphere in the fruit, chitosan coating decreases ethylene production and senescence rate and prevents cell wall degradation processes [44]. The positive effect of the ECNPC in retaining fruit firmness in this study demonstrates the effectiveness of this treatment in decreasing the activity of cell wall degrading enzymes probably by decreasing ethylene production and action. However, the effect of chitosan on soluble solids content is different for different types of fruit; for example, in papaya fruit, the coating did not have a significant effect on soluble solids but increased the soluble solids in mangos [45].

Weight loss of harvested crops is the consequence of water loss and the respiration process. With progress in senescence, the water of the fruit is lost due to transpiration; the weight loss is also related to the consumption of fruit carbohydrates and organic acid reserves. Water loss, in turn, accelerates senescence by enhancing the production of free radicals and ROS [46]. The amount of respiration and the consumption of sugar by the cell lead to more water loss, which is an important factor in the deterioration of the products [13]. Findings demonstrate that these films/coatings act as a barrier on the surface of fruit and vegetables, which causes higher moisture and water retention, creating favorable micro-environments by optimizing the concentration of gases and delaying the ripening process [47]. The role of chitosan-based edible coating is to restrict water vapor exchange between the fruit and the environment, decrease the respiration and metabolic activities, and activate the mechanisms of cuticle formation in the tissues. In fact, as an elicitor, chitosan can activate the PAL and polyphenol oxidase (PPO) enzymes and other enzymes playing roles in the biosynthesis of polyphenols and cuticle [48]. It is well demonstrated by researchers that these additives play an important role in extending the shelf life and maintaining the nutrient profile of numerous fruit and vegetables. These functional molecules play a synergistic role along with the chitosan and alginate-based edible coatings and retain the moisture and antioxidant potential, enhance the activity of antioxidant enzymes, reduce the activity of browning enzymes, and impart antimicrobial properties in fresh fruit and vegetables. All these properties help in sustaining the appearance, lowering the senescence, and extending the life of the coated fruit and vegetables [47]. In this study, the effects of chitosan nanoparticle-containing coating on enhancing PAL, glutathione peroxidase (GPOX), and different antioxidant fractions were demonstrated. Our results show that chitosan-based coating acts as a physical barrier for the gases and water vapor [49] and enhances the anti-stress and antioxidant properties of the fruit.

The role of chitosan nanoparticles in decreasing decay extension is related both to the direct antifungal activity of the chitosan and the activation of different resistance mechanisms and anti-stress systems [50]. The quality of fruit is defined by its characteristics, such as shape, size, color, and lack of defects, including cuts and decay [42]. The appearance of the product is the most important index affecting the marketability and the presence of any signs of contamination and decay and softening of fruit reduces the market demand of the product. The decay organisms can extend and grow on the surface of shrunken, wilted, injured, and softened fruit. Therefore, reducing the senescence and deterioration rate prevents the growth of decay symptoms and maintains the appearance and market demand of the product [51]. Our results indicate that ECNPC can decrease decay extension in raspberries during storage resulting in the safety and postharvest life enhancement. The results of this study showed that with an increase in the concentration of chitosan nanoemulsion, the decay extension rate was better controlled and fruit treated with 5 g L^{-1} chitosan nanoemulsion showed the lowest decays.

Chitosan acts as an active substance at the surface of fungal and bacterial cells, which makes them more permeable [52]. This interaction is mainly assumed to be electrostatic and occurs among the positive loads of chitosan amino acids and negative loads on the cell surface of microorganisms [53]. In general, the permeation of the cell surface causes the leakage of intracellular material and thereby causing cell death [54].

Phenolic compounds, including different polyphenols, flavonoids, and anthocyanins, are powerful antioxidants and antipathogen compounds, and a coating containing chitosan nanoparticles is able to restrict decay extension directly as a natural fungicide and indirectly as a resistance mediating elicitor [55]. Phenolic compounds are one of the most important secondary metabolites and are chemically very diverse; their biosynthesis is initiated with the phenylalanine, tyrosine, and tryptophan amino acids [56]. Anthocyanins also play the most important role in raspberry fruit color and overall quality. Phenolic compounds of raspberries inhibit liposome oxidation in the body. These compounds also have significantly shown a high capacity to eliminate singlet oxygen (free radical) or act as a hydrogen supplier [57]. The powerful antioxidative and anticancer activities of raspberries phenolic antioxidants in the human body have been well demonstrated [58]. After harvest, the concentrations of phenolic and flavonoids compounds are either substantially fixed or decreased [59]. As a novel and interesting finding, in this study, we found that chitosan nanoemulsion coating substantially enhances the biosynthesis and accumulation of different phenolic contents including total phenolics, flavonoids, and anthocyanin contents of the raspberry fruit during storage, resulting in an enhanced health-promoting capacity of the fruit. Moreover, these compounds, by detoxifying the free radicals and ROS, act as anti-stress and anti-senescence agents in the fruit resulting in enhanced postharvest life and quality maintenance.

PAL plays a key role in different phenolic biosynthesis pathways called the phenylpropanoid pathway [60]. It has been demonstrated that PAL activity decreases during the maturity and postharvest stages [61]. In addition, with the increase in the activity of the PPO activity during the senescence process, the consumption of polyphenols increases, and as a result, the content of total phenolic compounds decreases with the aging of the fruit tissue [62]. The results of this study show that the treatment enhanced the activity of PAL, resulting in the production of different phenolic compounds. On the other hand, edible coatings such as chitosan, by protecting the surface of the products, have been shown to reduce oxygen and thereby reducing the oxidation of phenolics [63].

Guaiacol peroxidase is one of the important antioxidant enzymes that were increased by an edible coating containing nanoparticles of chitosan in this study. This enzyme has been reported to show high activity in biological systems against hydrogen peroxide. Hydrogen peroxide is an important ROS-attacking cell membrane, and the increase in GPOX activity is pivotal for protecting the fruit cell membranes against peroxidation of lipids and DNA hydroperoxides [64].

Different enzymatic antioxidants are the cells' weapons against the oxidation of biological molecules such as lipids, proteins, carbohydrates, and deoxyribonucleic acid by free radicals and ROS [65,66]. The free radicals and ROS are produced during normal cell metabolism and attach the biomolecules of the cells called oxidative damage. To prevent oxidative stress, plant cells activate their antioxidant systems and respond to oxidative damage through the activation of different antioxidant systems [67]. Our results indicate that the antioxidant capacity, measured with both DPPH and FRAP methods, increases in the fruit treated with 5 g L^{-1} chitosan nanoemulsion. The use of edible coatings increases the capacity of the fruit antioxidant system and protects the cells against oxidative stress and pathogen attack [68,69].

5. Conclusions

In summary, our findings indicate that the use of chitosan nanoparticles can enhance different antioxidant fractions and the total antioxidant activity of raspberry fruit. Chitosan nanoemulsion treatments effectively prevented from the substantial loss in fruit taste and flavor during the storage period, while the control fruit exhibited off-flavors and bad taste. The taste and aroma during storage in the warehouse are significantly reduced due to increased respiration and enzymatic activity of the fruit. In the present study, the fruit treated with chitosan nanoemulsion was generally considered to be better for the sensory quality due to reduced respiration, metabolic activities, decay extension, weight loss and softening, and enhanced phytochemical contents. Competency of edible film/coatings and nanoforms in extending the shelf life without affecting the nutritional properties and safety aspects of fruit and vegetables still requires further attention.

Author Contributions: Conceptualization, H.S. and A.A.; writing—original draft preparation, S.R.I. and M.R.A.; writing—review and editing, H.S., A.A., M.P., and J.M.L. All authors have read and agreed to the published version of the manuscript.

Funding: This research received no external funding.

Institutional Review Board Statement: Not applicable.

Informed Consent Statement: Not applicable.

Data Availability Statement: The data presented in this study are available on request from the corresponding authors.

Conflicts of Interest: The authors declare no conflict of interest.

References

1. Alirezalu, K.; Pateiro, M.; Yaghoubi, M.; Alirezalu, A.; Peighambardoust, S.H.; Lorenzo, J.M. Phytochemical constituents, advanced extraction technologies and techno-functional properties of selected Mediterranean plants for use in meat products. A comprehensive review. *Trends Food Sci. Technol.* **2020**, *100*, 292–306. [CrossRef]
2. Lorenzo, J.M.; Pateiro, M.; Domínguez, R.; Barba, F.J.; Putnik, P.; Kovačević, D.B.; Shpigelman, A.; Granato, D.; Franco, D. Berries extracts as natural antioxidants in meat products: A review. *Food Res. Int.* **2018**, *106*, 1095–1104. [CrossRef]
3. Tulio, A.Z.; Reese, R.N.; Wyzgoski, F.J.; Rinaldi, P.L.; Fu, R.; Scheerens, J.C.; Miller, A.R. Cyanidin 3-rutinoside and cyanidin 3-xylosylrutinoside as primary phenolic antioxidants in black raspberry. *J. Agric. Food Chem.* **2008**, *56*, 1880–1888. [CrossRef]
4. Kanegusuku, M.; Sbors, D.; Bastos, E.S.; de Souza, M.M.; Cechinel-Filho, V.; Yunes, R.A.; Monache, F.D.; Niero, R. Phytochemical and analgesic activity of extract, fractions and a 19-hydroxyursane-type triterpenoid obtained from *Rubus rosaefolius* (Rosaceae). *Biol. Pharm. Bull.* **2007**, *30*, 999–1002. [CrossRef]
5. Ščetar, M.; Kurek, M.; Galić, K. Trends in fruit and vegetable packaging–A review. *Croat. J. Food Technol. Biotechnol. Nutr.* **2010**, *5*, 69–86.
6. Ponder, A.; Hallmann, E. The effects of organic and conventional farm management and harvest time on the polyphenol content in different raspberry cultivars. *Food Chem.* **2019**, *301*, 125295. [CrossRef] [PubMed]
7. Jianglian, D.; Shaoying, Z. Application of chitosan based coating in fruit and vegetable preservation: A review. *J. Food Process. Technol.* **2013**, *4*, 227. [CrossRef]
8. Yu, C.; Zeng, L.; Sheng, K.; Chen, F.; Zhou, T.; Zheng, X.; Yu, T. γ-Aminobutyric acid induces resistance against *Penicillium expansum* by priming of defence responses in pear fruit. *Food Chem.* **2014**, *159*, 29–37. [CrossRef]
9. Francis, G.A.; Gallone, A.; Nychas, G.J.; Sofos, J.N.; Colelli, G.; Amodio, M.L.; Spano, G. Factors Affecting quality and safety of fresh-cut produce. *Crit. Rev. Food Sci. Nutr.* **2012**, *52*, 595–610. [CrossRef]
10. Mastromatteo, M.; Conte, A.; Del Nobile, M.A. Combined use of modified atmosphere packaging and natural compounds for food preservation. *Food Eng. Rev.* **2010**, *2*, 28–38. [CrossRef]
11. Cota-Arriola, O.; Cortez-Rocha, M.O.; Burgos-Hernández, A.; Ezquerra-Brauer, J.M.; Plascencia-Jatomea, M. Controlled release matrices and micro/nanoparticles of chitosan with antimicrobial potential: Development of new strategies for microbial control in agriculture. *J. Sci. Food Agric.* **2013**, *93*, 1525–1536. [CrossRef]
12. Asghari, M.; Zahedipour, P. 24-epibrassinolide acts as a growth-promoting and resistance-mediating factor in strawberry plants. *J. Plant Growth Regul.* **2016**, *35*, 722–729. [CrossRef]
13. Dhakad, A.; Sonkar, P.; Bepari, A.; Kumar, U. Effect of pre-harvest application of plant growth regulators and calcium salts on biochemical and shelf life of acid lime (*Citrus aurantifolia* Swingle). *J. Pharmacogn. Phytochem.* **2020**, *9*, 1983–1985.
14. Katiyar, D.; Hemantaranjan, A.; Singh, B.; Bhanu, A.N. A future perspective in crop protection: Chitosan and its oligosaccharides. *Adv. Plants Agric. Res.* **2014**, *1*, 1–8.
15. Castelo Branco Melo, N.F.; de Mendonçasoares, B.L.; Diniz, K.M.; Leal, C.F.; Canto, D.; Flores, M.A.; Henrique da Costa Tavares-Filho, J.; Galembeck, A.; Stamford, T.L.M.; Stamford-Arnaud, T.M.; et al. Effects of fungal chitosan nanoparticles as eco-friendly edible coatings on the quality of postharvest table grapes. *Postharvest Biol. Technol.* **2018**, *139*, 56–66. [CrossRef]
16. Rinaudo, M. Chitin and chitosan: Properties and applications. *Prog. Polym. Sci.* **2006**, *31*, 603–632. [CrossRef]
17. Liu, J.; Zhang, J.; Xia, W. Hypocholesterolaemic effects of different chitosan samples in vitro and in vivo. *Food Chem.* **2008**, *107*, 419–425. [CrossRef]
18. Muzzarelli, R.A.; Boudrant, J.; Meyer, D.; Manno, N.; DeMarchis, M.; Paoletti, M.G. Current views on fungal chitin/chitosan, human chitinases, food preservation, glucans, pectins and inulin: A tribute to Henri Braconnot, precursor of the carbohydrate polymers science, on the chitin bicentennial. *Carbohydr. Polym.* **2012**, *87*, 995–1012. [CrossRef]

19. Eshghi, S.; Hashemi, M.; Mohammadi, A.; Badii, F.; Mohammadhoseini, Z.; Ahmadi, K. Effect of nanochitosan-based coating with and without copper loaded on physicochemical and bioactive components of fresh strawberry fruit (Fragaria x ananassa Duchesne) during storage. *Food Bioprocess Technol.* **2014**, *7*, 2397–2409. [CrossRef]
20. McClements, D.J.; Rao, J. Food-grade nanoemulsions: Formulation, fabrication, properties, performance, biological fate, and potential toxicity. *Crit. Rev. Food Sci. Nutr.* **2011**, *51*, 285–330. [CrossRef]
21. Acosta, E. Bioavailability of nanoparticles in nutrient and nutraceutical delivery. *Curr. Opin. Colloid Interface Sci.* **2009**, *14*, 3–15. [CrossRef]
22. Acevedo-Fani, A.; Soliva-Fortuny, R.; Martín-Belloso, O. Nanoemulsions as edible coatings. *Curr. Opin. Food Sci.* **2017**, *15*, 43–49. [CrossRef]
23. Chaudhary, S.; Kumar, S.; Kumar, V.; Sharma, R. Chitosan nanoemulsions as advanced edible coatings for fruits and vegetables: Composition, fabrication and developments in last decade. *Int. J. Biol. Macromol.* **2020**, *152*, 154–170. [CrossRef] [PubMed]
24. Bhargava, K.Y.; Aggarwal, S.; Kumar, T.; Bhargava, S. Comparative evaluation of the efficacy of three anti-oxidants vs. NaOCl and EDTA: Used for root canal irrigation in smear layer removal-SEM study. *Int. J. Pharm. Pharm. Sci.* **2015**, *7*, 366–371.
25. Zhang, L.; Li, J.; Hogan, S.; Chung, H.; Welbaum, G.E.; Zhou, K. Inhibitory effect of raspberries on starch digestive enzyme and their antioxidant properties and phenolic composition. *Food Chem.* **2010**, *119*, 592–599. [CrossRef]
26. Ponder, A.; Hallmann, E. The nutritional value and vitamin C content of different raspberry cultivars from organic and conventional production. *J. Food Compos. Anal.* **2020**, *87*, 103429. [CrossRef]
27. Bazzocco, S.; Mattila, I.; Guyot, S.; Renard, C.M.G.C.; Aura, A.-M. Factors affecting the conversion of apple polyphenols to phenolic acids and fruit matrix to short-chain fatty acids by human faecal microbiota in vitro. *Eur. J. Nutr.* **2008**, *47*, 442–452. [CrossRef]
28. Horvitz, S. Postharvest Handling of Berries. *Postharvest Handl.* **2017**, 107–123. [CrossRef]
29. Ayala-Zavala, J.F.; Wang, S.Y.; Wang, C.Y.; González-Aguilar, G.A. High oxygen treatment increases antioxidant capacity and postharvest life of strawberry fruit. *Food Technol. Biotechnol.* **2007**, *45*, 166–173.
30. Khademi, Z.; Ershadi, A. Postharvest application of salicylic acid improves storability of peach (*Prunuspersicacv. Elberta*) Fruits. *Int. J. Agric. Crop Sci.* **2013**, *5*, 651.
31. Vargas, M.; Albors, A.; Chiralt, A.; González-Martínez, C. Quality of cold-stored strawberries as affected by chitosan–oleic acid edible coatings. *Postharvest Biol. Technol.* **2006**, *41*, 164–171. [CrossRef]
32. Alirezalu, A.; Ahmadi, N.; Salehi, P.; Sonboli, A.; Alirezalu, K.; Khaneghah, A.M.; Barba, F.J.; Munekata, P.E.; Lorenzo, J.M. physicochemical characterization, antioxidant activity, and phenolic compounds of hawthorn (*Crataegus* spp.) fruits species for potential use in food applications. *Foods* **2020**, *9*, 436. [CrossRef] [PubMed]
33. Shin, Y.; Liu, R.H.; Nock, J.F.; Holliday, D.; Watkins, C.B. Temperature and relative humidity effects on quality, total ascorbic acid, phenolics and flavonoid concentrations, and antioxidant activity of strawberry. *Postharvest Biol. Technol.* **2007**, *45*, 349–357. [CrossRef]
34. Shameh, S.; Alirezalu, A.; Hosseini, B.; Maleki, R. Fruit phytochemical composition and color parameters of 21 accessions of five Rosa species grown in North West Iran. *J. Sci. Food Agric.* **2019**, *99*, 5740–5751. [CrossRef] [PubMed]
35. Nakajima, J.-I.; Tanaka, I.; Seo, S.; Yamazaki, M.; Saito, K. LC/PDA/ESI-MS Profiling and radical scavenging activity of anthocyanins in various berries. *J. Biomed. Biotechnol.* **2004**, *2004*, 241–247. [CrossRef] [PubMed]
36. Alirezalu, A.; Salehi, P.; Ahmadi, N.; Sonboli, A.; Aceto, S.; Maleki, H.H.; Ayyari, M. Flavonoids profile and antioxidant activity in flowers and leaves of hawthorn species (*Crataegus* spp.) from different regions of Iran. *Int. J. Food Prop.* **2018**, *21*, 452–470. [CrossRef]
37. Karthikeyan, M.; Radhika, K.; Mathiyazhagan, S.; Bhaskaran, R.; Samiyappan, R.; Velazhahan, R. Induction of phenolics and defense-related enzymes in coconut (*Cocos nucifera* L.) roots treated with biocontrol agents. *Braz. J. Plant Physiol.* **2006**, *18*, 367–377. [CrossRef]
38. Hassanpour, H. Effect of Aloe vera gel coating on antioxidant capacity, antioxidant enzyme activities and decay in raspberry fruit. *LWT* **2015**, *60*, 495–501. [CrossRef]
39. Gardesh, A.S.K.; Badii, F.; Hashemi, M.; Ardakani, A.Y.; Maftoonazad, N.; Gorji, A.M. Effect of nanochitosan based coating on climacteric behavior and postharvest shelf-life extension of apple cv. *Golab Kohanz. LWT* **2016**, *70*, 33–40. [CrossRef]
40. Lara, I.; Garcia, P.; Vendrell, M. Post-harvest heat treatments modify cell wall composition of strawberry (*Fragaria × ananassa* Duch.) fruit. *Sci. Hortic.* **2006**, *109*, 48–53. [CrossRef]
41. Gao, P.; Zhu, Z.; Zhang, P. Effects of chitosan–glucose complex coating on postharvest quality and shelf life of table grapes. *Carbohydr. Polym.* **2013**, *95*, 371–378. [CrossRef]
42. Salvia-Trujillo, L.; Rojas-Graü, M.A.; Soliva-Fortuny, R.; Martín-Belloso, O. Use of antimicrobial nanoemulsions as edible coatings: Impact on safety and quality attributes of fresh-cut Fuji apples. *Postharvest Biol. Technol.* **2015**, *105*, 8–16. [CrossRef]
43. Conforti, F.D.; Ball, J.A. A comparison of lipid and lipid/hydrocolloid based coatings to evaluate their effect on postharvest quality of green bell peppers. *J. Food Qual.* **2002**, *25*, 107–116. [CrossRef]
44. Hong, K.; Xie, J.; Zhang, L.; Sun, D.; Gong, D. Effects of chitosan coating on postharvest life and quality of guava (*Psidium guajava* L.) fruit during cold storage. *Sci. Hortic.* **2012**, *144*, 172–178. [CrossRef]
45. Bautista-Baños, S.; Hernández-López, M.; Bosquez-Molina, E.; Wilson, C. Effects of chitosan and plant extracts on growth of *Colletotrichum gloeosporioides*, anthracnose levels and quality of papaya fruit. *Crop Prot.* **2003**, *22*, 1087–1092. [CrossRef]

46. Treviño-Garza, M.Z.; García, S.; Flores-González, M.D.S.; Arévalo-Niño, K. Edible active coatings based on pectin, pullulan, and chitosan increase quality and shelf life of strawberries (*Fragaria ananassa*). *J. Food Sci.* **2015**, *80*, M1823–M1830. [CrossRef]
47. Nair, M.S.; Tomar, M.; Punia, S.; Kukula-Koch, W.; Kumar, M. Enhancing the functionality of chitosan- and alginate-based active edible coatings/films for the preservation of fruits and vegetables: A review. *Int. J. Biol. Macromol.* **2020**, *164*, 304–320. [CrossRef]
48. Orzali, L.; Forni, C.; Riccioni, L. Effect of chitosan seed treatment as elicitor of resistance to *Fusarium graminearum* in wheat. *Seed Sci. Technol.* **2014**, *42*, 132–149. [CrossRef]
49. Hasan, S.M.K.; Ferrentino, G.; Scampicchio, M. Nanoemulsion as advanced edible coatings to preserve the quality of fresh-cut fruits and vegetables: A review. *Int. J. Food Sci. Technol.* **2019**, *55*, 1–10. [CrossRef]
50. Xing, K.; Zhu, X.; Peng, X.; Qin, S. Chitosan antimicrobial and eliciting properties for pest control in agriculture: A review. *Agron. Sustain. Dev.* **2015**, *35*, 569–588. [CrossRef]
51. Sivakumar, D.; Bautista-Baños, S. A review on the use of essential oils for postharvest decay control and maintenance of fruit quality during storage. *Crop Prot.* **2014**, *64*, 27–37. [CrossRef]
52. Chung, Y.-C.; Chen, C.-Y. Antibacterial characteristics and activity of acid-soluble chitosan. *Bioresour. Technol.* **2008**, *99*, 2806–2814. [CrossRef]
53. Krajewska, B.; Wydro, P.; Jańczyk, A. Probing the modes of antibacterial activity of chitosan. Effects of ph and molecular weight on chitosan interactions with membrane lipids in langmuir films. *Biomacromolecules* **2011**, *12*, 4144–4152. [CrossRef]
54. Verlee, A.; Mincke, S.; Stevens, C.V. Recent developments in antibacterial and antifungal chitosan and its derivatives. *Carbohydr. Polym.* **2017**, *164*, 268–283. [CrossRef]
55. Thumula, P. Studies on Storage Behaviour of Tomatoes Coated with Chitosan-Lysozyme Films. Ph.D. Thesis, McGill University, Montreal, QC, Canada, 2006.
56. Edwards, J.E.; Brown, P.N.; Talent, N.; Dickinson, T.A.; Shipley, P.R. A review of the chemistry of the genus *Crataegus*. *Phytochemistry* **2012**, *79*, 5–26. [CrossRef]
57. Dai, J.; Gupte, A.; Gates, L.; Mumper, R.J. A comprehensive study of anthocyanin-containing extracts from selected black-berry cultivars: Extraction methods, stability, anticancer properties and mechanisms. *Food Chem. Toxicol.* **2009**, *47*, 837–847. [CrossRef]
58. Baby, B.; Antony, P.; Vijayan, R. Antioxidant and anticancer properties of berries. *Crit. Rev. Food Sci. Nutr.* **2017**, *58*, 2491–2507. [CrossRef] [PubMed]
59. Vámos-Vigyázó, L.; Haard, N.F. Polyphenol oxidases and peroxidases in fruits and vegetables. *Crit. Rev. Food Sci. Nutr.* **1981**, *15*, 49–127. [CrossRef]
60. Meng, X.; Li, B.; Liu, J.; Tian, S. Physiological responses and quality attributes of table grape fruit to chitosan preharvest spray and postharvest coating during storage. *Food Chem.* **2008**, *106*, 501–508. [CrossRef]
61. Sanchez-Gonzalez, L.; Pastor, C.; Vargas, M.; Chiralt, A.; Gonzalez-Martinez, C.; Chafer, M. Effect of hydroxypropylmethylcellulose and chitosan coatings with and without bergamot essential oil on quality and safety of cold-stored grapes. *Postharvest Biol. Technol.* **2011**, *60*, 57–63. [CrossRef]
62. Xu, W.T.; Peng, X.L.; Luo, Y.B.; Wang, J.A.; Guo, X.; Huang, K.L. Physiological and biochemical responses of grapefruit seed extract dip on "Redglobe" grape. *LWT Food Sci. Technol.* **2009**, *42*, 471–476. [CrossRef]
63. Jiang, Y.; Li, Y. Effects of chitosan coating on postharvest life and quality of longan fruit. *Food Chem.* **2001**, *73*, 139–143. [CrossRef]
64. Chaudière, J.; Ferrari-Iliou, R. Intracellular antioxidants: From chemical to biochemical mechanisms. *Food Chem. Toxicol.* **1999**, *37*, 949–962. [CrossRef]
65. Dar, T.A.; Uddin, M.; Khan, M.M.A.; Hakeem, K.; Jaleel, H. Jasmonates counter plant stress: A Review. *Environ. Exp. Bot.* **2015**, *115*, 49–57. [CrossRef]
66. Wang, J.; Wang, K.; Wang, Y.; Lin, S.; Zhao, P.; Jones, G. A novel application of pulsed electric field (PEF) processing for improving glutathione (GSH) antioxidant activity. *Food Chem.* **2014**, *161*, 361–366. [CrossRef]
67. Gill, S.S.; Tuteja, N. Reactive oxygen species and antioxidant machinery in abiotic stress tolerance in crop plants. *Plant Physiol. Biochem.* **2010**, *48*, 909–930. [CrossRef] [PubMed]
68. Hu, Q.; Hu, Y.; Xu, J. Free radical-scavenging activity of Aloe vera (*Aloe barbadensis Miller*) extracts by supercritical carbon dioxide extraction. *Food Chem.* **2005**, *91*, 85–90. [CrossRef]
69. Ishkeh, S.R.; Asghari, M.; Shirzad, H.; Alirezalu, A.; Ghasemi, G. Lemon verbena (*Lippia citrodora*) essential oil effects on antioxidant capacity and phytochemical content of raspberry (*Rubus ulmifolius* subsp. *sanctus*). *Sci. Hortic.* **2019**, *248*, 297–304. [CrossRef]

applied
sciences

MDPI

Article

Thermodynamic and Quality Performance Studies for Drying Kiwi in Hybrid Hot Air-Infrared Drying with Ultrasound Pretreatment

Ebrahim Taghinezhad [1,*], Mohammad Kaveh [2] and Antoni Szumny [3]

[1] Department of Agricultural Technology Engineering, Moghan College of Agriculture and Natural Resources, University of Mohaghegh Ardabili, Ardabil 56199-11367, Iran
[2] Faculty of Agriculture and Natural Resources, University of Mohaghegh Ardabili, Ardabil 56199-11367, Iran; sirwankaweh@uma.ac.ir
[3] Department of Chemistry, Wroclaw University of Environmental and Life Science, CK Norwida 25, 50-375 Wrocław, Poland; antoni.szumny@upwr.edu.pl
* Correspondence: e.taghinezhad@uma.ac.ir; Tel.: +98-4532715417

Abstract: The present study examined the effect of ultrasonic pretreatment at three time the levels of 10, 20 and 30 min on some thermodynamic (effective moisture diffusivity coefficient(D_{eff}), drying time, specific energy consumption (SEC), energy efficiency, drying efficiency, and thermal efficiency) and physical (color and shrinkage) properties of kiwifruit under hybrid hot air-infrared(HAI) dryer at different temperatures (50, 60 and 70 °C) and different thicknesses (4, 6 and 8 mm). A total of 11 mathematical models were applied to represent the moisture ratio (MR) during the drying of kiwifruit. The fitting of MR mathematical models to experimental data demonstrated that the logistic model can satisfactorily describe the MR curve of dried kiwifruit with a correlation coefficient (R^2) of 0.9997, root mean square error (RMSE) of 0.0177 and chi-square (χ^2) of 0.0007. The observed D_{eff} of dried samples ranged from 3.09×10^{-10} to 2.26×10^{-9} m^2/s. The lowest SEC, color changes and shrinkage were obtained as 36.57 kWh/kg, 13.29 and 25.25%, respectively. The highest drying efficiency, energy efficiency, and thermal efficiency were determined as 11.09%, 7.69% and 10.58%, respectively. The results revealed that increasing the temperature and ultrasonic pretreatment time and decreasing the sample thickness led to a significant increase ($p < 0.05$) in drying efficiency, thermal efficiency, and energy efficiency, while drying time, SEC and shrinkage significantly decreased ($p < 0.05$).

Keywords: dry; efficiency; energy; kiwifruit; quality; ultrasound

check for
updates

Citation: Taghinezhad, E.; Kaveh, M.; Szumny, A. Thermodynamic and Quality Performance Studies for Drying Kiwi in Hybrid Hot Air-Infrared Drying with Ultrasound Pretreatment. *Appl. Sci.* **2021**, *11*, 1297. https://doi.org/10.3390/app11031297

Academic Editor: Theodoros Varzakas

Received: 12 January 2021
Accepted: 27 January 2021
Published: 1 February 2021

Publisher's Note: MDPI stays neutral with regard to jurisdictional claims in published maps and institutional affiliations.

1. Introduction

Kiwifruit with the scientific name of *Actinidia deliciosa* is a subtropical fruit belonging to the Actinidiaceae family. This fruit is rich in vitamins E and C, phenolic substances, flavonoids, various pigments, and antioxidants [1]. According to the research by the Food and Agriculture Organization (FAO) of the United Nations, kiwifruit is the fourth most popular fruit in the world after banana, citrus fruits, and apple [2].

Fruits and vegetables with a moisture content (*MC*) of about 80% are classified as highly perishable materials. The high-potential drying is one of the most common methods to preserve and increase the productivity of food [3]. The most common method for drying food is to use hot air flow, which requires hot air to provide a simultaneous transfer of heat and moisture between hot air and food to evaporate water and to bring the moisture to the desired level [4]. In addition to having advantages, such as the possibility of accurate temperature control regardless of the size and shape of the product and not needing direct contact, the drying with hot air also has some disadvantages. These disadvantages include the long time and high temperature for drying the product during the period of descending speed. In addition, high temperatures reduce the nutritional value and increase energy consumption [5]. The infrared dryers have been commonly used in recent years and have received more attention in developing countries due to the low cost, simple equipment, and low price. Infrared drying is the process in which wet materials are heated by infrared radiation. Infrared radiation is a type of electromagnetic radiation with a wavelength between 0.76 and 400 μm [6]. In the hybrid method, the use of hot air and infrared waves coincide in all stages of product drying. The reports show that the hybrid method at mild air speeds can improve energy consumption and drying time while increasing product quality [7]. The application of the pretreatments that reduce the initial moisture of the food and also modify the food texture to accelerate the transfer of moisture and reduce drying time will be very useful.

Ultrasonic pretreatment is performed by immersing the fruit in water and applying ultrasound waves. Ultrasonic waves cause a series of rapid, intermittent contractions and expansions of the cell wall (sponge effect), which creates microscopic channels in the structure of the material and increases the possibility of moisture transfer. As a result, an ultrasound can be used as a pretreatment in drying heat-sensitive food products without adversely affecting the qualitative characteristics of the food, since it improves the speed and decreases the temperature required for drying [8].

Mierzwa et al. examined the effect of ultrasonic pretreatment with two types of microwaves and hot air dryers on some drying parameters (kinetics, shrinkage, specific energy consumption, and rehydration ratio) of raspberry, and stated that the use of ultrasonic pretreatment reduced specific energy consumption and shrinkage in both types of dryers [9]. Abbaspour et al. investigated the drying kinetics, SEC, color, shrinkage, and rehydration of carrots using three types of hot air, microwave and infrared dryers with ultrasonic pretreatment [10]. They reported that the use of ultrasonic pretreatment reduced the drying time of the sample compared to the control treatment. The use of ultrasonic pretreatment in all three types of hot air, microwave and infrared dryers also increased the D_{eff} and reduced the SEC, color changes, and shrinkage. Dehghannya et al. studied the effect of using ultrasonic pretreatment on the drying of potato slices in hot air microwave dryers and measured the MC, D_{eff}, SEC, rehydration percentage, and shrinkage [11]. They stated that the application of ultrasonic waves reduced the drying time. Additionally, the samples under ultrasonic pretreatment had lower shrinkage and SEC and a higher moisture diffusivity coefficient and rehydration percentage than the control sample. Ghanbarian et al. dried peppermint using the hot air dryer (at three temperature levels) with ultrasonic pretreatment (four levels of ultrasonic power) [12]. They examined MC, drying time, SEC, and energy efficiency of peppermint. The results showed that the use of ultrasonic pretreatment reduced the drying time and SEC and increased the energy efficiency of peppermint. Thus far, various studies have been conducted on the drying of kiwifruit using hot air dryer [13,14], hot air dryer with recirculation under heat pump [15], microwave [16], infrared-vacuum dryer [17], vacuum-hot air dryer [18], hot air, vacuum, freezer, hot air-microwave-vacuum [19], hot air, microwave and freezer [20], HAI [21], freezer [22], and short and medium-wave infrared radiation with/without osmotic dehydration [23].

Due to the importance of kiwifruit in the human diet and the effect of the drying process on the maintenance of this product with high qualitative properties and low energy consumption, and given the lack of a comprehensive study on the effect of ultrasound on the qualitative and thermodynamic properties of kiwifruit, this research aimed to evaluate the effect of ultrasonic pretreatment on the changes in D_{eff}, SEC, energy efficiency, drying efficiency, color, shrinkage, and drying time of kiwifruit slices during drying with the hybrid HAI dryer.

2. Materials and Methods

Fresh Hayward kiwifruit was bought from the local market of ParsAbad, Ardabil province, Iran and stored at 4 °C. One hour before the test, the samples were removed from the refrigerator and kept at 24 °C. Using an electric slicer (Celme GE300, Montebello Vicentino, Italy), kiwifruits were cut into 4, 6 and 8 mm thick slices and the slices were then cut into 24 mm discs by a circular slicer. The initial MC of kiwifruit samples was obtained in an oven at 105 °C for 24 h. The initial *MC* of kiwifruit (6.85% (d.b.)) was determined and the drying continued until reaching an *MC* of 0.2% (d.b.).

2.1. Ultrasonic

The kiwifruit samples selected for ultrasonic pretreatment were transferred to an ultrasonic bath. All samples were placed in a 6 Lultrasonic bath (Parsonic 7500S, Tehran, Iran), which was filled with distilled water at an ambient temperature of 25 °C in a ratio of 1:4. The samples were then exposed to ultrasonic waves for 0, 10, 20 and 30 min. The frequency and power of this test were 28 kHz and 70 W, respectively.

2.2. Hybrid Hot Air-Infrared (HAI) Dryer

A hybrid *HAI* dryer (GC 400, Grouc, Tehran, Iran) was used for the experiments. The dryer consisted of two infrared lamps (Philips, Flemish, Belgium), each with a rated power of 250 W (500 W in total), which was installed in the upper part of the drying chamber at a height of 30 cm. The flow rate of the air entering the dryer was set to 1 m/s. The kiwifruits under study were placed in the middle of the channel on the containers made from wire mesh on digital scales (AND, GF–6000, A&D Company Ltd, Tokyo, Japan) with a precision of 0.01 g embedded below and outside the channel. The drying of kiwifruit was performed at three air temperature levels of 50, 60 and 70 °C. To start the test, the machine was put into operation for 15 min to fix the temperature and flow rate of the dryer's air. In each experiment, 55 g of kiwifruit was placed in a single layer on a drying tray. During the drying experiments, the mean range of changes in ambient temperature and the relative humidity were 20 ± 4 °C and 15 ± 5%, respectively.

2.3. Moisture Ratio (MR) and Mathematical Modeling

To model the drying process, the *MR* parameter during the kiwifruit drying was obtained using the following Equation (1) [24]:

$$MR = \frac{M_t - M_e}{M_o - M_e} \tag{1}$$

MR: moisture ratio (dimensionless), M_t: MC of samples at any given time on dry basis (d.b.), M_e: equilibrium moisture content (d.b.), and M_o: initial moisture of samples (d.b.). According to Equation (1), the moisture ratio depends on the initial moisture, balanced moisture, and moisture of samples at any time during the drying. For long drying times, the M_e values are very small compared to the M_o and M_t values. Therefore, the equation of *MR* during the drying can be simplified, as in Equation (2), and to calculate the *MR*, it is not necessary to measure the balanced moisture [25].

$$MR = \frac{M_t}{M_o} \tag{2}$$

2.4. Modeling

In this study, a number of experimental and proposed models were used to model the drying kinetics (fitting of *MR* versus drying time) of green kiwifruit samples (Table 1). To select the most appropriate model to describe the drying kinetics of green kiwifruit, the criteria of determination coefficient (R^2), root mean square error (*RMSE*), and chi-square (χ^2) were calculated by each model and compared with other models. The higher the R^2 value and the lower the *RMSE* and χ^2 values, the better the model fitting to the experimental data and it is chosen as the best model. The R^2, *RMSE* and χ^2 values were calculated according to Equations (3)–(5):

$$R^2 = \frac{\sum\limits_{i=1}^{N} \left(MR_{\mathrm{exp},i} - MR_{pre,i}\right)^2}{\sum\limits_{i=1}^{N} \left(\dfrac{\sum\limits_{i=1}^{N} MR_{pre,i}}{N} - MR_{pre,i}\right)^2} \tag{3}$$

$$\chi^2 = \frac{\sum\limits_{i=1}^{N} \left(MR_{\mathrm{exp},i} - MR_{pre,i}\right)^2}{N - z} \tag{4}$$

$$RMSE = \left[\frac{1}{N}\sum\limits_{i=1}^{N} \left(MR_{pre,i} - MR_{\mathrm{exp},i}\right)^2\right]^{\frac{1}{2}} \tag{5}$$

N, z, $MR_{pre,i}$ and $MR_{\mathrm{exp},i}$ are the number of observations, number of dryer constants, predicted MR_i, and experimental MR_i, respectively [20].

Table 1. Applied models to fit the experimental data $MR = 1 + at + bt^2$.

References	Equations	Models
Two-term	$MR = a\exp(-kt0) + b\exp(-k_1 t)$	[24]
Two-term exponential	$MR = a\exp(-kt) + (1-a)\exp(-kat)$	[20]
Verma	$MR = a\exp(-kt) + (1-a)\exp(-gt)$	[17]
Demir et al.	$MR = a\exp\left(-kt\right)^n + b$	[26]
Midilli et al.	$MR = a\exp(-kt^n) + bt$	[19]
Newton (Lewis)	$MR = a\exp(-kt)$	[27]
Page	$MR = a\exp(-kt^n)$	[28]
Wang and Singh	$MR = 1 + at + bt^2$	[13]
Henderson and Pabis	$MR = a\exp(-kt)$	[16]
Logistic	$MR = a/(1 + b\exp(kt))$	[29]

a, b, c, k, k_0, k_1 and n are drying constants.

2.5. Determination of Effective Moisture Diffusivity Coefficient (D_{eff})

The number of moisture transfer mechanisms is extensive and often complex. Transfer phenomena are usually classified into pressure diffusion, forced diffusion, and ordinary diffusion (net transfer of matter without fluid motion). According to Equation (6), the Fick's second law for unstable conditions can describe the transfer of moisture in the descending stage of the drying process [29,30]:

$$\frac{\partial X}{\partial t} = D_{eff}\frac{\partial^2 X}{\partial^2 x} \tag{6}$$

where X is the local MC on dry basis, t is the time and x is the spatial index. The study of the Fick's second law of diffusion indicates the mass transfer during the period of the declining drying rate of agricultural products [26]. To apply Fick's law, it is assumed that the food product is one-dimensional, the initial MC is uniform, and it has an internal movement of moisture such as the major resistance to moisture transfer. The Fick's equation for a plane sheet is solved as follows [31,32]:

$$MR = \frac{8}{\pi^2}\sum\limits_{n=1}^{\infty}\frac{1}{(2n+1)}\exp\left(\frac{-D_{eff}(2n+1)^2\pi^2 t}{4L^2}\right) \tag{7}$$

From 1 to infinity, t is the drying time (s), and D_{eff} is the effective diffusivity coefficient (m^2/s). The effective diffusivity coefficient is obtained by calculating the slope in Equation (8) [11]:

$$MR = \frac{8}{\pi^2} \exp\left(\frac{-D_{eff}\pi^2 t}{4L^2}\right) \tag{8}$$

The diffusivity coefficient is usually determined by plotting the experimental drying data versus LnMR over time. When plotting the LnMR value over time, place the resulting line slope in Equation (9) to obtain the D_{eff}:

$$K = \left(\frac{D_{eff}\pi^2}{4L^2}\right) \tag{9}$$

where K is the line slope.

2.6. Calculation of Some Thermodynamic Parameters

2.6.1. Energy Consumption

The thermal energy consumed by the hybrid *HAI* dryer for drying kiwifruit at infrared temperature and power at constant air flow rate is obtained using Equation (10) [33]:

$$EU_{ter} = A.v.\rho_a.C_a.\Delta t.t + K.t \tag{10}$$

where EU_{ter} is the total thermal energy consumption per drying period (kWh), A is the area of the container in which the test sample is placed (m^2), v is the wind speed (m/s), ρ_a is the air density (kg/m^3), t is the total drying time of each sample (h), ΔT is the temperature difference (°C), and C_a is the specific heat (KJ/kg °C).

The mechanical energy consumed by the blower in the dryer during the drying process is obtained from Equation (11) [34]:

$$EU_{mec} = \Delta P.W_{air}.t \tag{11}$$

EU_{mec} is the total mechanical energy consumption per period (kJ), ΔP is the pressure drop (Pa), and W_{air} is the airflow rate (m^3/s).

The energy consumption for ultrasonic pretreatment can be obtained using the following equation [10]:

$$EU_{ult} = UI \cos \Phi \tag{12}$$

where EU_{ult} is the ultrasonic power (kw), U (V) and I (A) are the input voltage and current in the ultrasonic machine, respectively, and $\cos\Phi$ is the power factor equal to 0.8.

The energy consumption required for 1 kg of moisture to be removed from the drying product is calculated by Equation (13):

$$SEC_{total} = \frac{EU_{(mec+ter+ult)}}{M_W} \tag{13}$$

where *SEC* is the specific energy consumption (kWh/kg), $EU_{(mec+ter+ult)}$ is the total energy consumption during the drying process (kWh), and M_w is the amount of moisture removed from the product (kg).

2.6.2. Drying Efficiency

Drying efficiency is the division of the total energy required to heat the drying product and the energy required to evaporate moisture by the total energy consumption during the drying process (D_e):

$$D_e = \left(\frac{E_{evap} + E_{heating}}{SEC}\right) \tag{14}$$

where D_e is the drying efficiency (%), $E_{heating}$ is the energy required to increase the temperature of the dried product (kJ), E_{evap} is the energy needed to evaporate moisture (kJ), and SEC is the total energy consumption (kJ).

2.6.3. Energy Efficiency

Energy efficiency is obtained by dividing the energy required to evaporate moisture from the drying product by the total energy SEC during the drying process.

$$\eta_e = \left(\frac{E_{evap}}{SEC} \right) \times 100 \tag{15}$$

where η_e is the energy efficiency and E_{eva} is the energy required to evaporate moisture (kJ).

2.6.4. Thermal Efficiency

Thermal efficiency is the division of the moisture evaporated from the product by the heat used to dry the product, which can be calculated from Equation (16):

$$TE = \frac{M_{evap}}{H_{total}} \tag{16}$$

where TE is the thermal efficiency (kg/kJ), M_{evap} is the weight of moisture evaporated from the product (kg), and H_{total} is the total heat consumption (kg).

2.6.5. Energy Required to Evaporate Moisture from Product

The energy required to evaporate moisture from the product during the kiwifruit drying process in the hybrid HAI dryer with ultrasonic pretreatment is calculated from Equation (17):

$$Q_w = h_{f.g}.M_w \tag{17}$$

where Q_w is the energy required to evaporate moisture (kJ), h_{fg} is the latent heat of vaporization (kJ/kg), and M_w is the moisture evaporated from the product (kg).

The latent heat of vaporization can be obtained from Equation (18):

$$h_{f.g} = (7.33 \times 10^6 - 16T_{abs}^2)^{0.5}$$
$$273.16 \leq T_{abs} \leq 338.72$$
$$h_{f.g} = 2.503 \times 10^3 - 2.386(T_{abs} - 273.16) \tag{18}$$
$$337.72 \leq T_{abs} \leq 533.16$$

2.6.6. Energy Required to Increase Product Temperature

The energy required to increase the product temperature from the initial temperature (temperature before entering the drying chamber) to the latent temperature (highest product temperature) can be calculated from Equations (19)–(21):

$$E_{heating} = W_d C_m (T_2 - T_1) \tag{19}$$

$$C_m = 1465 + 3560(\frac{M_P}{1 + M_P}) \tag{20}$$

$$M_P = \frac{W_w - W_d}{W_d} \tag{21}$$

where $E_{heating}$ is the energy required to increase the product temperature (kJ), W_d is the dry matter weight (kg), C_m is the specific heat of the product (kJ/kg K), T_1 is the initial product temperature (K), T_2 is the latent product temperature (K), M_p is the MC of the product on dry basis (kg water/kg solid), and M_w is the initial weight of the product (kg).

2.7. Shrinkage

Shrinkage is the ratio between the initial volume of fresh kiwifruit and the final volume of dried kiwifruit, which is expressed by the following equation:

$$S_v = 1 - \frac{V_t}{V_o} \times 100 \tag{22}$$

where S_v is the percentage of shrinkage (dimensionless), V_t is the volume at the desired time (cm^3) and V is the initial sample volume (cm^3). The sample volume is obtained by the toluene displacement method according to Dehghannya et al. [11].

2.8. Color

The color of the samples prior to and following the drying was measured by a color meter (Portable colorimeter, HP–200, China). To describe the color changes during the drying, the ΔE index (color difference of total dried samples from fresh samples) was used. This index is defined by Equation (8) where L represents the brightness, b represents the yellow-blue color and a indicates the red-green color [25,35]:

$$\Delta E = \sqrt{(L^* - L_0^*)^2 + (a^* - a_0^*)^2 + (b^* - b_0^*)^2} \tag{23}$$

2.9. Data Analysis

In this study, the factorial experiment was used in a completely randomized design. The treatments were considered at three drying temperatures (50, 60 and 70 °C), three levels of thickness (4, 6 and 8 mm) and three levels of ultrasonic radiation time (10, 20 and 30 min). Additionally, the effect of treatments on the qualitative (color changes and shrinkage) and thermodynamic parameters (D_{eff}, SEC, thermal efficiency, energy efficiency, and drying efficiency) was investigated for the drying of kiwifruit slices. To examine the significance of the factors, the data analysis of variance was performed using SPSS software and the mean comparison of treatments was made at the 5% probability level.

3. Results and Discussion

3.1. Mathematical Modeling

Table 2 shows the statistical parameters obtained from the fitting of the experimental data of the drying experiments to the 11 models listed in Table 1. According to the tables, it can be seen that, in the method of hybrid *HAI* drying with ultrasonic pretreatment, the logistic model has the highest value of determination coefficient (0.9997) and the lowest values of chi-square ($\chi^2 = 0.0007$) and root mean square error ($RMSE = 0.0177$) compared to other models.

Table 2. Statistical analysis of mathematical modeling of kiwifruit slices drying kinetics.

Models	*RMSE*	χ^2	R^2
Two-term	0.0501	0.0167	0.9979
Two-term exponential	0.0648	0.0221	0.9968
Verma	0.0402	0.0092	0.9985
Demir et al.	0.0367	0.0074	0.9988
Midilli et al.	0.0211	0.0017	0.9995
Newton (Lewis)	0.0761	0.0272	0.9958
Page	0.0285	0.0037	0.9991
Wang and Singh	0.0473	0.0132	0.9981
Henderson and Pabis	0.0533	0.0192	0.9977
Logistic	0.0177	0.0007	0.9997

3.2. Drying Time

Figure 1 shows the effect of drying air temperature, thickness and ultrasonic pretreatment time on the drying time of kiwifruit slices. As shown in Figure 1, the maximum drying time was obtained at 50 °C with a thickness of 8 mm (control sample) for 420 min. Also, the minimum drying time was obtained at 70 °C, 4 mm thickness and ultrasonic radiation time of 30 min for 70 min. As expected, the drying time is significantly reduced ($p < 0.05$) by raising the temperature due to the increased thermal gradient (as a result of higher mass and heat transfer) and greater and quicker removal of moisture from the product [36].

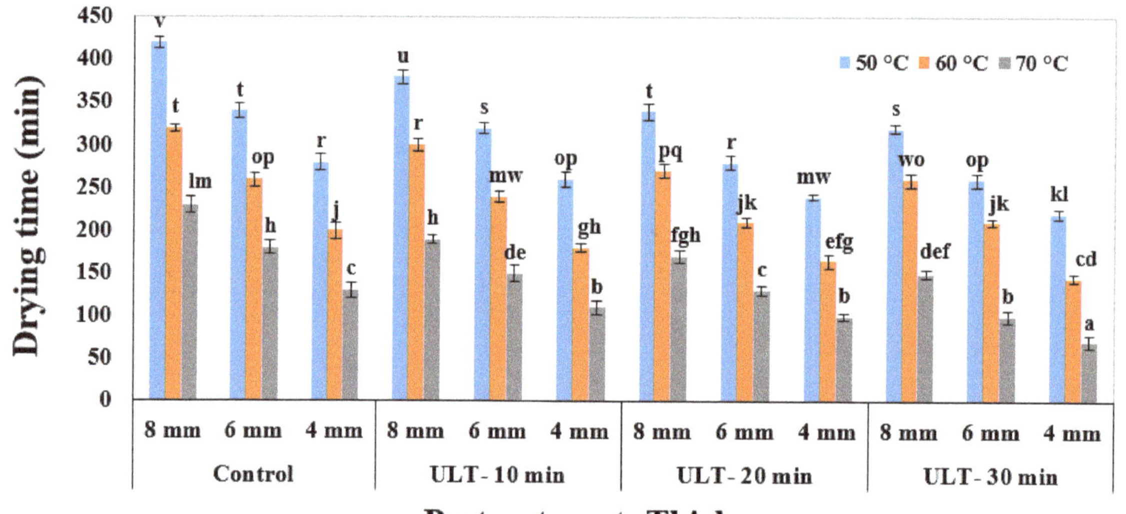

Figure 1. Drying time under different conditions for drying kiwifruit slices.

The results obtained in Figure 1 also show that the highest effect of using ultrasonic pretreatment on the drying time of kiwifruit is seen for 30 min of ultrasonic application and the lowest one is seen for the control sample. By applying the cavitation process (asymmetric bursting of bubbles) near the foodsurface, ultrasonic pretreatment causes the rapid and eruptive flow of sound waves to the surface and forms microscopic channels in the samples by creating successive contractions and expansions. In addition, by increasing the duration of applying this energy (ultrasound), the channels expands and the product has a spongy texture, facilitating the flow of water out of the product during the drying through the created channels [37]. Similar results were observed for carrot drying [24], banana drying [31], and blackberry drying [29]. They showed that, by increasing the ultrasonic radiation time, the drying time decreases significantly ($p < 0.05$). As shown in Figure 1, the drying time was longer as the thickness of the samples increased. The reason for this issue is the hardening mechanism of the cutting surface, increased amount of water, and also a longer path of vaporremoval, which made it difficult to transfer moisture through the texture and ultimately increased the drying time [38]. Doymaz and Ozdemir for drying tomatoes and Onwude et al. for drying potatoes showed that the drying time increases by enlarging the thickness of the samples [28,36].

3.3. D_{eff}

Different values of D_{eff} for different treatments of this study are reported in Figure 2. By increasing the drying temperature, the sample thickness and ultrasonic radiation time during the kiwifruit drying process D_{eff} had a significantly ($p < 0.05$) increasing trend. By increasing the air temperature from 50 to 70 °C, it was observed that the moisture diffusivity coefficient increased from 3.09×10^{-10} to 2.26×10^{-9} m^2/s. The reason for this issue is the significant effect of air temperature in creating more molecular movement and surface suction, and increasing the diffusivity coefficient. Additionally, increasing the temperature increases the enthalpy of the entering air, and increasing the enthalpy increases the mass and heat transfer, which increases D_{eff}. Further, by increasing the thickness of the samples, the amount of D_{eff} increased. Nguyen and Price reported that superficial hardening generally occurs much faster in thinner samples than in thicker samples, while the rate of moisture evaporation in thinner samples is much higher [39]. Therefore, rapid superficial hardening in the samples with less thickness causes a restriction in moisture transfer compared to thicker samples and thus leads to a decrease in moisture diffusivity coefficient in thinner samples [40,41]. Similar observations on the effect of thickness on moisture diffusivity coefficient have been reported by other researchers, including Doymaz and Gol, Azimi-Nejadian and Hoseini and Onwude et al. [28,40,42]. The D_{eff} increased by prolonging the ultrasonic pretreatment time from 0 to 30 min. Ultrasonic pretreatment causes the opening of the capillary tubes due to the dispersion of surface compounds and the formation of longer microscopic channels due to the deformation of the cells, followed by the formation ofmore open capillary tubes [43]. Therefore, ultrasonic pretreatment accelerates the removal of moisture from the product by deforming the cells and destroying the cell wall.

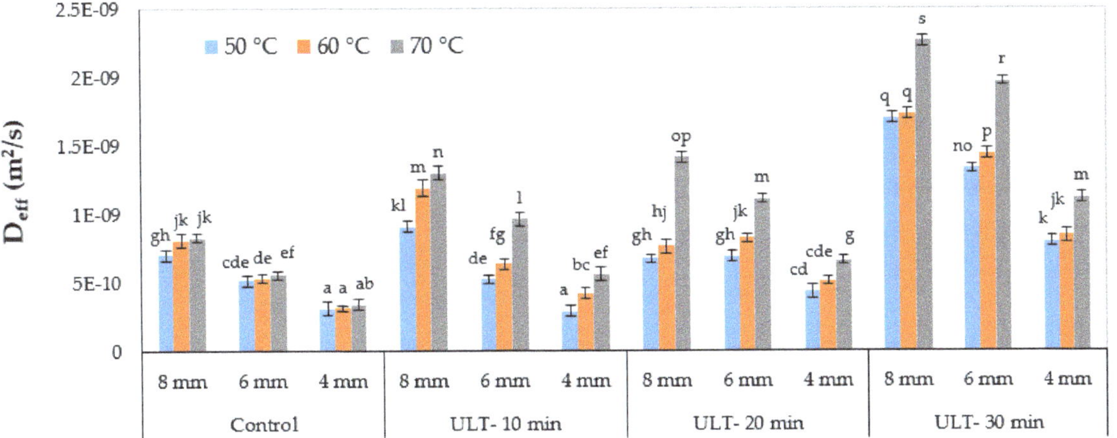

Figure 2. The value of D_{eff} under different conditions for drying kiwifruit slices.

Many researchers have reported D_{eff} for other products with ultrasonic pretreatment. For instance, the obtained moisture diffusivity coefficient values for potatoes in the hybrid microwave-hot air dryer with ultrasonic pretreatment ranged from 1.07×10^{-7} to 1.283×10^{-7} m²/s [11]. Vallespir et al. obtained the D_{eff} for kiwifruit in a hot air dryer with ultrasonic pretreatment ranging from 3.67×10^{-11} to 9.45×10^{-11} m²/s [5]. Abbaspour-Gilandeh et al. calculated the diffusivity coefficient for the drying of walnut in the microwave and hot air dryers at different ultrasonic pretreatment times between 3.12×10^{-9} and 8.99×10^{-9} m²/s and between 2.77×10^{-9} and 5.56×10^{-9} m²/s [43].

3.4. Thermodynamic Parameters

3.4.1. SEC

According to Figure 3, the lowest *SEC* (36.57 kWh/kg) was obtained with a significant difference ($p < 0.05$) at an air temperature of 70 °C, ultrasonic radiation time of 30 min, and thickness of 4 mm. Additionally, the control sample with a thickness of 8 mm for drying kiwifruit significantly ($p < 0.05$) had the highest *SEC* (181.97 kWh/kg). This can be due to the low moisture removal from the drying product and the increase in drying time. As the air temperature grows from 50 to 70 °C, the thermal gradient within the drying product increases, resulting in an increase in the moisture evaporation rate (the downward slope of the *MR* variation diagram becomes steeper as the temperature increases). Therefore, with the increasing temperature, the total energy consumption decreases [44]. Further, by enhancing the drying temperature from 50 to 70 °C, the difference between the drying temperature and the ambient temperature increased. As a result, increasing the temperature difference significantly decreased the drying time of the product, which is a strong reason for the reduced energy consumption. Motevali et al. and Kaveh et al., for drying *Roman chamomile* and *Pistacia Atlantica*, respectively, showed decreased energy consumption with increasing temperatures [33,44].

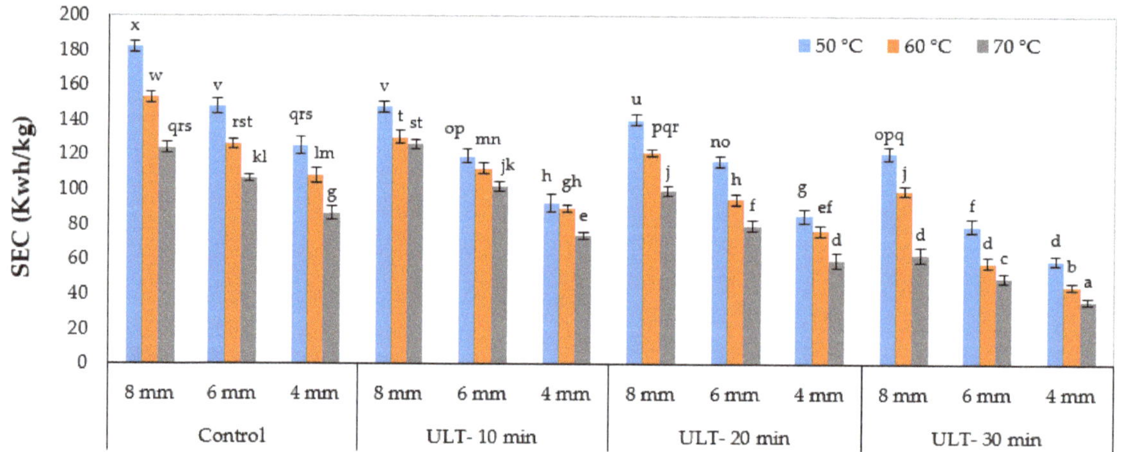

Figure 3. Amounts of *SEC* under different conditions for kiwifruit slices.

Increasing the ultrasonic radiation time from 0 (control sample) to 30 min significantly decreased the *SEC* ($p < 0.05$), because with increasing the ultrasonic pretreatment time, the product texture is more destroyed and the hard layer does not form in the product during the drying process, and thus, the product dries faster, and consequently, the *SEC* in the dryer is reduced [45]. The researchers showed that, with an increasing ultrasonic radiation time, the *SEC* decreases during the drying of blackberries [29].

The study of the effect of thickness on energy consumption revealed that reducing the thickness during the drying of kiwi reduces the drying time due to the increased thermal gradient and accelerated the removal of moisture from the product, thus reducing the total energy consumption. Darvishi et al., for drying kiwifruit in different thicknesses using the microwave dryer, showed that by reducing the thickness of the samples, the *SEC* is significantly reduced ($p < 0.05$) [16].

3.4.2. Energy Efficiency and Drying Efficiency

Figures 4 and 5 show the effect of different treatments (temperature, thickness and ultrasonic radiation time) on energy efficiency and drying efficiency, respectively. The highest energy efficiency and drying efficiency values (7.69% and 11.09%) were obtained at 70 °C, 4 mm thickness, and 30 min ultrasonic radiation time, respectively. Increasing the temperature from 50 to 70 °C increases the energy efficiency and the drying efficiency. Drying temperature increases the rate of moisture removal from the product and reduces the drying time, and consequently, both efficiency values will increase with increasing temperatures. However, comparing the application times of ultrasonic pretreatment shows that the highest values of both efficiencies occur when the 30 min ultrasonic pretreatment is used. According to Figures 4 and 5, it can be seen that reducing the thickness of kiwifruit samples will significantly ($p < 0.05$) increase energy efficiency and drying efficiency, because less thickness will increase the gradient of the drying product and more moisture is removed from the product, and as a result, the drying time of kiwifruit at low thicknesses is reduced. The lowest energy efficiency and drying efficiency values can be seen in the control treatment with a thickness of 8 mm (1.54% and 1.88%), respectively. The results show that increasing the ultrasonic pretreatment time improves both energy efficiency and drying efficiency. This is because, by increasing the ultrasonic pretreatment time, the changes in the product texture increases and the hard layer does not form in the process of drying this pretreated product and the product dries faster.

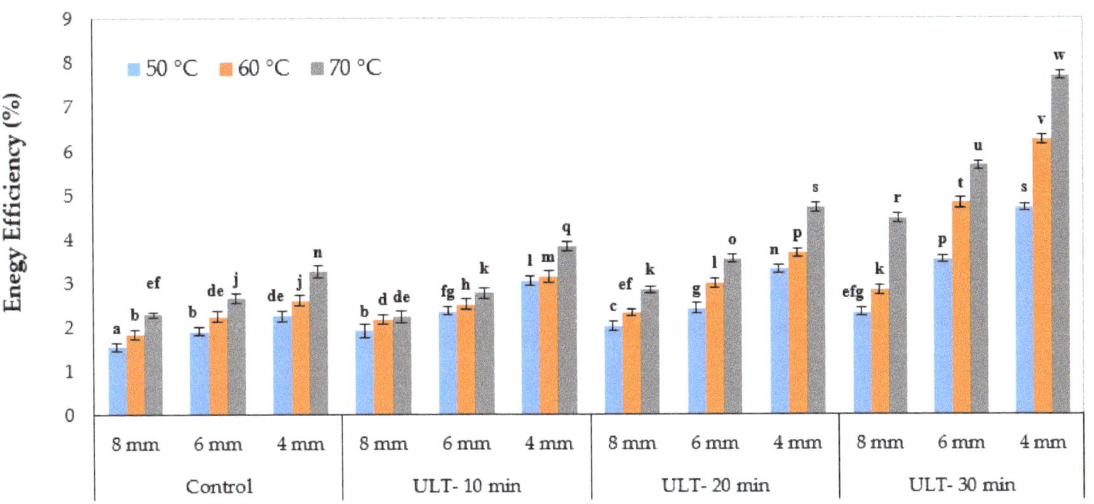

Figure 4. Energy efficiency under different conditions for kiwifruit slices drying.

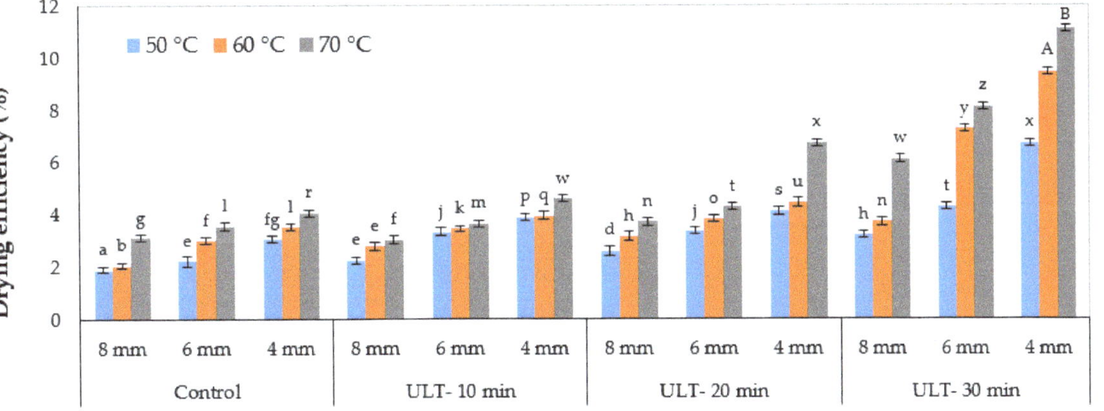

Figure 5. Drying efficiency under different conditions for drying kiwifruit slices.

The results of this study are consistent with the results published by other researchers. The energy efficiency values for drying chamomile in hot air dryers ranged between 1.91% and 6.76% [44], for apple slices between 2.87% and 9.11% [46], for *Pistacia atlantica* in a hot air dryer equal to 5.65% [47], and for peppermint leaves in a hot air dryer with ultrasonic pretreatment varying between 1.41% and 3.69% [12]. Albini et al. obtained the drying efficiency and energy efficiency for barley drying in a hot air dryer ranging 8%–17.4% and 24%–34.2%, respectively [48]. In another study, the drying efficiency was calculated between 3.42% and 12.29% for apple drying and showed that the drying efficiency increases by increasing the temperature [46].

3.4.3. Thermal Efficiency

Figure 6 depicts the effect of different treatments on thermal efficiency. It shows that the highest amount of thermal efficiency (10.58%) was achieved at 70 °C, ultrasonic radiation time of 30 min, and thickness of 4 mm, while the lowest value of this efficiency (1.54%) was obtained at 50 °C, control sample, and thickness of 8 mm. The results show that, with increasing temperatures, the changes in thermal efficiency have a significantly ($p < 0.05$) upward trend. The increase in thermal efficiency is due to the fact that, with increasing temperatures, the difference between the temperature of the drying product and the temperature of the drying air increases and leads to the accelerated removal of moisture from the product and thus reduces the time and amount of energy required.

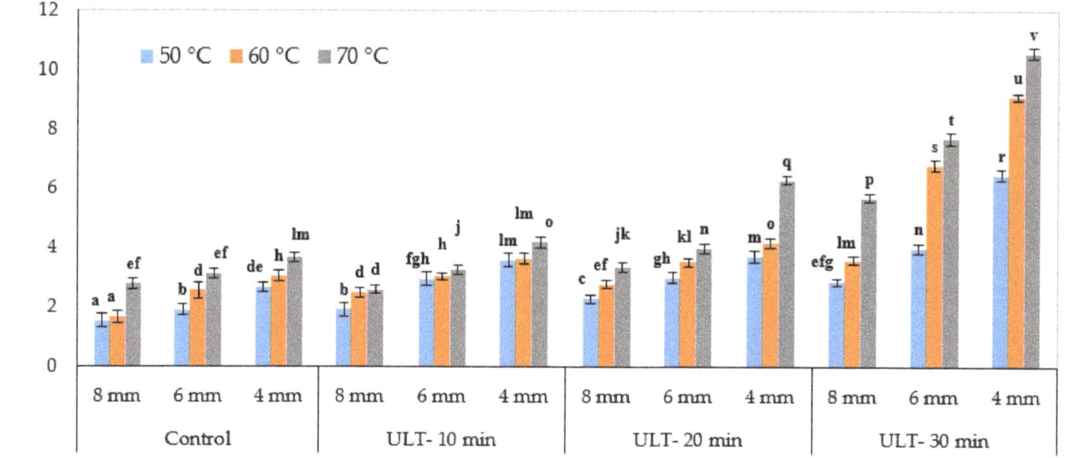

Figure 6. Thermal efficiency under different conditions for drying kiwifruit slices.

Thermal efficiency had an upward trend when using ultrasonic pretreatment. In the case of ultrasonic pretreatment, due to the reduced relative humidity of the drying air, the drying time is significantly ($p < 0.05$) shorter, and as a result, the share of energy consumption is smaller. The effect of thickness on the changes in thermal efficiency shows that, with decreasing thicknesses, the amount of thermal efficiency increases significantly ($p < 0.05$). This can be due to the increased drying rate as a result of the decreased thickness, because with decreasing thicknesses, the thermal gradient increases, the removal of moisture from within the product to the surface increases, and the drying time decreases.

Torki-Harchegani et al. showed that the thermal efficiency of peppermint leaf drying was 38.36%–49.33% for the microwave dryer and 3.69%–5.47% for the hot air dryer [49]. In another study by Beigi for drying apple, the thermal efficiency was calculated as ranging

3.70%–9.44%, and it was shown that, with increasing temperatures, the thermal efficiency increases [46]. Motevali et al. calculated the thermal efficiency for drying chamomile in a hot air dryer between 2.12% and 6.87% and stated that the thermal efficiency increases with increasing drying temperatures [44].

3.5. Qualitative Properties

3.5.1. Color

As can be seen in Figure 7, by raising the air temperature from 50 to 70 °C, the overall color change significantly increased ($p < 0.05$). One of the reasons is that, during the process of food drying, the pigments are decomposed due to slow heating and will cause discoloration in the products [18]. Islam et al. showed that increasing the temperature causes papaya color changes in a hot air dryer [35]. Sufer and Palazoglu achieved similar results for drying pomegranate [50]. For drying kiwifruit in a hot air dryer at different temperatures, Izli et al. showed that the color changes of kiwifruit increased by increasing the temperature [20]. Additionally, by increasing the thickness of the sample, the overall color changes increased, which was due to the fact that at high thicknesses, as the drying time increases, the kiwifruits slices were exposed to hot air for a longer period of time, and the high heat exposure to kiwifruit slices intensified the decomposition of pigments [26]. Therefore, at a thickness of 4 mm, the least amount of overall color changes was seen. The use of ultrasonic pretreatment also reduced color changes. Increasing the time of ultrasonic application reduced the rate of color changes in the ultrasonic pretreatment. Additionally, applying ultrasonic pretreatment time had less color changes than the control sample. The color changes during kiwifruit drying are due to various factors such as thermal degradation of chlorophyll, oxidation, enzymatic reactions and non-enzymatic browning reactions [51]. Taghinezhad et al. studied the drying of blackberries and concluded that the least amount of overall color change was observed at 30 min ultrasonic radiation time [37]. In another study, Dehghannya et al. examined the color changes of *Mirabelle plum* in the hot air dryer with ultrasonic pretreatment [52]. They showed that the amount of color changes decreases with increasing ultrasonic radiation time. Further, the results obtained by Wang et al. study, which was gathered on kiwifruit slices in the hot air dryer with ultrasonic pretreatment, showed that the highest amount of overall color change was 32.45 in the control sample and the lowest color change amount of 9.59 was observed in the ultrasonic radiation time of 30 min [27].

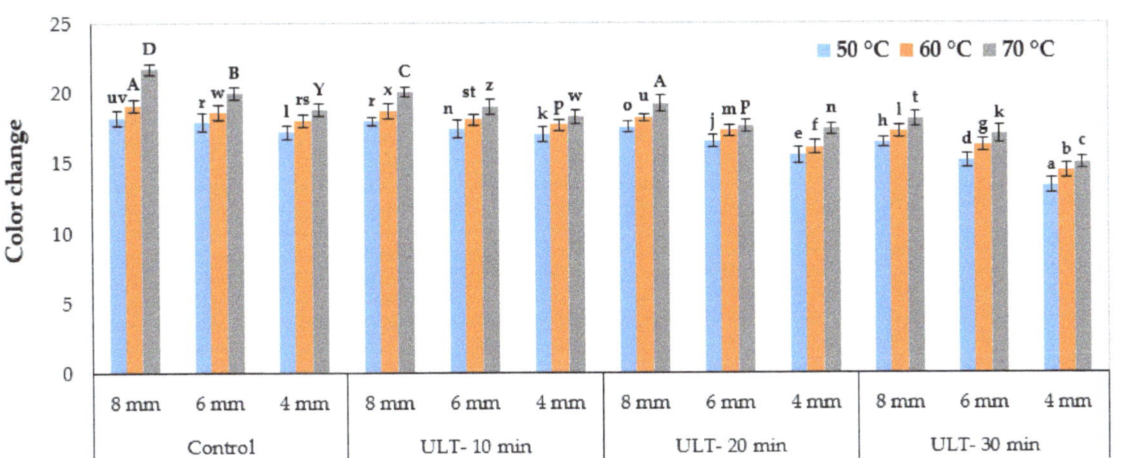

Figure 7. Color changes of kiwifruit slices under different drying conditions.

3.5.2. Shrinkage

Figure 8 shows the percentage of shrinkage changes in the *HAI* dryer with ultrasonic pretreatment at different thicknesses of kiwifruit. The results show that, at high temperatures, high ultrasonic radiation time and less thickness, the amount of shrinkage is significantly reduced ($p < 0.05$). As kiwifruit has a high initial moisture, due to the shrinkage, which is also significant, some changes occur in the initial structure during the drying process. In this way, the fluid in the cell wall exerts pressure to the cell wall and the fluid inside the cell is put under pressure [51]. During the drying process, the removal of water from the cell reduces the stress induced by the liquid on the cell wall. This reduction in stress causes the shrinkage of the constituent texture of the substance [26]. However, air temperature, high ultrasonic radiation time and less thickness accelerate the water removal rate and do not allow the product to deform. The results of this study corroborate those obtained by Kaveh et al. [29]. They investigated the effect of ultrasonic radiation time on blackberry shrinkage in the hot air dryer, and demonstrated that with increasing air temperature and ultrasonic radiation time, the amount of shrinkage decreased. Dehghannya et al. also achieved similar results for drying potato slices [11].

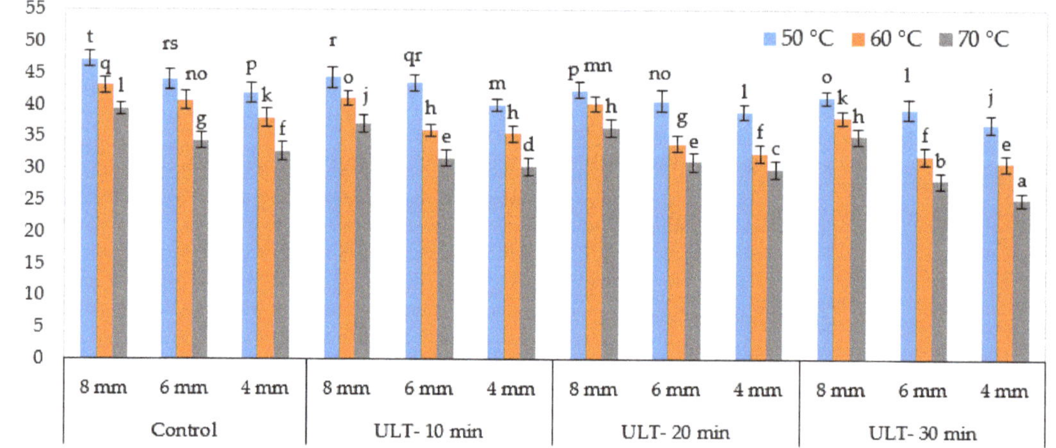

Pretreatment—Thicknes

Figure 8. Shrinkage of kiwifruit slices under different drying conditions.

4. Conclusions

In this study, drying of kiwifruit was performed at thicknesses of 4, 6 and 8 mm with a hybrid *HAI* dryer at air temperatures of 50, 60 and 70 °C and ultrasonic times of 0 (control sample), 10, 20 and 30 min. The results showed that, by increasing the temperature, ultrasonic time and decreasing the thickness, the drying time and *SEC* decreased. The logistic model was able to predict the MR with the highest value of $R^2 = 0.9997$ and the lowest values of *RMSE* = 0.0117 and $\chi^2 = 0.0007$ were chosen as the best model. Additionally, by increasing the thickness of the samples, ultrasonic pretreatment time and by prolonging the air temperature, the amount of D_{eff} increased from 3.09×10^{-10} to 2.26×10^{-9} m^2/s. The *SEC* value decreased significantly ($p < 0.05$) by increasing the temperature, ultrasonic time and decreasing the thickness. The lowest *SEC* value was obtained at 70 °C, ultrasonic time of 30 min, and thickness of 4 mm. The highest values of drying efficiency, energy efficiency, and thermal efficiency were 11.09%, 7.69% and 10.58%, respectively, at air an temperature of 70 °C, an ultrasonic time of 30 min, and sample thickness of 4 mm. The lowest color changes and shrinkage values were 13.2%9 and 25.25%, respectively, at 50 °C air temperature, 30 min ultrasonic time, and 4 mm thickness.

Author Contributions: Conceptualization, E.T. and M.K.; methodology, E.T. and M.K.; validation, E.T. and M.K.; formal analysis, E.T.; investigation, E.T. and M.K.; resources, E.T. and A.S.; data curation, E.T. and M.K.; writing—original draft preparation, E.T. and A.S.; writing—review and editing, E.T., A.S.; and M.K.; visualization, E.T. and A.S.; funding acquisition, E.T. and A.S. All authors have read and agreed to the published version of the manuscript.

Funding: This research was funded by the office of vice chancellor for research at Mohaghegh Ardabili University, and Wrocław University of Environmental and Life Sciences.

Institutional Review Board Statement: Not applicable.

Informed Consent Statement: Not applicable.

Data Availability Statement: Data for this research will available.

Acknowledgments: The authors are highly thankful to Department of Agricultural Technology Engineering, Mohaghegh Ardabili University, Ardabil, Iran for providing facilities to conduct this research work.

Conflicts of Interest: The authors declare no conflict of interest.

References

1. Liu, Y.; Zeng, Y.; Hu, X.; Sun, X. Effect of Ultrasonic Power on Water Removal Kinetics and Moisture Migration of Kiwifruit Slices During Contact Ultrasound Intensified Heat Pump Drying. *Food Bioprocess Technol.* **2020**, *13*, 430–441. [CrossRef]
2. Zakipour-Molkabadi, E.; Hamidi-Esfahani, Z.; Abbasi, S. Formulation of leather from kiwi fruit losses. *Iranian Food Sci. Technol. Res. J.* **2011**, *6*, 263–271.
3. Kumar, C.; Karim, A.; Saha, S.C.; Joardder, M.U.H.; Brown, R.J.; Biswas, D. Multiphysics modelling of convective drying of food materials. In Proceedings of the Global Engineering, Science and Technology Conference, Marrakech, Morocco, 17–20 April 2012.
4. Taghinezhad, E.; Szumny, A.; Kaveh, M.; Rasooli Sharabiani, V.; Kumar, A.; Shimizu, N. Parboiled Paddy Drying with Different Dryers: Thermodynamic and Quality Properties, Mathematical Modeling Using ANNs Assessment. *Foods* **2020**, *9*, 86. [CrossRef] [PubMed]
5. Vallespir, F.; Rodríguez, Ó.; Cárcel, J.A.; Rosselló, C.; Simal, S. Ultrasound assisted low-temperature drying of kiwifruit: Effects on drying kinetics, bioactive compounds and antioxidant activity. *J. Sci. Food Agric.* **2019**, *99*, 2901–2909. [CrossRef]
6. Salehi, F. Recent applications and potential of infrared dryer systems for drying various agricultural products: A review. *Int. J. Fruit Sci.* **2020**, *20*, 586–602. [CrossRef]
7. Barzegar, M.; Zare, D.; Stroshine, R.L. An integrated energy and quality approach to optimization of green peas drying in a hot air infrared-assisted vibratory bed dryer. *J. Food Eng.* **2015**, *166*, 302–315. [CrossRef]
8. Charoux, C.M.; Ojha, K.S.; O'Donnell, C.P.; Cardoni, A.; Tiwari, B.K. Applications of airborne ultrasonic technology in the food industry. *J. Food Eng.* **2017**, *208*, 28–36. [CrossRef]
9. Mierzwa, D.; Szadzińska, J.; Pawłowski, A.; Pashminehazar, R.; Kharaghani, A. Nonstationary convective drying of raspberries, assisted by microwaves and ultrasound. *Dry. Technol.* **2019**, *37*, 988–1001. [CrossRef]
10. Abbaspour-Gilandeh, Y.; Kaveh, M.; Aziz, M. Ultrasonic-Microwave and Infrared Assisted Convective Drying of Carrot: Drying Kinetic, Quality and Energy Consumption. *Appl. Sci.* **2020**, *10*, 6309. [CrossRef]
11. Dehghannya, J.; Kadkhodaei, S.; Heshmati, M.K.; Ghanbarzadeh, B. Ultrasound-assisted intensification of a hybrid intermittent microwave-hot air drying process of potato: Quality aspects and energy consumption. *Ultrasonics* **2019**, *96*, 104–122. [CrossRef]
12. Ghanbarian, D.; Torki-Harchegani, M.; Sadeghi, M.; Pirbalouti, A.G. Ultrasonically improved convective drying of peppermint leaves: Influence on the process time and energetic indices. *Renew. Energy* **2020**, *153*, 67–73. [CrossRef]
13. Darıcı, S.; Şen, S. Experimental investigation of convective drying kinetics of kiwi under different conditions. *Heat Mass Transfer.* **2015**, *51*, 1167–1176. [CrossRef]
14. Mahjoorian, A.; Mokhtarian, M.; Fayyaz, N.; Rahmati, F.; Sayyadi, S.; Ariaii, P. Modeling of drying kiwi slices and its sensory evaluation. *Food Sci. Nutr.* **2017**, *5*, 466–473. [CrossRef] [PubMed]
15. Mohammadi, I.; Tabatabaekoloor, R.; Motevali, A. Effect of air recirculation and heat pump on mass transfer and energy parameters in drying of kiwifruit slices. *Energy* **2019**, *170*, 149–158. [CrossRef]
16. Darvishi, H.; Zarein, M.; Farhudi, Z. Energetic and exergetic performance analysis and modeling of drying kinetics of kiwi slices. *J. Food Sci. Technol.* **2016**, *53*, 2317–2333. [CrossRef] [PubMed]
17. Aidani, E.; Hadadkhodaparast, M.; Kashaninejad, M. Experimental and modeling investigation of mass transfer during combined infrared-vacuum drying of Hayward kiwifruits. *Food Sci. Nutr.* **2017**, *5*, 596–601. [CrossRef]
18. Orikasa, T.; Koide, S.; Okamoto, S.; Imaizumi, T.; Muramatsu, Y.; Takeda, J.-I.; Shiina, T.; Tagawa, A. Impacts of hot air and vacuum drying on the quality attributes of kiwifruit slices. *J. Food Eng.* **2014**, *125*, 51–58. [CrossRef]
19. Akar, G.; Barutçu Mazı, I. Color change, ascorbic acid degradation kinetics, and rehydration behavior of kiwifruit as affected by different drying methods. *J. Food Process Eng.* **2019**, *42*, e13011. [CrossRef]

20. Izli, N.; Izli, G.; Taskin, O. Drying kinetics, colour, total phenolic content and antioxidant capacity properties of kiwi dried by different methods. *J. Food Meas. Charact.* **2017**, *11*, 64–74. [CrossRef]
21. Nadian, M.H.; Abbaspour-Fard, M.H.; Martynenko, A.; Golzarian, M.R. An intelligent integrated control of hybrid hot air-infrared dryer based on fuzzy logic and computer vision system. *Comput. Electron. Agric.* **2017**, *137*, 138–149. [CrossRef]
22. Ergün, K.; Çalışkan, G.; Dirim, S.N. Determination of the drying and rehydration kinetics of freeze dried kiwi (Actinidia deliciosa) slices. *Heat Mass Transfer.* **2016**, *52*, 2697–2705. [CrossRef]
23. Lyu, J.; Chen, Q.; Bi, J.; Zeng, M.; Wu, X. Drying characteristics and quality of kiwifruit slices with/without osmotic dehydration under short-and medium-wave infrared radiation drying. *Inter. J. Food Eng.* **2017**, *13*. [CrossRef]
24. Ricce, C.; Rojas, M.L.; Miano, A.C.; Siche, R.; Augusto, P.E.D. Ultrasound pre-treatment enhances the carrot drying and rehydration. *Food Res. Int.* **2016**, *89*, 701–708. [CrossRef] [PubMed]
25. Nadian, M.H.; Abbaspour-Fard, M.H.; Sadrnia, H.; Golzarian, M.R.; Tabasizadeh, M. Optimal pretreatment determination of kiwifruit drying via online monitoring. *J. Sci. Food Agric.* **2016**, *96*, 4785–4796. [CrossRef]
26. Jahanbakhshi, A.; Kaveh, M.; Taghinezhad, E.; Rasooli Sharabiani, V. Assessment of kinetics, effective moisture diffusivity, specific energy consumption, shrinkage, and color in the pistachio kernel drying process in microwave drying with ultrasonic pretreatment. *J. Food Process. Preserv.* **2020**, *44*, e14449. [CrossRef]
27. Wang, J.; Xiao, H.-W.; Ye, J.-H.; Wang, J.; Raghavan, V. Ultrasound pretreatment to enhance drying kinetics of Kiwifruit (Actinidia deliciosa) slices: Pros and cons. *Food Bioprocess Technol.* **2019**, *12*, 865–876. [CrossRef]
28. Onwude, D.I.; Hashim, N.; Abdan, K.; Janius, R.; Chen, G. Modelling the mid-infrared drying of sweet potato: Kinetics, mass and heat transfer parameters, and energy consumption. *Heat Mass Transfer.* **2018**, *54*, 2917–2933. [CrossRef]
29. Kaveh, M.; Taghinezhad, E.; Aziz, M. Effects of physical and chemical pretreatments on drying and quality properties of blackberry (*Rubus* spp.) in hot air dryer. *Food Sci. Nutr.* **2020**, *8*, 3843–3856. [CrossRef]
30. Varzakas, T.H.; Leach, G.C.; Israilides, C.J.; Arapoglou, D. Theoretical and experimental approaches towards the determination of solute effective diffusivities in foods. *Enzyme Microb. Technol.* **2005**, *37*, 29–41. [CrossRef]
31. La Fuente, C.I.; Tadini, C.C. Ultrasound pre-treatment prior to unripe banana air-drying: Effect of the ultrasonic volumetric power on the kinetic parameters. *J. Food Sci. Technol.* **2018**, *55*, 5098–5105. [CrossRef]
32. Varzakas, T.; Escudero, I.; Economou, I.G. Estimation of endoglucanase and lysozyme effective diffusion coefficients in polysulphone membranes. *J. Biotechnol.* **1999**, *72*, 77–83. [CrossRef]
33. Kaveh, M.; Chayjan, R.A.; Taghinezhad, E.; Sharabiani, V.R.; Motevali, A. Evaluation of specific energy consumption and GHG emissions for different drying methods (Case study: Pistacia Atlantica). *J. Clean. Prod.* **2020**, *259*, 120963. [CrossRef]
34. Motevali, A.; Koloor, R.T. A comparison between pollutants and greenhouse gas emissions from operation of different dryers based on energy consumption of power plants. *J. Clean. Prod.* **2017**, *154*, 445–461. [CrossRef]
35. Islam, M.; Saha, T.; Monalisa, K.; Hoque, M. Effect of starch edible coating on drying characteristics and antioxidant properties of papaya. *J. Food Meas. Charact.* **2019**, *13*, 2951–2960. [CrossRef]
36. Doymaz, İ.; Özdemir, Ö. Effect of air temperature, slice thickness and pretreatment on drying and rehydration of tomato. *Int. J. Food Sci. Tech.* **2014**, *49*, 558–564. [CrossRef]
37. Taghinezhad, E.; Kaveh, M.; Khalife, E.; Chen, G. Drying of organic blackberry in combined hot air-infrared dryer with ultrasound pretreatment. *Dry. Technol.* **2020**. [CrossRef]
38. Doymaz, İ. Drying kinetics, rehydration and colour characteristics of convective hot-air drying of carrot slices. *Heat Mass Transfer.* **2017**, *53*, 25–35. [CrossRef]
39. Nguyen, M.-H.; Price, W.E. Air-drying of banana: Influence of experimental parameters, slab thickness, banana maturity and harvesting season. *J. Food Eng.* **2007**, *79*, 200–207. [CrossRef]
40. Azimi-Nejadian, H.; Hoseini, S.S. Study the effect of microwave power and slices thickness on drying characteristics of potato. *Heat Mass Transfer.* **2019**, *55*, 2921–2930. [CrossRef]
41. Varzakas, T.; Arapoglou, D.; Israilides, C. Kinetics of endoglucanase and endoxylanase uptake by soybean seeds. *J. Biosci. Bioeng.* **2006**, *101*, 111–119. [CrossRef]
42. Doymaz, I.; Göl, E. Convective drying characteristics of eggplant slices. *J. Food Process. Eng.* **2011**, *34*, 1234–1252. [CrossRef]
43. Abbaspour-Gilandeh, Y.; Kaveh, M.; Jahanbakhshi, A. The effect of microwave and convective dryer with ultrasound pretreatment on drying and quality properties of walnut kernel. *J. Food Process. Preserv.* **2019**, *43*, e14178. [CrossRef]
44. Motevali, A.; Minaei, S.; Banakar, A.; Ghobadian, B.; Khoshtaghaza, M.H. Comparison of energy parameters in various dryers. *Energy Convers. Manag.* **2014**, *87*, 711–725. [CrossRef]
45. Taghinezhad, E.; Kaveh, M.; Jahanbakhshi, A.; Golpour, I. Use of artificial intelligence for the estimation of effective moisture diffusivity, specific energy consumption, color and shrinkage in quince drying. *J. Food Process Eng.* **2020**, *43*, e13358. [CrossRef]
46. Beigi, M. Energy efficiency and moisture diffusivity of apple slices during convective drying. *Food Sci. Technol.* **2016**, *36*, 145–150. [CrossRef]
47. Abbaspour-Gilandeh, Y.; Kaveh, M.; Fatemi, H.; Hernández-Hernández, J.L.; Fuentes-Penna, A.; Hernández-Hernández, M. Evaluation of the Changes in Thermal, Qualitative, and Antioxidant Properties of Terebinth (Pistacia atlantica) Fruit under Different Drying Methods. *Agronomy* **2020**, *10*, 1378. [CrossRef]
48. Albini, G.; Freire, F.B.; Freire, J.T. Barley: Effect of airflow reversal on fixed bed drying. *Chem. Eng. Process. Process. Intensif.* **2018**, *134*, 97–104. [CrossRef]

49. Torki-Harchegani, M.; Ghanbarian, D.; Pirbalouti, A.G.; Sadeghi, M. Dehydration behaviour, mathematical modelling, energy efficiency and essential oil yield of peppermint leaves undergoing microwave and hot air treatments. *Renew. Sustain. Energy Rev.* **2016**, *58*, 407–418. [CrossRef]
50. Sufer, O.; Palazoglu, T.K. A study on hot-air drying of pomegranate Kinetics of dehydration, rehydration and effects on bioactive compounds. *J. Therm. Anal. Calorim.* **2019**, *137*, 1981–1990. [CrossRef]
51. Motevali, A.; Zabihnia, F. Effect of the Different Pre-Treatments Thermal, Pulse, Chemical and Mechanical on the External Mass Transfer Coefficient Changes, Moisture Diffusion Coefficient and Activation Energy. *J. Res. Innov. Food Sci. Technol.* **2017**, *6*, 277–290.
52. Dehghannya, J.; Gorbani, R.; Ghanbarzadeh, B. Influence of combined pretreatments on color parameters during convective drying of Mirabelle plum (Prunus domestica subsp. syriaca). *Heat Mass Transfer.* **2017**, *53*, 2425–2433. [CrossRef]

Article

Two-Stage Production System Pondering upon Corporate Social Responsibility in Food Supply Chain: A Case Study

Ten-Suz Chen [1], Yung-Fu Huang [2], Ming-Wei Weng [2] and Manh-Hoang Do [2,3,*]

1   Department of Business Administration, Chaoyang University of Technology, Taichung 41349, Taiwan; s10837906@cyut.edu.tw
2   Department of Marketing and Logistics Management, Chaoyang University of Technology, Taichung 41349, Taiwan; huf@cyut.edu.tw (Y.-F.H.); mwweng@cyut.edu.tw (M.-W.W.)
3   Faculty of Economics, Tay Nguyen University, Buon Ma Thuot 63000, Vietnam
*   Correspondence: s10537912@gm.cyut.edu.tw; Tel.: +886-4-2332-3000

Abstract: Corporate social responsibility (CSR) has witnessed remarkable attention in academic studies as well as being widely conducted in different industries globally. This specific case was chosen as one of the biggest dairy companies that may be represented for Vietnam dairy supply chain management. This research aims to integrate CSR initiatives into food supply chain management to clarify the optimal replenishment policy, paying close attention to the relationship between midstream manufacturers and final customers. The classical economic production quantity model has been employed, relying on the two-stage assembly production system. The three parameters that contribute to the total profit formulation that have been considered consist of the social charity amount for per unit selling, the unit wholesale price of the manufacturer, and the return rate of used goods from the customer. The study has stressed that there is a significant impact from implementing CSR initiatives on the enterprise's inventory policy that leads to enhance the firm's financial performance.

Keywords: Two-stage production system; corporate social responsibility; supply chain; dairy industry; social charity; Vietnam

Citation: Chen, T.-S.; Huang, Y.-F.; Weng, M.-W.; Do, M.-H. Two-Stage Production System Pondering upon Corporate Social Responsibility in Food Supply Chain: A Case Study. *Appl. Sci.* **2021**, *11*, 1088. https://doi.org/10.3390/app11031088

Academic Editor: Theodoros Varzakas

Received: 24 December 2020
Accepted: 21 January 2021
Published: 25 January 2021

Publisher's Note: MDPI stays neutral with regard to jurisdictional claims in published maps and institutional affiliations.

1. Introduction

The idea of conducting corporate social responsibility (CSR) has undergone phenomenal growth in recent decades and has been rolled out in diverse sectors globally, such as the European banking industry [1], Vietnamese coffee sector [2], Jordanian pharmaceutical industry [3], the U.S. restaurant industries [4], Chinese food industry [5], and so on. Indeed, this positive signal has attracted increasing interest from corporations due to the advantages of adopting CSR into their business strategy [6–8]. Some recent studies have stated the critical role of CSR activities; for example, the positive effects of CSR on the firm's financial performance [9,10], the value perception by consumers is generated [11], corporate reputation, consumer's satisfaction, or competitive advantage of the company [12–14]. Thus, two major characteristics contribute to CSR simply understanding as follows: firstly, the relationship between corporate responsibility and the firm's stakeholders, including internal and external. Secondly, toward sustainable development, which obligation has divided into three major issues: enhancing the economy, protecting the environment, and contributing to local society. Hence, CSR initiatives have been conducted and considered an excellent tool to attain sustainability [15–17].

In terms of supply chain management (SCM), the structure starts from the first stage known as upstream suppliers to the last step known as downstream customers; thus, there are many relationships among stakeholders in the supply chain that must be solved and predicted [18–20]. Additionally, the positive signal from CSR initiatives can enhance SCM performance and may obtain a sustainable supply chain; thus, the number of studies integrating CSR initiatives into SCM has also been gradually becoming popular and diverse in many different industries worldwide [14,21,22]. Hence, adopting CSR initiatives into SCM may have been affected by various factors or barriers [18,19]. The customer is at the last stage of the supply chain but also plays a decisive role in the sales of products and impacts the economic responsibility of the firms. However, there are only a few studies concerning the relationship between customers and manufacturers, which have an influence on CSR ideas, especially the food supply chain [16,17]. The food supply chain has several particular characteristics compared to other sectors; thereby, research integration of CSR initiatives into the food SCM is attaining much attention due to its contribution to sustainability, particularly in transitional economies [23]. From some of the existing CSR-SCM studies about stakeholder theory, they have demonstrated that businesses' efforts to adopt CSR activities into corporate governance policies to satisfy the interests of shareholders and stakeholders because SCM plays a role as the critical interface with suppliers or agency services, for instance, logistics services, transport providers, service of yards and warehouses [24,25]. Unfortunately, only a few studies consider the optimal model related to SCM stakeholders; meanwhile, the excellent role of CSR programs in SCM sustainability has not been considered properly, particularly in transitional economies [21,22]. Thus, this research aims to propose an optimal model for the food supply chain, which can adopt CSR initiatives into the food SCM. In this way, supporting the food company has a social contribution as well as obtaining superior financial performance. This study has some research questions as follows:

RQ1. What is the optimal social charity amount for per unit product selling?
RQ2. What is the optimal production run time of the manufacturer?
RQ3. What is the optimal return rate of used products from the customer?

This study makes several contributions to the published literature regarding CSR and SCM. Firstly, this research develops the classical economic production quantity (EPQ) model to address the stakeholders' satisfaction under the two-stage assembly production system. An analysis between two supply chain members will be conducted, including midstream manufacturers and downstream buyers, whose members are witnessing the CSR adoption. Secondly, the proposed EPQ model's efficacy has been assessed through the Vietnamese dairy supply chain case study, which is among the critical sectors contributing to the sustainable development of a developing nation like Vietnam [26]. This is the unique research considering the relationship between CSR practices and food supply chain management in Vietnam; thus, the research findings provide useful information to SCM's managers. Moreover, this model's scope can also be expanded to various sectors to determine the proposed model's efficacy. The structure of this study as follows. The authors discussed the introduction in Section 1. The literature review and the methodology of this study will be described in Sections 2 and 3, respectively. The next section will outline the notation and model formulation for this research. A numerical example of this study will be shown in Sections 5 and 6. In the last part, Section 7 outlines the research implications and limitations of the study, as well as suggestions for new study directions in the future.

2. Literature Review

To date, the definition of CSR is still controversial about consistency, and there are many different perspectives on CSR definition [7,11]. In studies of the stakeholders' theory, the scholars agreed that businesses would enhance their profit; however, they must also assume responsibilities to the environment, community, and society where the company is operating [27]. Therefore, the CSR concept has been understood under three responsibilities dimensions: economic, social, and environmental responsibility [2,27]. Regarding supply chain management, this includes many different stages with several complex relationships between the company and other stakeholders. The role of stakeholders in sustainable supply chain management (SSCM) has been described in several previous studies, such as empirical research in India of Das (2018) [12]; a review research from Sodhi and Tang (2018) [18]; or Sarkar et al. (2018) who suggested a production model of the automobile SSCM related to environment protection [20]. Consequently, corporatations need to determine their responsibilities with goods from the first stage as forming ideas [28]. CSR initiatives have been employed as useful tools to incorporate with SCM to improve corporate decision-making effectiveness, as well as to accomplish SSCM targets [21,24,29]. In fact, the findings obtained have recently shown that various studies incorporating CSR and SCM have been increasing rapidly. For instance, Feng (2017) systematically reviewed the relationship between CSR and SCM among previous studies and pointed out that these studies are heavily concentrated in developed countries; meanwhile, just a few studies have been conducted in emerging economies [21]; Das (2018) evaluated the model to enhance SSCM performance of the Indian enterprises through competitiveness measurement that had been influenced by CSR practices [12]; Khalid et al. (2015) suggested employing three-aspect obligations to attain SSCM, however, they found a lack of attention on the social dimension of the SCM compared to other responsibilities [30]; Wu et al. (2017) indicated that supplier's CSR misconduct directly remarkably affects the economic performance in the SCM [25].

In terms of the food supply chain sector, which is always a difference between the food industry and other industries since the characteristics of food products such as food quality, safety, storage condition, and shelf life as well as the process of transporting between warehouses, and stores to end customers [5,31]. These are factors that can easily occur in incidents related to food goods. Moreover, it also may impact food supply chain management and make them more difficult and complicated, although the companies have applied risk mitigation strategies in the supply chain [5,31,32]. Evaluation of the correlation between CSR issues and the food SCM has been indicated as a hot topic and given much attention by some scholars in recent years [21]. For instance, Chkanikova and Mont (2015) highlighted some drivers and barriers for food retailers in conducting CSR in their supply chains [33]; Prakash (2018) indicated the rise of CSR and SCM studies; however, it is not really significant and worthy of research potential [34]. Hence, there has been a gap in sustainability efforts and outcomes between developed nations and emerging nations in attaining a sustainable food supply chain. The comparison research between two types of countries has been conducted and claims that in developed countries' conditions where mature supply chains already exist, the industry can adopt risk mitigation strategies and achieve the effectiveness. They also suggested that the future of research should clarify which conditions in transitional economies are needed to achieve a sustainable supply chain [35]. However, their similar findings suggested that companies should pay attention to the different stages and their roles from up-stream to down-stream [28,36].

Regarding SCM in the developing country context, especially in the perishable food supply chain, no company wants to suffer the risk of damage because of perishable products such as meat, seafood, or other perishable goods. The relationship between manufacturers and retailers is very complicated and mainly focused on their economic benefits. There are two aspects, including the price markdown costs and it is related to some parameters such as potential customer quantity and market price fluctuations. These two factors affect the optimal choice of pricing strategy of the enterprises in the perishable food supply chain [37]. Meanwhile, the perceived consumer in developing countries has changed significantly in recent years. Huang et al. (2019) stressed that Vietnamese consumers have more interest and are willing to spend more money on products from socially responsible companies [11]. In terms of the dairy SCM, the dairy quality was the top priority of all stakeholders as well as stages in the SCM; thus, the biggest issue is dairy quality assurance across the entire supply chain because the specificity of dairy products is highly perishable and it is very costly to preserve refrigeration and collect it continuously. Previous studies have shown that manufacturers and stakeholders such as farmers, logistic services, and suppliers have mutual responsibility related to dairy quality [23,38–41]. Hence, studying the optimal model of integrating CSR into SCM in the food industry context will enrich the CSR and SCM literature and suggest more new promising ideas for other scholars in the future.

3. The Research Methodology

3.1. The EPQ Model

The classical economic production quantity model is developed based on the single product and single-stage production system; furthermore, it has been widely employed in various SCM studies [42–45]. In fact, the manufacturing process's complexity recently has witnessed more involvement with different components, from raw materials procurement to distribution systems for finished products. Therefore, several articles have been interested in studying how the end product is manufactured through a multi-stage production system [46–48]. Several studies have noted that multi-component production strategies adopt opportunistic maintenance policies [49–52]. However, only a little literature has been published on the impact of CSR in the inventory field. Those recommendations have mentioned that the impact of CSR on the manufacturing process as well as the supply chain would be given new opportunities to explore. For instance, Modak et al. (2019) have considered the effect of selling price and social work donation on demand for three-channel structures [53].

The authors will illustrate the optimal model to enrich the findings of Chang et al. (2012), who have conducted that model in the automation industry. They have also suggested that other scholars may evaluate their model in various industries or another country context [47]. Consequently, many scholars have been conducting the two-stage production system approach to point out the most suitable EPQ model for their examples. This example can be demonstrated by Gupta and Mohanty (2015), Dey et al. (2019), or Sabbaghnia and Taleizadeh (2020); they have argued the two-stage production system perspective through their research [48,54,55], particularly SCM integration with CSR similarly to Nematollahi et al. (2017), and Jokar and Hosseini-Motlagh (2020) [56,57]. Furthermore, they have recommended evaluating this approach's applicability to different sectors in the real world. From these suggestions include relying on different characteristics related to the dairy industry and the emerging economy; thus, the author has employed the proposed model of Chang et al. (2012) to evaluate concordance through a case study in the Vietnamese dairy industry. Moreover, it has also contributed to filling the literature gap in theoretical inventory management in the dairy supply chain.

3.2. The Research Context

Vietnam is a transitional economy with agriculture as a majority development sector [58,59]. The Vietnamese dairy industry is one of the priority sectors for sustainable development targets and receives attention from government policies and enterprises, and the community [26]; furthermore, the Vietnamese dairy industry represents a typical case study in the Vietnamese supply chain (be presented in Figure 1). However, only a few dairy firms are involved in CSR programs and attempt to commit to stakeholders, the community, and society. Hence, their CSR programs' undeniable efforts over the years related to shareholders, employees, local-residents, environment protection, and others, have had a phenomenal contribution to the sustainability of a transitional economy like Vietnam [11]. Furthermore, the perception of Vietnamese consumers for products is also increased when they pay more attention to the origin of the products as well as the responsibilities of those companies to society instead of only being interested in price and quality [11]; therefore, it has given a positive sign and become the dynamic for Vietnamese dairy firms to CSR adoption.

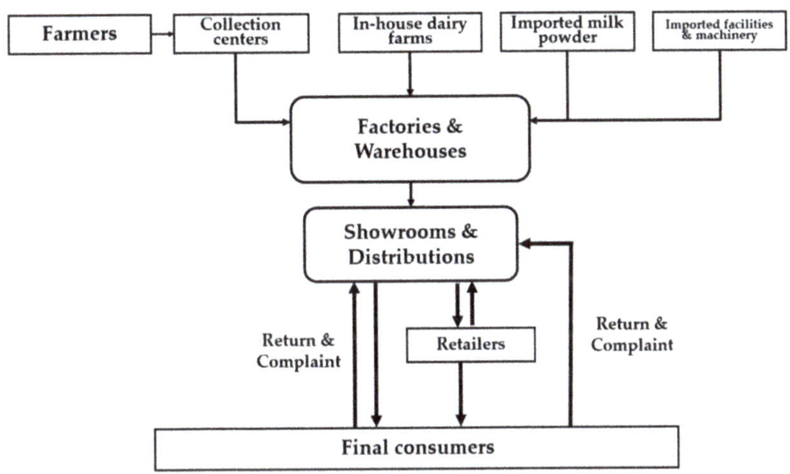

Figure 1. Simplified Vietnam's Dairy Supply Chain. (Source: The authors).

4. Model Formulation

4.1. Notation

In this study, the authors proposed an EPQ model for a two-stage assembly production system throughout one cycle. The system illustrates the relationship for all returned products in Stage 1 (manufacturing process) and end product in Stage 2 (assembly process). The manufacturer also directly collects the returned goods through a closed-loop supply chain with a reverse dual-channel. To ease readability, the authors will adopt the same notation in the production system and formulation from Chang et al. (2012) for this paper. The notation used in this paper, including system parameters and decision variables, has been listed in Tables 1 and 2, respectively.

Table 1. System parameters of this study.

m	The number of return points at retailer stage.
n	The number of raw material sourcing at the manufacturing stage.
p_i	Production rate of the return source i at retailer stage, where $i = 1, 2, \ldots, n$ and $p_1 > p_2 > \ldots > p_n$
D	Demand rate.
k	Setup cost.
c_1	The unit price of a returned product that the manufacturer paid to the customer.
c_0	The unit cost of the manufacturer paid to customers for a returned unit.
r	The return rate of used goods from the customer ($0 \leq r \leq 1$)
h_i	Holding cost of the source i for raw materials at the manufacturer stage, with $i = 1, 2, \ldots, n$.
h_e	Holding cost of an end product at retailer stage.
θ_i	The defective rate of the source i for raw material at the manufacturer stage, with $i = 1, 2, \ldots, n$.
θ_e	The defective rate of an end product at retailer stage.
d_i	Disposal cost for a defective source i for raw materials at the manufacturer stage, with $i = 1, 2, \ldots, n$.
d_e	Disposal cost for a defective end product at retailer stage.
d_{re}	Disposal cost for a returned product at retailer stage.
t_{id}	The time period when the inventory of source i for raw materials depletes, with $i = 1, 2, \ldots, n$.
t_{rdi}	Return period for an inventory of the used product at i retailer, with $i = 1, 2, \ldots, m$.
I_r	Investment in collection activities at the retailer stage.
I_s	Investment in CSR activity at the manufacturing stage.
T	The amount of cycle time.
Z_i	The maximum inventory level of source i for raw material at the manufacturer stage, with $i = 1, 2, \ldots, n$.
Z_e	The maximum inventory level of the end product at the retailer stage.
Z_{re}	The maximum inventory level of the returned product at the retailer stage.
p_w	The unit wholesale price of the manufacturer.

Table 2. Decision variables.

s	Social charity (SC) amount for per unit selling.
t_i	Production run time for the raw material of resource i, where $i = 1, 2, \ldots, n$.
r	The return rate of used goods from the customer.

4.2. Assumptions

The aim is to determine the effects of an optimal model for this study, therefore, the authors have adopted four assumptions as below.

(1) The production cycle repeats infinitely.

(2) Following the line of Savaskan et al. (2004), the total collection cost : $C(r) = I_r + c_0 r D = \alpha r^2 + c_0 r D$, where rD was the total amount of goods which will be returned from customers; besides, $D = a - bp_w + \delta s$; moreover, $a > 0$, is the entire size of the market, b is the price sensitivity of demand, δ is the social charity of demand.

(3) The return rate for used products from the customers to the manufacturer in the reverse channel, where $r = \sqrt{I_r/\alpha}$ and α is a scaling parameter.

(4) Based on Ma et al. (2013) and Hosseini-Motlagh et al. (2018), MI is the investment function of the manufacturer which is illustrated as follows $MI = \sqrt{I_s/\varepsilon}$, where ε is a scaling parameter [46,60].

(5) The authors consider forward and reverse dual-channel in reverse logistics with a manufacturer and a retailer. In the forward dual-channel, the manufacturer sells a product on the market through the retailer through reverse logistics. In reverse dual-channel, the manufacturer collects the returned products; additionally, they also conduct CSR activities.

4.3. Model Formulation

From the notation and assumptions illustrated above, the graph of production system during the time period from 0 to T [0,T] will be demonstrated in the two-stage production system (reference in Figure A1 of Appendix A). The total profit function will be established; the model of research is a perfect cyclic process. The demands of customers with amounts of dairy products in a cycle will be met by the right number of goods from manufacturers (i.e., $p_i t_i = p_n t_n = DT$).

Therefore, t_i and T can be presented in turn as follows:

$$t_i = \frac{p_n t_n}{p_i}, \tag{1}$$

with $i = 1, 2, \ldots, n$ and

$$T = \frac{p_n t_n}{D}, \tag{2}$$

Hence, the maximum level of inventory for raw material i at retailer stage is

$$Z_i = (p_i - p_e)t_i, \tag{3}$$

with $i = 1, 2, \ldots, n$.

Therefore, the period when the inventory of raw material i depletes can be determined as:

$$t_{id} = \frac{Z_i}{p_e} = \frac{(p_i - p_e)t_i}{p_e} = \left(\frac{p_i}{p_e} - 1\right)t_i, \tag{4}$$

with $i = 1, 2, \ldots, n$.

$$H_e = (p_e - D)(t_n + t_{nd})$$

$$= (p_e - D)\left[t_n + \left(\frac{p_n}{p_e} - 1\right)t_n\right] \tag{5}$$

$$= (p_e - D)\frac{p_n}{p_e}t_n.$$

However, the maximum inventory level of return used product as

$$Z_{re} = rt_{rdi} \tag{6}$$

where $i = 1, 2, \ldots, m$.

Based on six Equations (1)–(6), with the influence of these components, the total profit function will be established as follows:

(1) Sales revenue (SR)

$$SR = (p_w - s)$$

(2) Setup cost

$$C_s = k$$

(3) Holding the cost of the end product (HC_e)

$$HC_e = \frac{h_e Z_e T}{2} = \frac{h_e(p_e - D)}{2Dp_e}p_n{}^2 t_n{}^2$$

(4) Holding the cost of all materials (HC_c)

$$HC_c = \frac{p_n{}^2 t_n{}^2}{2}\left[\sum_{i=1}^{n} h_i\left(\frac{1}{p_e} - \frac{1}{p_i}\right)\right]$$

(5) Disposal costs of the defective end product/all raw materials per cycle (DC)

$$DC = \left(d_e\theta_e + \sum_{i=1}^{n} d_i\theta_i\right)DT + \left(T - \sum_{i=1}^{m} t_{rdi}\right)rd_{re}$$

(6) Return costs for used products at retailer stage (RC).

$$RC = c_1 r \sum_{i=1}^{m} t_{rdi}.$$

(7) Based on Giri's proposed model in 2005 [61], the production cost (PC) as:

$$PC = \left(\beta_0 + \frac{\beta_1}{p_e} + \beta_2 p_e\right)p_n t_n.$$

In the integrated supply chain system, all upstream manufacturers are willing to conglomerate resources. In an organizational decision-making process involving the manufacturer and the retailer, a product is manufactured in a single batch. Thus, making capital investment decisions on CSR activities is given by I_s and collecting the used products for recycling is given by I_r. Based on Savaskan et al. (2004), both investment costs are $I_s = \beta s^2$ and $I_r = \alpha r^2$, where α and β are scaling parameters, respectively [62]. Therefore, the objective function of the proposed model consisting of seven parts to maximize the total profit per unit of time is given by optimizing s, t_n, and r.

Thus, the total profit per unit time (denoted by $TP(s, t_n, r)$) is given by

$$
\begin{aligned}
TP(s, t_n, r) \;=\;& (SR - SC - HC_e - HC_c - DC - RC - PC) \\
=\;& (p_w - s + c_1 r)(a - bp_w + \delta s) \\
& - \left[\frac{h_e[p_e - (a - bp_w + \delta s)]}{(a - bp_w + \delta s)p_e} + \left[\sum_{i=1}^{n} h_i \left(\frac{1}{p_e} - \frac{1}{p_i} \right) \right] \right] \frac{p_n^2 t_n^2}{2} \\
& - \left[\left(d_e \theta_e + \sum_{i=1}^{n} d_i \theta_i \right) + \left(\beta_0 + \frac{\beta_1}{p_e} + \beta_2 p_e \right) + \frac{r d_{re}}{a - bp_w + \delta s} \right] p_n t_n \\
& + (d_{re} - c_1) r \sum_{i=1}^{m} t_{rdi} \\
& - \beta s^2 - \alpha r^2 - k.
\end{aligned}
\tag{7}
$$

Aiming to address this nonlinear programming issue, the authors first ignore the restriction and take the first-order derivation of $TP(s, t_n, r)$ with respect to s, t_n, r, respectively. We obtain

$$
\begin{aligned}
\frac{\partial TP(s, t_n, r)}{\partial s} \;=\;& \left[\delta c_1 r + (b - \beta)p_w + (\beta + \delta)(p_w - 2s) - a \right] \\
& - (p_n t_n h_e + 2 r d_{re}) \left[\frac{\delta}{(a - bp_w + \delta s)^2} \right] \frac{p_n t_n}{2},
\end{aligned}
\tag{8}
$$

$$
\begin{aligned}
\frac{\partial TP(s, t_n, r)}{\partial t_n} \;=\;& - \left\{ \frac{h_e[p_e - (a - bp_w + \delta s)]}{(a - bp_w + \delta s)p_e} + \left[\sum_{i=1}^{n} h_i \left(\frac{1}{p_e} - \frac{1}{p_i} \right) \right] \right\} p_n^2 t_n \\
& - \left[\left(d_e \theta_e + \sum_{i=1}^{n} d_i \theta_i \right) + \left(\beta_0 + \frac{\beta_1}{p_e} + \beta_2 p_e \right) + \frac{r d_{re}}{a - bp_w + \delta s} \right] p_n,
\end{aligned}
\tag{9}
$$

and

$$
\frac{\partial TP(s, t_n, r)}{\partial r} = c_1(a - bp_w + \delta s) + (d_{re} - c_1) \sum_{i=1}^{m} t_{rdi} - 2\alpha r.
\tag{10}
$$

To find the optimal solution of (s, t_n, r), let $\partial TP(s, t_n, r)/\partial s = 0$, $\partial TP(s, t_n, r)/\partial t_n = 0$, and $\partial TP(s, t_n, r)/\partial r = 0$, simultaneously. Solving these three equations, we obtain

$$
\delta c_1 r + (b + \delta)p_w = \left[\frac{p_n t_n h_e + 2 r d_{re}}{2(a - bp_w + \delta s)^2} \right] \delta p_n t_n + a + 2s(\delta + \beta),
\tag{11}
$$

$$
\begin{aligned}
& \left\{ \frac{h_e[p_e - (a - bp_w + \delta s)]}{(a - bp_w + \delta s)p_e} + \left[\sum_{i=1}^{n} h_i \left(\frac{1}{p_e} - \frac{1}{p_i} \right) \right] \right\} p_n^2 t_n + \frac{r d_{re}}{a - bp_w + \delta s} p_n \\
& = - \left[\left(d_e \theta_e + \sum_{i=1}^{n} d_i \theta_i \right) - \left(\beta_0 + \frac{\beta_1}{p_e} + \beta_2 p_e \right) \right] p_n,
\end{aligned}
\tag{12}
$$

and

$$
r = \frac{c_1(a - bp_w + \delta s) + (d_{re} - c_1) \sum_{i=1}^{m} t_{rdi}}{2\alpha}.
\tag{13}
$$

From Equations (11)–(13), it is clear that s and r can be uniquely determined as functions of t_n. To analyze the inventory problem, we will show that for any given feasible (s^*, r^*), the optimal production run time also exists and is unique. For given s^* and r^*, the first-order necessary condition for $TP(t_n | s^*, r^*)$ to be maximum is

$$\frac{dTP(t_n|s^*,r^*)}{dt_n} = -\left\{ \frac{h_e[p_e-(a-bp_w+\delta s)]}{(a-bp_w+\delta s)p_e} - \left[\sum_{i=1}^{n} h_i\left(\frac{1}{p_e} - \frac{1}{p_i}\right) \right] \right\} p_n^2 t_n$$

$$- \left[\left(d_e\theta_e + \sum_{i=1}^{n} d_i\theta_i \right) + \left(\beta_0 + \frac{\beta_1}{p_e} + \beta_2 p_e \right) + \frac{rd_{re}}{a-bp_w+\delta s} \right] p_n \tag{14}$$

$$= 0.$$

Let $G(t_n)$ be the left-hand side of Equation (14), i.e.,

$$G(t_n) = -\left\{ \frac{h_e[p_e-(a-bp_w+\delta s^*)]}{(a-bp_w+\delta s^*)p_e} - \left[\sum_{i=1}^{n} h_i\left(\frac{1}{p_e} - \frac{1}{p_i}\right) \right] \right\} p_n^2 t_n$$

$$- \left[\left(d_e\theta_e + \sum_{i=1}^{n} d_i\theta_i \right) + \left(\beta_0 + \frac{\beta_1}{p_e} + \beta_2 p_e \right) + \frac{r^* d_{re}}{a-bp_w+\delta s^*} \right] p_n. \tag{15}$$

We first rewrite Equation (14) and have

$$\left[\left(d_e\theta_e + \sum_{i=1}^{n} d_i\theta_i \right) + \left(\beta_0 + \frac{\beta_1}{p_e} + \beta_2 p_e \right) + \frac{r^* d_{re}}{a-bp_w+\delta s^*} \right] p_n$$

$$= -\left\{ \frac{h_e[p_e-(a-bp_w+\delta s^*)]}{p_e(a-bp_w+\delta s^*)} - \left[\sum_{i=1}^{n} h_i\left(\frac{1}{p_e} - \frac{1}{p_i}\right) \right] \right\} p_n^2 t_n. \tag{16}$$

Because of the left-hand side of Equation (16)

$$\left[\left(d_e\theta_e + \sum_{i=1}^{n} d_i\theta_i \right) + \left(\beta_0 + \frac{\beta_1}{p_e} + \beta_2 p_e \right) + \frac{r^* d_{re}}{a-bp_w+\delta s^*} \right] p_n > 0,$$

then we have $\Delta > 0$, where

$$\Delta \equiv \left\{ \left[\sum_{i=1}^{n} h_i\left(\frac{1}{p_e} - \frac{1}{p_i}\right) \right] - \frac{h_e[p_e-(a-bp_w+\delta s^*)]}{p_e(a-bp_w+\delta s^*)} \right\} p_n^2 t_n.$$

Next, taking the first-order derivative of $G(t_n)$ with respect to t_n, we obtain

$$\frac{dG(t_n)}{dt_n} = \left\{ \left[\sum_{i=1}^{n} h_i\left(\frac{1}{p_e} - \frac{1}{p_i}\right) \right] - \frac{h_e[p_e-(a-bp_w+\delta s^*)]}{(a-bp_w+\delta s^*)p_e} \right\} p_n^2 = \frac{\Delta}{t_n} > 0. \tag{17}$$

Therefore, $G(t_n)$ is a strictly increasing function $t_n \in (0, \infty)$.

Theorem 1. *For any given $t_n \geq 0$, we consider the interval $t_n \in (0, \infty)$,*
(a) If $G(t_n) < 0$, then the solution (s^*, t_n^*, r^*) which maximizes $TP(s, t_n, r)$ not only exists but also is unique, and $t_n^* \in (0, \infty)$.
(b) If $G(t_n) \geq 0$, then the optimal value of t_n is $t_n^* \to 0$. The production system should not be opened.

Proof.
(a) Firstly, we consider the interval $t_n \in (0, \infty)$. Because $\lim_{t_n \to \infty} G(t_n) = \infty$ and $\lim_{t_n \to 0} G(t_n) < 0$, from the Intermediate Value Theorem, we can find a unique solution $t_n^* \in (0, \infty)$ such that $G(t_n^*) = 0$. Substituting t_n^* into Equations (8)–(10), the corresponding s^* and r^* can be determined. Furthermore, we also calculate that

$$\mathbf{H}_{11} = \frac{\partial^2 TP(s, r|t_n)}{\partial s^2}\bigg|_{(s,r)=(s^*,r^*)} = 2\left\{ \left[\frac{\delta^2 p_n t_n(h_e p_n t_n + rd_{re})}{(a-bp_w+\delta s)^3} \right] - (\beta+\delta) \right\},$$

$$\mathbf{H}_{22} = \left.\frac{\partial^2 TP(s,r|t_n)}{\partial r^2}\right|_{(s,r)=(s^*,r^*)} = -2\alpha < 0,$$

and

$$\mathbf{H}_{12} = \mathbf{H}_{21} = \left.\frac{\partial^2 TP(s,\ r|t_n)}{\partial s \partial r}\right|_{(s,r)=(s^*,r^*)} = 0.$$

Therefore, the determinant of the Hessian matrix at the stationary point (s^*, r^*) is

$$\det(H) = \mathbf{H}_{11} \times \mathbf{H}_{22} - \mathbf{H}_{12} \times \mathbf{H}_{21}$$
$$= -4\alpha \left\{ \left[\frac{\delta^2 p_n t_n (h_e p_n t_n + r d_{re})}{(a - b p_w + \delta s)^3} \right] - (\beta + \delta) \right\} < 0.$$

(b) From Equation (9), we obtain that $\frac{\partial TP(s,t_n,r)}{\partial t_n} < 0$, which implies that t_n causes a higher value $TP(s, t_n, r)$. Hence, the maximum value $TP(s, t_n, r)$ occurs at the point $t_n^* \to 0$. It seems reasonable to conclude that the production system will not be opened. This completes the proof. $\square$

Summarizing the above results, the Algorithm 1 was used to obtain the optimal solution to our problem.

Algorithm 1 Calculating the optimal solution

Step 1:	Start with $i = 1$ and $t_i \to t_n$.
Step 2:	Put t_i into Equation (12) to obtain the corresponding value of t_n, i.e., t_n'.
Step 3:	If $G(t_n') < 0$, go to Step 4. Otherwise, set the optimal solutions $r^* = 0, s^* = 0$, and $t_n^* = 0$ (i.e., the manufacturer will not adopt CSR initiatives), then stop the Algorithm.
Step 4:	Find the optimal value t_n' such that $G(t_n') = 0$.
Step 5:	If the difference between t_i and t_{i+1} is sufficiently small, set $t_n^* = t_{i+1}$, then go to Step 6.
Step 6:	The maximum total profit per unit time $TP(s^*, t_n^*, r^*)$ can be obtained by substituting s^*, t_n^*, and r^* into Equation (7).

---

5. Application Example

The empirical research context is Vietnam's dairy supply chain to demonstrate the utility and feasibility of the proposed model. ABC dairy corporation was chosen, which is currently the largest dairy company in Vietnam [63]. The ABC company currently has a complete supply chain with several farms, factories, and its own distribution system; moreover, their products have been distributed to nearly all supermarkets and retail points all over Vietnam. Towards sustainable development, ABC dairy firm's orientation would invest heavily in CSR initiatives and be directed to the five major objectives, including community development support; environment and energy; responsibility for employees; responsibility for products; and local economic development. Regarding the ABC dairy company's commitment to the stakeholders, they are committed to providing safe products to consumers with top quality and appropriate prices. Moreover, this company also declares complying with environmental regulations and integrating social responsibility into business strategies. Food products' characteristics are easily broken for many different reasons before reaching customers, combined with the ABC company's product quality assurance policy. Thus, the case study selected as the ABC dairy corporation would help the authors better assess the proposed model in the empirical research.

The authors considered an application example of CSR initiatives to demonstrate the proposed model and verify the obtained analytical results. The real data were collected from ABC dairy firm and employed for the numerical analysis as following: $k = \$80/cycle$, $\beta_0 = 0.1$, $\beta_1 = 0.03$, $\beta_2 = 0.025$, $p_w = \$10$, $a = 15$, $b = 9$, $\alpha = 7$, $\beta = 2$, $\delta = 4$.

Raw material source 1: p_1 = 100/per unit time, h_1 = \$0.01/per unit, θ_1 = 0.03, d_1 = \$0.01/per unit.

Raw material source 2: p_2 = 200/per unit time, h_2 = \$0.02/per unit, θ_2 = 0.02, d_2 = \$0.02/per unit.

Raw material source 3: p_3 = 300/per unit time, h_3 = \$0.375/per unit/per unit time, θ_3 = 0.01, d_3 = \$0.03/per unit.

End product in showrooms: p_e = 60/per unit time, h_e = \$2/per unit/per unit time, θ_e = 0.05, d_e = \$0.04/per unit.

Then, we find the outcome as s^* = 9.58943; t_n^* = 0.495389; r^* = 0.0595798 and $TP(s^*, t_n{}^*, r^*)$ = 287.392.

In the finding of this section, the authors evaluate the impact of implementing CSR practices on the company's inventory policy. Thus, the effect of changes in various parameters of the model, for example (in Table A1 of Appendix B), was indicated that the effects of h_e, β_2, and θ_e on total profit are significant. These imply that the quality of the end product is essential for the ABC dairy enterprise.

Thus, maintaining high-quality products secures a high level of demand by end customers whereas poor quality products affect the customer's confidence, reputation, and sales of the company. Aim to create a positive brand image through CSR activities of the enterprise towards the community. Hence, the dairy company was aware of the importance of the public, especially their target customers, having a positive perception of them. As Figure 2 indicated, the concave function of the total economic profit based on per unit time.

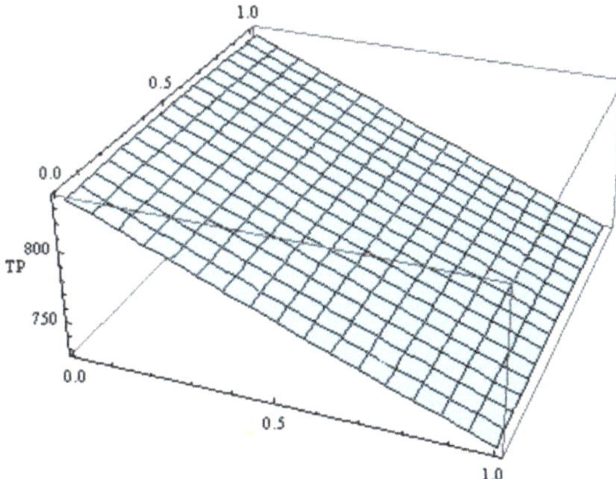

Figure 2. The total profit per unit time $TP(s^*, t_n^*, r)$ for application example.

6. Discussion of Research Findings

Toward sustainable development targets that lead to ABC dairy company's CSR programs being conducted to solve the comprehensive enterprise issue, which has a profit-sharing obligation to society beyond making a profit. In fact, to contribute the profit-sharing to the nation's target through charitable activities or community development, the ABC dairy firm has an obligation that implements financial responsibilities before other duties. Therefore, it requires the ABC dairy firm to ensure their accountability to stakeholders, especially shareholders' interests. A numerical example in the above subsection is considered to study the effects of changes in the system parameters (h_1, h_2, h_3, h_e, θ_1, θ_2, θ_3, θ_e, d_1, d_2, d_3, d_e, β_0, β_1, and β_2) on the optimal values of s^*, t_n^*, r, and $TP(s^*, t_n^*, r)$. The result indicated that the effects of h_e, β_2 and θ_e on total profit are significant. These imply that the quality of the end product is essential for the ABC dairy enterprise. Thus, maintaining high-quality products aims to secure a high demand by end customers while struggling with the risk that poor quality products affect customers' confidence, reputation, and sales. Brand image, reputation, as well as financial performance can be significantly enhanced through CSR activities of the enterprise towards the community; hence, ABC dairy company must be aware of the local community's importance, especially their target customers. For the present, it may be useful to look more closely at some of the more important features of the sensitivity analysis performed by changing each of the parameters by +50%, +25%, −25%, and −50%; taking one parameter at a time and keeping the remaining parameters unchanged. The analytical results were clearly demonstrated in Table A1 of Appendix B.

From the Table A1 in Appendix B, the following the management implications have been very positive as follows:

(1) When the values of parameters ($h_1, h_2, h_3, h_e, \theta_1, \theta_2, \theta_3$ and θ_e) are decreased, it leads to the growth for $TP(s^*, t_n^*, r)$. It indicates that the firm cooperation with manufacturers is based on these factors, including inventory turnover and the quality of raw material or end products. The total profit has been enhanced, leading to the firm's social donation fund also increasing. Thus, it positively directs on the ABC dairy enterprise's brand and its reputation from the local community.

(2) With decreases in the value of parameter β_0, then $TP(s^*, t_n^*, r)$ increases. If the processing cost of raw materials from different sources could be reduced effectively, the total profit would be enhanced. This implies that the ABC dairy firm should invest more in new employee orientation to improve productivity.

(3) With decreases in the value of parameter β_1, then $TP(s^*, t_n^*, r)$ increases. This implies that the ABC dairy firm should decrease the labor costs per unit of time (i.e., wage or salary) to increase total profit.

(4) With decreases in the value of parameter β_2, then $TP(s^*, t_n^*, r)$ increases. This implies that the ABC dairy firm should decrease the tool or idle cost per unit of time to increase total profit. Furthermore, the β_2 parameter will have an influence rather than β_0 and β_1 on the total profit per unit time.

(5) With decreases in the value of parameters as d_1, d_2, d_3, then $TP(s^*, t_n^*, r)$ can be increased. This implies that the firm should decrease scrape costs per unit of time to increase total profit.

Apart from complying with laws related to the environment and workers, improving the quality of end products and reducing their holding costs to enhance financial performance are the firms' priority targets. Based on the proposed model of Chang et al. (2012), the authors have developed and evaluated this suggestion to the Vietnamese food supply chain via collected data from the ABC dairy firm case study; thereby, the suitability and positive results have been clarified. These research findings are consistent with other studies related to evaluating the EPQ model under the two-stage assembly production system, such as Dey et al. (2019) and Sabbaghnia and Taleizadeh (2020). They indicated that relationships between SCM members could be comprehensively evaluated to point out the best decision-making; hence, an application example along with sensitivity analysis has been conducted to support this argument. In particular, the relationship between CSR initiatives and SCM of some studies has supported our findings, and the research examples can be seen by Nematollahi et al. (2017); Modak et al. (2019); and Jokar and Hosseini-Motlagh (2020). They have similar conclusions in terms of the vital role as well as the CSR initiatives that could affect decision-making and various SCM performances. Furthermore, to increase donations to local society, the economic performance would be impacted by these activities; however, there is a positive signal from the final customers, whose willingness to buy green products [11,53], or think they are making a small contribution to society generates the enhancing of economic obligation for corporations.

Although the implementation of social activities affects the economic benefits of ABC dairy enterprise, its total profits still increase, which ensures its commitment to its economic responsibility with shareholders and employees. Thus, ABC dairy corporation is a particular example for the EPQ model's suitability, which can point out some practical implications for the top managers in decision-making related to CSR programs under SCM. They need to establish a risk management mechanism to control any potential risks that may affect their operations and profits. With such an arrangement, they can significantly lower the company's operational risk, damage, and impact.

7. Conclusions

This study is unique considering integrating CSR initiatives into the food supply chain in Vietnam, which considered some critical decision parameters together and linked the ideas and conclusions based on a particular case study. The noteworthy contributions of Vietnam's dairy firms and their CSR programs are significant to the sustainable development goal of an agricultural country like Vietnam. Aiming to demonstrate the efficiency of the proposed EPQ model, ABC dairy corporation was chosen, which is a typical example that can represent the Vietnam dairy industry in both the main issues of this study as CSR and SCM. Thus, the authors demonstrated the proposed model and clarified how to integrate CSR initiatives into the food SCM. A two-stage assembly production system involving the return rate of used goods from the customer was suggested, which has impacted the CSR policy and top manager decision-making in the dairy supply chain. The research findings are consistent with several prior studies in terms of the CSR initiatives' role, which is considered an excellent tool that may obtain a sustainable supply chain by enhancing social responsibility and the firm's financial performance. Furthermore, total profits per unit of time have been influenced by these critical parameters leading to the following research implications.

(1) Based on theories regarding the relationship between CSR and SCM, these results have contributed to the existing literature on developing the classical economic production quantity model under the two-stage assembly production system.

(2) The findings will help the top managers of dairy corporations in Vietnam better understand the integrating approach and the efficiency of adopting CSR programs into SCM.

(3) The production costs will be effectively controlled, such as material and operating costs. Hence, the firms have more dynamics to devote more money to charitable donations, which positively impacts consumer perception, and then economic performance will be enhanced.

The supply chain management includes all relationships from upstream suppliers to downstream consumers; unfortunately, in this study, the author has just focused on evaluating the relationship between producer ABC and their final customer through social activities, total profits of the dairy company, and the return of the goods issue of the customer. Thus, the authors suggested that further research might extend to different relationships between manufacturers and stakeholders based on CSR activities and SCM nexus. Considering the relationship between manufacturers and suppliers, the linkage effect between the dairy company and retailers in the dairy supply chain of an emerging economy is a promising idea for other scholars. Moreover, aiming to reduce their holding costs and enhance financial benefits, this proposed model could be conducted in various sectors to broaden and enrich the knowledge of the CSR initiatives' usefulness.

Author Contributions: M.-H.D. and Y.-F.H. developed conceptualization; methodology was developed by M.-W.W. and T.-S.C.; data curation and resources were undertaken by M.-H.D. and T.-S.C.; Y.-F.H. and M.-W.W. analyzed the data; wrote the manuscript, M.-H.D., M.-W.W., and T.-S.C.; writing—review and editing, T.-S.C., M.-H.D., and M.-W.W.; visualization, Y.-F.H.; supervision, Y.-F.H.; project administration, Y.-F.H. All authors have read and agreed to the published version of the manuscript.

Funding: The author's research was partially supported by the Ministry of Science and Technology of the Republic of China under Grant MOST 109-2622-H-324-001. Besides, this research was supported by the Ministry of Education, Taiwan, R. O. C. (Intelligence commerce and marketing craft the social practice model for the Prionailurus bengalensis and agricultural optimized regional revitalization in Chungliao).

Institutional Review Board Statement: Not applicable.

Informed Consent Statement: Not applicable.

Data Availability Statement: Due to its proprietary nature <or ethical concerns>, supporting data cannot be made openly available.

Acknowledgments: The authors would like to thank the editor and anonymous reviewers for their valuable and constructive comments, which have led to a significant improvement in the manuscript.

Conflicts of Interest: The authors declare no conflict of interest.

Appendix A

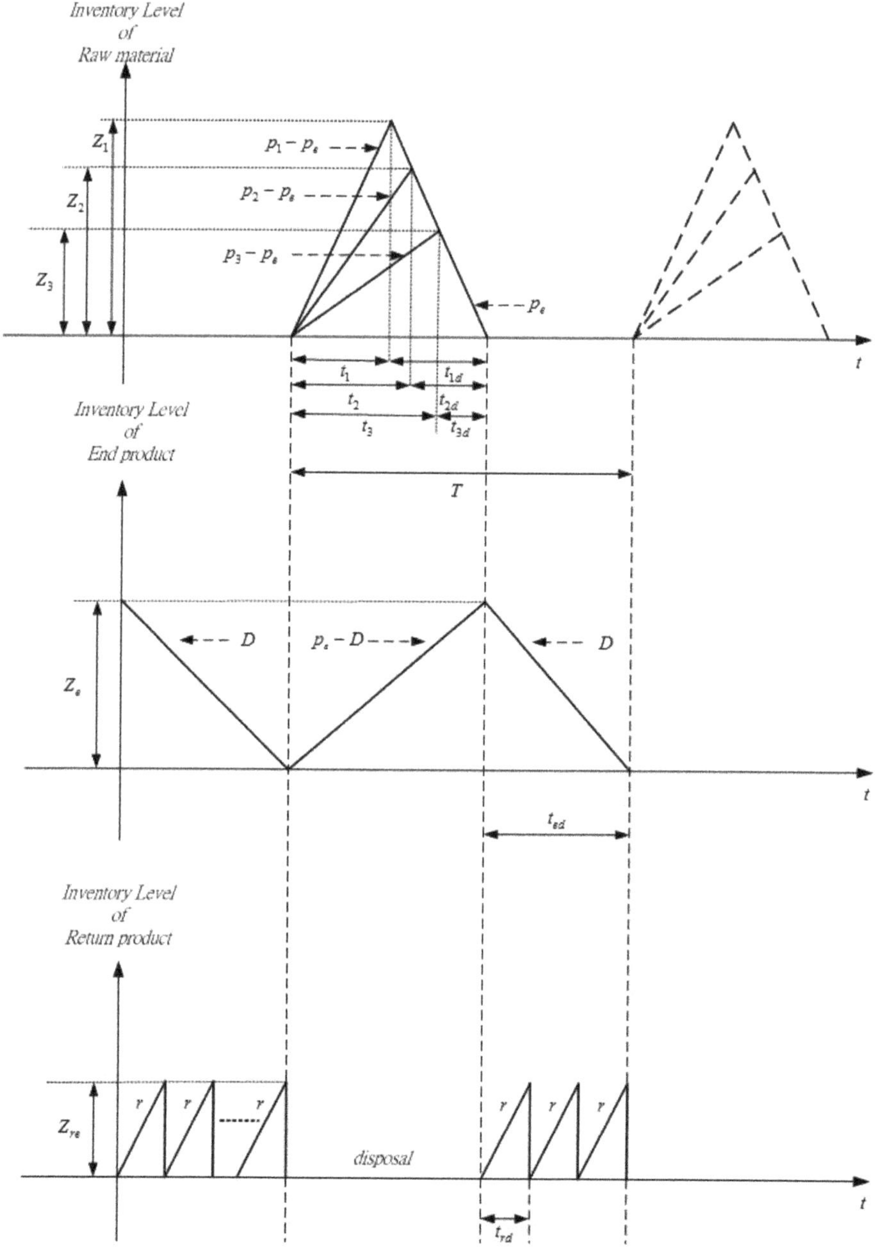

Figure A1. The graph of production system during time period $[0, T]$.

Appendix B

Table A1. Effect of changes in various parameters of the model for application example.

Parameter	Change (%)	Optimal Solutions			
		s^*	t_n^*	r^*	$TP(s^*,t_n^*,r^*)$
h_1	−50%	0.85486	0.15500	0.02282	287.395
	−25%	9.58942	0.49506	0.05962	287.395
	0	9.58942	0.49538	0.05957	287.392
	25%	9.58943	0.49572	0.05954	287.389
	50%	9.58944	0.49605	0.05950	287.386
h_2	−50%	9.58939	0.49401	0.05975	287.404
	−25%	9.58941	0.49470	0.05966	287.398
	0	9.58942	0.49538	0.05957	287.392
	25%	9.58944	0.49608	0.05950	287.386
	50%	9.58946	0.49678	0.05941	287.380
h_3	−50%	9.58883	0.47072	0.06271	287.627
	−25%	9.58977	0.50912	0.06115	287.506
	0	9.58942	0.49538	0.05957	287.392
	25%	9.58977	0.50912	0.05797	287.283
	50%	9.59015	0.52394	0.05633	287.181
h_e	−50%	9.59251	0.85894	0.03436	289.676
	−25%	9.59027	0.61039	0.04835	288.211
	0	9.58942	0.49538	0.05957	287.392
	25%	9.58898	0.42676	0.06916	286.836
	50%	9.58872	0.38024	0.07762	286.421
θ_1	−50%	9.58942	0.49535	0.05959	287.393
	−25%	9.58943	0.49537	0.05958	287.392
	0	9.58943	0.49538	0.05957	287.392
	25%	9.58944	0.49541	0.05957	287.391
	50%	9.58944	0.49543	0.05957	287.391
θ_2	−50%	9.58942	0.49533	0.05958	287.393
	−25%	9.58943	0.49536	0.05958	287.392
	0	9.58943	0.49538	0.05957	287.392
	25%	9.58943	0.49542	0.05957	287.391
	50%	9.58943	0.49545	0.05957	287.391
θ_3	−50%	9.58942	0.49535	0.05959	287.393
	−25%	9.58943	0.49537	0.05958	287.392
	0	9.58943	0.49538	0.05957	287.392
	25%	9.58943	0.49541	0.05957	287.391
	50%	9.58943	0.49543	0.05957	287.391
θ_e	−50%	9.58907	0.48082	0.04808	287.643
	−25%	9.58925	0.48806	0.04880	287.517
	0	9.58943	0.49538	0.05957	287.392
	25%	9.58961	0.51027	0.05870	287.269
	50%	9.58981	0.51028	0.05784	287.148
d_1	−50%	9.58942	0.49535	0.05959	287.391
	−25%	9.58943	0.49536	0.05958	287.391
	0	9.58943	0.49538	0.05957	287.392
	25%	9.58943	0.49541	0.05957	287.392
	50%	9.58943	0.49543	0.05957	287.393
d_2	−50%	9.58942	0.49533	0.05959	287.391
	−25%	9.58943	0.49536	0.05958	287.391
	0	9.58943	0.49538	0.05957	287.391
	25%	9.58943	0.49542	0.05957	287.392
	50%	9.58943	0.49545	0.05957	287.393

Table A1. *Cont.*

Parameter	Change (%)	Optimal Solutions			
		s^*	t_n^*	r^*	$TP(s^*, t_n^*, r^*)$
d_3	−50%	9.58942	0.49535	0.05959	287.391
	−25%	9.58943	0.49536	0.05958	287.391
	0	9.58943	0.49538	0.05957	287.391
	25%	9.58943	0.49541	0.05957	287.392
	50%	9.58943	0.49543	0.05957	287.393
d_e	−50%	9.58943	0.49535	0.05959	287.391
	−25%	9.58943	0.49537	0.05958	287.391
	0	9.58943	0.49538	0.05958	287.391
	25%	9.58943	0.49541	0.05958	287.391
	50%	9.58943	0.49543	0.05957	287.391
β_0	−50%	9.58907	0.48082	0.02282	287.643
	−25%	9.58925	0.48806	0.05962	287.517
	0	9.58943	0.49538	0.05958	287.391
	25%	9.58961	0.50279	0.06047	287.269
	50%	9.58981	0.51027	0.06138	287.148
β_1	−50%	9.58942	0.49495	0.05963	287.399
	−25%	9.58942	0.49516	0.05960	287 396
	0	9.58943	0.49538	0.05958	287.391
	25%	9.58943	0.49561	0.05955	287.388
	50%	9.58944	0.49583	0.05953	287.384
β_2	−50%	9.58858	0.45958	0.06422	288.035
	−25%	9.58899	0.47723	0.06185	287.707
	0	9.58943	0.49538	0.05958	287.391
	25%	9.58989	0.51405	0.05742	287.088
	50%	9.59039	0.53319	0.05536	286.795

References

1. Gangi, F.; Mustilli, M.; Varrone, N. The impact of corporate social responsibility (CSR) knowledge on corporate financial performance: Evidence from the European banking industry. *J. Knowl. Manag.* **2019**, *23*, 110–134. [CrossRef]
2. Do, M.-H.; Huang, Y.-F.; Do, T.-N. The effect of total quality management-enabling factors on corporate social responsibility and business performance: Evidence from Vietnamese coffee firms. *Benchmarking Ann. Int. J.* **2020**. [CrossRef]
3. Sharabati, A.A.A. Effect of corporate social responsibility on Jordan pharmaceutical industry's business performance. *Soc. Responsib. J.* **2018**, *14*, 566–583. [CrossRef]
4. Rhou, Y.; Singal, M.; Koh, Y. CSR and financial performance: The role of CSR awareness in the restaurant industry. *Int. J. Hosp. Manag.* **2016**, *57*, 30–39. [CrossRef]
5. Kong, D.; Shi, L.; Yang, Z. Product recalls, corporate social responsibility, and firm value: Evidence from the Chinese food industry. *Food Policy* **2019**, *83*, 60–69. [CrossRef]
6. Jamali, D.; Karam, C. Corporate Social Responsibility in Developing Countries as an Emerging Field of Study. *Int. J. Manag. Rev.* **2016**, *20*, 32–61. [CrossRef]
7. Carroll, A.B. Carroll's pyramid of CSR: Taking another look. *Int. J. Corp. Soc. Responsib.* **2016**, *1*, 1–8. [CrossRef]
8. Hategan, C.D.; Sirghi, N.; Curea-Pitorac, R.I.; Hategan, V.P. Doing well or doing good: The relationship between corporate social responsibility and profit in Romanian companies. *Sustainability* **2018**, *10*, 1041. [CrossRef]
9. Maqbool, S.; Zameer, M.N. Corporate social responsibility and financial performance: An empirical analysis of Indian banks. *Futur. Bus. J.* **2018**, *4*, 84–93. [CrossRef]
10. Nollet, J.; Filis, G.; Mitrokostas, E. Corporate social responsibility and financial performance: A non-linear and disaggregated approach. *Econ. Model.* **2016**, *52*, 400–407. [CrossRef]
11. Huang, Y.F.; Do, M.H.; Kumar, V. Consumers' perception on corporate social responsibility: Evidence from Vietnam. *Corp. Soc. Responsib. Environ. Manag.* **2019**, *26*, 1272–1284. [CrossRef]
12. Das, D. Sustainable supply chain management in Indian organisations: An empirical investigation. *Int. J. Prod. Res.* **2018**, *56*, 5776–5794. [CrossRef]
13. Heinberg, M.; Ozkaya, H.E.; Taube, M. Do corporate image and reputation drive brand equity in India and China?-Similarities and differences. *J. Bus. Res.* **2018**, *86*, 259–268. [CrossRef]

14. Khosroshahi, H.; Rasti-Barzoki, M.; Hejazi, S.R. A game theoretic approach for pricing decisions considering CSR and a new consumer satisfaction index using transparency-dependent demand in sustainable supply chains. *J. Clean. Prod.* **2019**, *208*, 1065–1080. [CrossRef]
15. Bisogno, M. Corporate Social Responsibility and Supply Chains: Contribution to the Sustainability of Well-being. *Agric. Agric. Sci. Procedia* **2016**, *8*, 441–448. [CrossRef]
16. Hsu, S.Y.; Chang, C.C.; Lin, T.T. Triple bottom line model and food safety in organic food and conventional food in affecting perceived value and purchase intentions. *Br. Food J.* **2019**, *121*, 333–346. [CrossRef]
17. Pino, G.; Amatulli, C.; De Angelis, M.; Peluso, A.M. The influence of corporate social responsibility on consumers' attitudes and intentions toward genetically modified foods: Evidence from Italy. *J. Clean. Prod.* **2016**, *112*, 2861–2869. [CrossRef]
18. Sodhi, M.M.S.; Tang, C.S. Corporate social sustainability in supply chains: A thematic analysis of the literature. *Int. J. Prod. Res.* **2018**, *56*, 882–901. [CrossRef]
19. Zhang, M.; Tse, Y.K.; Doherty, B.; Li, S.; Akhtar, P. Sustainable supply chain management: Confirmation of a higher-order model. *Resour. Conserv. Recycl.* **2018**, *128*, 206–221. [CrossRef]
20. Sarkar, B.; Omair, M.; Choi, S.B. A multi-objective optimization of energy, economic, and carbon emission in a production model under sustainable supply chain management. *Appl. Sci.* **2018**, *8*, 1744. [CrossRef]
21. Feng, Y.; Zhu, Q.; Lai, K.H. Corporate social responsibility for supply chain management: A literature review and bibliometric analysis. *J. Clean. Prod.* **2017**, *158*, 296–307. [CrossRef]
22. Liu, Y.; Quan, B.T.; Xu, Q.; Forrest, J.Y.L. Corporate social responsibility and decision analysis in a supply chain through government subsidy. *J. Clean. Prod.* **2019**, *208*, 436–447. [CrossRef]
23. Ding, H.; Fu, Y.; Zheng, L.; Yan, Z. Determinants of the competitive advantage of dairy supply chains: Evidence from the Chinese dairy industry. *Int. J. Prod. Econ.* **2019**, *209*, 360–373. [CrossRef]
24. Chen, Y.S.; Chiu, S.C.; Lin, S.; Wu, K.H. Corporate social responsibility and income smoothing: Supply chain perspectives. *J. Bus. Res.* **2019**, *97*, 76–93. [CrossRef]
25. Wu, Y.; Li, H.; Gou, Q.; Gu, J. Supply chain models with corporate social responsibility. *Int. J. Prod. Res.* **2017**, *55*, 6732–6759. [CrossRef]
26. Tan, A.; Ngan, P.T. A proposed framework model for dairy supply chain traceability. *Sustain. Futur.* **2020**, *2*, 100034. [CrossRef]
27. Svensson, G.; Ferro, C.; Høgevold, N.; Padin, C.; Carlos Sosa Varela, J.; Sarstedt, M. Framing the triple bottom line approach: Direct and mediation effects between economic, social and environmental elements. *J. Clean. Prod.* **2018**, *197*, 972–991. [CrossRef]
28. Gómez-Luciano, C.A.; Rondón Domínguez, F.R.; González-Andrés, F.; Urbano López De Meneses, B. Sustainable supply chain management: Contributions of supplies markets. *J. Clean. Prod.* **2018**, *184*, 311–320. [CrossRef]
29. Ghadimi, P.; Azadnia, A.H.; Heavey, C.; Dolgui, A.; Can, B. A review on the buyer-supplier dyad relationships in sustainable procurement context: Past, present and future. *Int. J. Prod. Res.* **2016**, *54*, 1443–1462. [CrossRef]
30. Khalid, R.U.; Seuring, S.; Beske, P.; Land, A.; Yawar, S.A.; Wagner, R. Putting sustainable supply chain management into base of the pyramid research. *Supply Chain Manag.* **2015**, *20*, 681–696. [CrossRef]
31. Zhong, R.; Xu, X.; Wang, L. Food supply chain management: Systems, implementations, and future research. *Ind. Manag. Data Syst.* **2017**, *117*, 2085–2114. [CrossRef]
32. Scalia, G.L.; Settanni, L.; Micale, R.; Enea, M. Predictive shelf life model based on RF technology for improving the management of food supply chain: A case study. *Int. J. RF Technol. Res. Appl.* **2016**, *7*, 31–42. [CrossRef]
33. Chkanikova, O.; Mont, O. Corporate supply chain responsibility: Drivers and barriers for sustainable food retailing. *Corp. Soc. Responsib. Environ. Manag.* **2015**, *22*, 65–82. [CrossRef]
34. Prakash, G. Review of the food processing supply chain literature: A UK, India bilateral context. *J. Adv. Manag. Res.* **2018**, *15*, 457–479. [CrossRef]
35. Gouda, S.K.; Saranga, H. Sustainable supply chains for supply chain sustainability: Impact of sustainability efforts on supply chain risk. *Int. J. Prod. Res.* **2018**, *56*, 5820–5835. [CrossRef]
36. Golini, R.; Moretto, A.; Caniato, F.; Caridi, M.; Kalchschmidt, M. Developing sustainability in the Italian meat supply chain: An empirical investigation. *Int. J. Prod. Res.* **2017**, *55*, 1183–1209. [CrossRef]
37. Chen, X.; Wu, S.; Wang, X.; Li, D. Optimal pricing strategy for the perishable food supply chain. *Int. J. Prod. Res.* **2019**, *57*, 2755–2768. [CrossRef]
38. Dries, L.; Swinnen, J.F.M. The impact of interfirm relationships on investment: Evidence from the Polish dairy sector. *Food Policy* **2010**, *35*, 121–129. [CrossRef]
39. Derville, M.; Allaire, G. Change of competition regime and regional innovative capacities: Evidence from dairy restructuring in France. *Food Policy* **2014**, *49*, 347–360. [CrossRef]
40. Glover, J.L.; Champion, D.; Daniels, K.J.; Dainty, A.J.D. An Institutional Theory perspective on sustainable practices across the dairy supply chain. *Int. J. Prod. Econ.* **2014**, *152*, 102–111. [CrossRef]
41. Augustin, M.A.; Udabage, P.; Juliano, P.; Clarke, P.T. Towards a more sustainable dairy industry: Integration across the farm-factory interface and the dairy factory of the future. *Int. Dairy J.* **2013**, *31*, 2–11. [CrossRef]
42. Krishnamoorthi, C.; Panayappan, S. An EPQ model for an imperfect production system with rework and shortages. *Int. J. Oper. Res.* **2013**, *17*, 104–123. [CrossRef]

43. Wee, H.M.; Wang, W.T.; Cárdenas-Barrón, L.E. An alternative analysis and solution procedure for the EPQ. model with rework process at a single-stage manufacturing system with planned backorders. *Comput. Ind. Eng.* **2013**, *64*, 748–755. [CrossRef]
44. Teng, J.T.; Lou, K.R.; Wang, L. Optimal trade credit and lot size policies in economic production quantity models with learning curve production costs. *Int. J. Prod. Econ.* **2014**, *155*, 318–323. [CrossRef]
45. Wu, J.; Teng, J.T.; Skouri, K. Optimal inventory policies for deteriorating items with trapezoidal-type demand patterns and maximum lifetimes under upstream and downstream trade credits. *Ann. Oper. Res.* **2018**, *264*, 459–476. [CrossRef]
46. Ma, P.; Wang, H.; Shang, J. Contract design for two-stage supply chain coordination: Integrating manufacturer-quality and retailer-marketing efforts. *Int. J. Prod. Econ.* **2013**, *146*, 745–755. [CrossRef]
47. Chang, H.J.; Su, R.H.; Yang, C.-T.; Weng, M.W. An economic manufacturing quantity model for a two-stage assembly system with imperfect processes and variable production rate. *Comput. Ind. Eng.* **2012**, *63*, 285–293. [CrossRef]
48. Sabbaghnia, A.; Taleizadeh, A.A. Quality, buyback and technology licensing considerations in a two-period manufacturing–remanufacturing system: A closed-loop and sustainable supply chain. *Int. J. Syst. Sci. Oper. Logist.* **2020**, 1–19. [CrossRef]
49. Wensing, T.; Kuhn, H. Analysis of a multi-component periodic review inventory system in an assembly environment. *OR Spectr.* **2013**, *35*, 107–126. [CrossRef]
50. Yao, Z.; Lee, L.H.; Chew, E.P.; Hsu, V.N.; Jaruphongsa, W. Dual-channel component replenishment problem in an assemble-to-order system. *IIE Trans.* **2013**, *45*, 229–243. [CrossRef]
51. Keizer, M.C.A.O.; Teunter, R.H.; Veldman, J. Joint condition-based maintenance and inventory optimization for systems with multiple components. *Eur. J. Oper. Res.* **2017**, *257*, 209–222. [CrossRef]
52. Cheng, G.Q.; Zhou, B.H.; Li, L. Joint optimization of lot sizing and condition-based maintenance for multi-component production systems. *Comput. Ind. Eng.* **2017**, *110*, 538–549. [CrossRef]
53. Modak, N.M.; Kazemi, N.; Cárdenas-Barrón, L.E. Investigating structure of a two-echelon closed-loop supply chain using social work donation as a Corporate Social Responsibility practice. *Int. J. Prod. Econ.* **2019**, *207*, 19–33. [CrossRef]
54. Gupta, M.; Mohanty, B.K. Multi-stage multi-objective production planning using linguistic and numeric data-a fuzzy integer programming model. *Comput. Ind. Eng.* **2015**, *87*, 454–464. [CrossRef]
55. Dey, B.K.; Sarkar, B.; Pareek, S. A two-echelon supply chain management with setup time and cost reduction, quality improvement and variable production rate. *Mathematics* **2019**, *7*, 328. [CrossRef]
56. Nematollahi, M.; Hosseini-Motlagh, S.M.; Heydari, J. Coordination of social responsibility and order quantity in a two-echelon supply chain: A collaborative decision-making perspective. *Int. J. Prod. Econ.* **2017**, *184*, 107–121. [CrossRef]
57. Jokar, A.; Hosseini-Motlagh, S.M. Simultaneous coordination of order quantity and corporate social responsibility in a two-Echelon supply chain: A combined contract approach. *J. Oper. Res. Soc.* **2020**, *71*, 69–84. [CrossRef]
58. The World Bank Group (WB) Overview Vietnam. Available online: https://www.worldbank.org/en/country/vietnam/overview (accessed on 18 December 2020).
59. Do, T.N.; Kumar, V.; Do, M.H. Prioritize the key parameters of Vietnamese coffee industries for sustainability. *Int. J. Product. Perform. Manag.* **2020**, *69*, 1153–1176. [CrossRef]
60. Hosseini-Motlagh, S.-M.; Nematollahi, M.; Nouri, M. Coordination of green quality and green warranty decisions in a two-echelon competitive supply chain with substitutable products. *J. Clean. Prod.* **2018**, *196*, 961–984. [CrossRef]
61. Giri, B.C.; Yun, W.Y.; Dohi, T. Optimal design of unreliable production-inventory systems with variable production rate. *Eur. J. Oper. Res.* **2005**, *162*, 372–386. [CrossRef]
62. Savaskan, R.C.; Bhattacharya, S.; Wassenhove, L.N. van Closed-Loop Supply Chain Models with Product Remanufacturing. *Manage. Sci.* **2004**, *50*, 239–252. [CrossRef]
63. VNR500 Vietnam Report top 500 Company. Available online: http://www.vnr500.com.vn/Nganh-nghe/San-xuat-kinh-doanh-thuc-pham-che-bien-Post/26.html (accessed on 12 March 2020).

Article

New Retailing Problem for an Integrated Food Supply Chain in the Baking Industry

Ning Xu [1,2], Yung-Fu Huang [3], Ming-Wei Weng [3,*] and Manh-Hoang Do [3,4]

1   Department of Business Administration, Chaoyang University of Technology, Taichung 41349, Taiwan; xn@hzu.edu.cn
2   School of Economics and Management, Huizhou University, Huizhou 516007, China
3   Department of Marketing and Logistics Management, Chaoyang University of Technology, Taichung 41349, Taiwan; huf@cyut.edu.tw (Y.-F.H.); s10537912@gm.cyut.edu.tw (M.-H.D.)
4   Faculty of Economics, Tay Nguyen University, Buon Ma Thuot 630000, Vietnam
*   Correspondence: mwweng@cyut.edu.tw; Tel.: +886-4-23323000

Abstract: As coronavirus disease 2019 (COVID-19) continues to spread, online consumption habits in China have changed significantly. Thus, the booming online-to-offline (O2O) food ordering and delivery industry via the online bakery have been changing customers' food shopping behavior. This article proposes a comparison relying on advanced O2O strategy for a single-vendor-single-retailer integrated system. Three coordination mechanisms consist of revenue-sharing, buy-back, and quantity flexibility contracts have been employed for optimizing the order quantity. Replenishment strategies and temperature for the supply chain members are considered based on the new retailing framework. Herein, the authors suggest an algorithm for the computation of the optimal solution. Lastly, numerical examples and sensitivity analysis are also conducted to clarify the usefulness of the proposed model in the food supply chain. Sensitivity analysis revealed a number of managerial insights. For example, the results obtained under O2O operations can be compared with those obtained under online/offline operations (under various parameters settings) to determine an opportune moment for three coordination mechanisms.

Keywords: inventory; new retailing; baking industrial; food supply chain coordination

1. Introduction

Several recent studies have appeared to tackle the new retailing (NR) issue, which is the phrase coined by Jack Ma, Alibaba Executive Chairman [1]. Alibaba was significantly implementing NR and online, offline, logistics, and data across the individual value chain to be involved. New retailing is one sector undergoing an Internet of Things (IoT)-driven revolution, as traditional offline retail stores such as convenience stores and supermarkets have become the point of convergence for online and online integration. According to a research report [2], on "China's Online Shopping Market Research Report" published in 2012 by CNNIC, internet shopping market accounted for about RMB1.26 trillion trading volume. The online-to-offline (O2O) business model will revolutionize the customer experience in retailing, allowing companies to introduce innovative customer services. The study is still at an early stage in evaluating NR, not to mention a paucity of literature on this subject. Du and Jiang [3] evaluated that the offline service quality problems of O2O model in China. E-commerce's development in the food supply chain has played an essential role in supporting small and medium enterprises (SMEs) to deal with challenges, stimulating domestic demand for bakery, and creating jobs; thus, the term NR is considered as a new concept based on the development of O2O model. Regarding the history line of the O2O term development, which witnessed the booming of the O2O food market in China in recent years, as recommended by Maimaiti et al. [4], more than twenty percent out of China's total population has already become the customers of the O2O food delivery market. The O2O model's impact on the optimal power structure of a retail service supply chain was studied by Chen et al. [5]. On the other hand, He et al. [6] identified the role of the competitive O2O model to adapt to multi-store service firms. In the early 20th century, Smith and Sparks [7] stressed that product quality is the most vital factor in the food supply chain. Van der Vorst et al. [8] explained that organizations can gain a competitive edge by incorporating food quality into food supply chain. In understanding quality assurance with food security, Trienekens and Zuurbier [9] presented the process of production and distribution is the major interest under food supply chain. Product quality, costs, and sustainability have been highlighted as important factors that influence the food supply chain model was dealt with by Zanoni and Zavanella [10]. In line with that, numerous studies have also investigated those issues in terms of the food supply chain [11–13]. Due to globalization, the complexity of food supply chains is increasingly becoming more complex. The concept of food supply chain traceability has become not just a way to help food manufacturers and retailers ensure food security (ex: temperature controlled), but also creating a linkage among business floors each other. Furthermore, the adoption of the IoT technology into food supply chain management (SCM) to assist with the information management process has gradually become more popular [14–17]. Early theorization of a single-vendor single-retailer integrated inventory model can be traced back to Goyal [18]. Moreover, to struggle with the rapidly changing market conditions, the supply chain system must have smooth coordination, which is considered the vital key to attaining the flexibility necessary as well as to meet superior logistics processes. Besides, several prior studies also stressed that a sounder theoretical basis for vendor-retailer integrated inventory model [19–22]. Consequently, these articles have been published that focus on a wide range of the coordination of the supply chain. Unfortunately, there is no existing article published comparing different food supply chain contracts under the new retailing framework. The channel coordination is considering as the critical key parameter of the SCM, which has been obtained by employ the contract to address related problems. The contract is a major concern in a wide variety of applications; examples of contract are wholesale price, revenue sharing, buy-back, or quantity flexibility.

1.1. Revenue-Sharing Contract

The contracting mechanism has been adopted by supply chain members and well-known as a popular method to mitigate risks. Based on the revenue-sharing approach, which has been widely employed in the e-commerce and the manufacturing sector [23–28], Govindan and Malomfalean [29] have employed the Stackelberg game theory to propose the framework for comparing those coordination mechanisms.

1.2. Buy-Back Contract

Recently, there is an increase remarkably in the number of companies that apply a buy-back contract into supply chain management. Some firms believed that buy-back contracts can improve overall SC coordination in the manufacturing sector [30–35] and retailing sector [36,37].

1.3. Quantity-Flexibility Contract

The current market is witnessing fierce competition among the firms to obtain acceptance from consumers. Companies are making an effort to offer new goods for sale while customers struggle with so many brands for their needs. One of the roles of the new products coordinator is to track the status of all products submittals. Manufacturers expected retailers to set valuable orders for them, such as early time or large quantities due to the fact that they must spend a long time for production preparation. Quantity-flexibility contracts are efficient coordination schemes to balance the risk of each side, especially in the manufacturing sector [38–44] and retailing sector [45]. Three kinds of contracts have been more widely available and conducted in various supply chains. Based on existing literature, the authors illustrated a taxonomy that reflects accepted categories (as represented in Table 1). On the whole, there has been relatively little progress in comparing the three coordination mechanisms until recently.

Table 1. The main classification of inventory models from relevant studies.

Reference	Model	Demand	Revenue-Sharing Contract	Buy-Back Contract	Quantity-Flexibility Contract
Pang et al. [23]	Supply chain	Sales effort dependent	V		
Tsao & Lee [24]	Supply chain	Uncertain and promotion-sensitive	V		
Zhao & Wu [25]		Stochastic	V		
Cachon & Lariviere [26]	Supply chain	Price-dependent	V		
Sang [27]	Supply chain	Fuzzy demand	V		
Yao et al. [28]	Supply chain	Price-dependent demand	V		
Govindan & Malomfalean [29]	Supply chain	Constant	V	V	V
Gong [28]	Supply chain	Stochastic		V	
Wang & Zipkin [30]	Supply chain	Constant		V	
Hou et al. [32]	Supply chain	Constant		V	
Ding et al. [33]	EOQ	Price-dependent demand		V	
Tibrewala et al. [34]	Supply chain	Stochastic		V	
Zhang et al. [35]	Supply chain	Stochastic		V	
Dai et al. [36]	Supply chain	Stochastic		V	
Sainathan & Groenevelt [37]	Supply chain	Constant	V	V	V
Tsay & Lovejoy [38]	Supply chain	Stochastic			V
Li & Kouvelis [39]	Supply chain	Constant	V		V
Sethi et al. [40]	Supply chain	Stochastic			V
Wu [41]	Supply chain	Stochastic			V
Milner & Kouvelis [42]	Supply chain	Stochastic			V
Bag et al. [43]	EPQ	Stochastic			V
Li et al. [44]	Supply chain	Constant			V
Lian & Deshmukh [45]	Supply chain	Stochastic			V
Present article	Food supply chain	Constant	V	V	V

Table 1 demonstrated the comparison among existing relevant studies under different supply chain. It is hoped that this brief and necessarily oversimplified review will lead to a better understanding of three contracts in supply chain, in order to address these areas of concern by shedding light on the nature of the problems and to maximize profits of each member of a food chain made up of one manufacturer and one retailer with three contracts. Hence, the contribution of this article was to investigate the decision-making of the temperature, order quantity, number of shipments in food supply chain between manufacturer and retailer. It is important to recognize the efforts of three coordination mechanisms, but we lack empirical support.

1.4. Food Security

Temperature control is a major concern in a wide variety of applications such as preserving food (or food security), distribution food, and storage food. It is evident that a key element within these chains is temperature, also in its different forms, as it is a necessary source to guarantee quality-based processes. Shock freezing is widely used in the food industry. However, it is not popular with catering companies that produce their own bread and bakery products. Many companies prefer to use ready-made frozen semi-finished products, rather than to bake them on their own. Center kitchen (CK), also known as the distribution center, is set up as a modern food production base, equipped with complete environmental protection equipment and food processing equipment. From ordering raw materials, food production, to selling food, the center kitchen carries out unified management. CK is a new trend in the food industry where centralized preparation and processing of fresh foods occurs in the catering chains. It is the basic function of CK to process raw food materials purchased in bulk into semi-products and/or products under the dictation of unified requirement on product qualities. However, the proper control and management of temperature is crucial in delivering perishables to consumers and ensuring that those perishables are in good condition and safe to eat. Aung and Chang [46] addressed the approach used to obtain the optimal temperature for multi-commodity refrigerated storage. Lorite et al. [47] developed a critical temperature indicator (CTI) to monitor the supply chain and storage conditions of perishable food products. Kuo and Chen [48] developed an advanced multi-temperature joint distribution for continuously temperature-controlled logistics. For the food manufacturer, positive food security will protect the company's brand and goodwill. However, "loss of goodwill" has been touched upon in management science textbooks of various business school curricula. Today it is widely accepted that a goodwill cost may be difficult to estimate. Lawrence and Pasternack [49] argued that many businesses do not have any idea what the long-term goodwill cost of an unsatisfied customer might be. Aksen [50] examined the role of loss of customer goodwill on the incapacitated lot-sizing problem over a finite planning horizon. Finn et al. [51] investigated whether the cost of food contamination can result in indirect losses such as lost reputation and goodwill. Mitreva and Gjurevska [52] applied the TQM tool to a bread and baking company in the quality system.

2. Problem Description

New business models have been relentlessly launched in the market due to the rapid advancement of technology and social changes. While internet shopping becomes more efficient and effective, traditional retailers need new digital innovation. An outbreak of coronavirus infection in Wuhan, China, has led to a global epidemic declared a Public Health Emergency of International Concern by the World Health Organization (WHO). COVID-19 is resistant to environment factors, has high penetration ability, and may be lethal. The food supply chain has needed to adjust rapidly to demand-sides shocks, including cake buying and changes in food purchasing patterns as well as planning for any supply-side disruptions due to potential labor shortages and disruptions to transportation and supply networks. However, few empirical studies have been done on this issue. Hobbs [53] proposed an assessment of implication of the COVID-19 pandemic for the food supply chain regarding the growth of the online grocery delivery sector and the extent to which consumers will prioritize "local" food supply chains. Singh et al. [54] investigated an economic model in helping to simulate a responsive food supply chain to match the varying demand and provided decision-making support for rerouting the vehicles during the COVID-19 pandemic. Rizou et al. [55] provided an excellent review of the possible transmission ways of COVID-19 through the food supply chain and environment before developing COVID-19 diagnostic tests. Therefore, we develop an integrated food supply chain system as well as discussing the optimal decisions found to maximize the total profit based on the new retailing framework through COVID-19 pandemic. Transportation issues caused another concern about food quality that concerned products' temperature control in customers or distribution centers from COVID hot spots. On the other hand, the traditional retailers recognize that COVID-19 will have a significant impact on their business. Online technology can support customer service staff, engage offline-to-online ordering, and gives sales assistants more opportunities to help customers explore their options. As fresh food deteriorates easily, there has been increasing consumer demand for logistics that ensure the timeliness of deliveries. He et al. [56] proposed an agent-based O2O food ordering model (AOFOM) that consists of customers, restaurants, and the online food ordering platform. Wang et al. [57] made a comparison of food shopping with four e-commerce modes: Business-to-consumer (B2C), online-to-offline delivery (O2O Delivery), online-to-offline in store (O2O In-store), and New Retail. Li et al. [58] provided an excellent review of online food delivery platforms. A fairly large body of literature exists on the supply chain. However, within that literature, there is a surprising lack of information on O2O model. Online food delivery provided a critical lifeline during 2020 COVID-19 pandemic for the tens of millions of people quarantined at home. Such research is still in its infancy, but it may have a contribution to make to comparison the understanding of the above three contracts. Next, the empirical evidence is provided in support of the claim. In order to face the highly competitive environment, the following issues were posed by the manufacturer/retailer:

RQ1. What is the optimal shipment size from manufacturer to the retailer?

RQ2. What is the optimal temperature in the food supply chain?

RQ3. What is the optimal order quantity for the retailer?

RQ4. Which of three coordinated agreements attains the best outcome under new retailing framework?

The main aim of the present article is to use the inventory model to discuss its properties in comparisons with the overall effects of three coordination mechanisms: Revenue sharing, buy-back, or quantity flexibility. The supply chain coordination proposed by Govindan and Malomfalean [29] was used the basis for replicating consumer behavior under O2O environment described above. We addressed the basic two-echelon supply chain proposed by Yang et al. [59]. They investigated single-supplier and single-retailer production and inventory models in both non-cooperative and cooperative situations. Thus, the contribution of this article is in understanding of three coordination mechanisms in the food supply chain. In actuality, however, coordination mechanism is a common problem in supply chain. Little is known about the comparison of the three coordination mechanism in the literature, with most research concentrating on single/two contracts. Moreover, the author suggests that manufacturers should focus on the effect of three coordination mechanisms in the food supply chain. Notations and assumptions will be demonstrated in Section 3, followed by Section 4, which presents the model formulation. Sections 5 and 6 will show the suggestion in terms of solutions and application example, respectively. Next, in Section 7, a numerical example will be presented as proof for the findings of Sections 5 and 6 as mentioned. Finally, conclusions are provided in Section 8.

3. Notations and Assumptions

In addition, the following assumptions are made in developing our mathematical model.

Assumptions

1. The article considers a supply chain system comprising of one manufacturer (supplier) and one retailer.
2. The retailer has remaining inventory at the end of a season; the whole amount will be sold back to the central kitchen.
3. The manufacturer considers the differential pricing strategy under O2O environment, $p_1 \geq p_2$.
4. The manufacturer considers the basics of price elasticity strategy, $\beta \geq \gamma$.
5. The buy-back price has to be lower than the wholesale price, $\beta' < w$.
6. The items expected sales at a varying rate of quantity $S(Q)$, where $S'(Q) > 0$ and $S''(Q) < 0$. Here $S'(Q)$ denotes the first derivative of $S'(Q)$ with respect to Q. Note that $S'(Q) > 0$ means that expected sales are increasing over time.

4. Model Formulation

The article considers a two-echelon supply chain with one manufacturer (central kitchen, CK) and one retailer (branch warehouse). The structure is developed under a coordinated case. The purpose of this study was to evaluate the effect of revenue-sharing, buy-back, and quantity- flexibility under new retailing framework.

- Temperature control

One of the major preoccupations of food processing in the past decade has been investigating the nature of temperature control. Temperature is one of the important factors to consider during the bulk production of food items. Storage temperature and time is discussed by Peleg et al. [60], as follows.

$$q(T, t) = q_0 e^{-b(T)t^{n(T)}},\tag{1}$$

where $b(T)$ and $n(T)$ are temperature-dependent coefficients. In particular, $b(T)$ can be described by the empirical model as follows.

$$b(T) = \ln[1 + e^{m(T - T_C)}],\tag{2}$$

where T is the temperature ($^\circ K$), where m and T_C are constants. It is possible to empirically obtain the values of parameters $b(T)$, $n(T)$, m, and T_C so as to use the equations above to calculate the expected quality of food products after storage, at given time periods and temperature levels.

- Temperature coefficient

The cost for storage and transportation depends on the temperature in central kitchen. The coefficient of performance (COP) for refrigeration is calculated as follows.

$$COP = \frac{T_L}{T_H - T_L},\tag{3}$$

where T_H and T_L are higher (environment) and lower (cooling) temperature measured in Kelvin. For example, if $T_H = 293$ K (20 $^\circ$C), and $T_L = 273.15$ K(1 $^\circ$C), then $COP = 13.76$. It means that for each unit of energy drawn from an electrical source, the coolant can absorb as much as 13.76 units of heat from the inside of the refrigerator.

We assume the cost of electrical energy at 1 $^\circ$C is 1, then relative cost at higher temperatures can be calculated by multiplying the cost with the ratio of COP values (where we use an environment temperature T_H of 20 $^\circ$C = 293 K).

- The quality degradation cost (C_{QD}) at central kitchen

The loss value of a batch of size Q is

$$p_2\left[Q - \int_0^Q e^{-b(T)(Q-q)/D} dq\right] = p_2\left[Q - \frac{D}{b(T)}(1 - e^{-b(T)Q/D})\right],\tag{4}$$

where $n(T) = 1$.

4.1. Retailer's Total Profit per Unit Time

The retailer's total profit per replenishment cycle is calculated as follows.

1. Sales revenue (SR): The sales revenue per replenishment cycle is expressed as $SR = p_2 S(Q)$.
2. Ordering cost (OC): The retailer's ordering cost per replenishment cycle is $OC = A/n$.
3. Purchasing cost (PC): The retailer's purchasing cost per replenishment cycle is $PC = w'Q$.
4. Freight cost (FC): Fixed cost of shipment F, and various transportation costs. Namely, the retailer's freight cost per replenishment is $FC = F + rQ$.
5. Goodwill cost (GC): Goodwill cost of the retailer is $GC = g_r[D - (\beta - \gamma)(p_1 + p_2) - S(Q)]$.
6. Marginal cost (MC): The retailer's marginal cost per replenishment cycle is $MC = c_r Q$.
7. Holding cost (HC): Based on Yang et al. [59], the retailer's inventory level in a replenishment cycle, the retailer's holding cost is calculated as $HC = \frac{1}{2}\frac{Q^2}{D}$.

Base on Govindan and Malomfalean [29], considering the relationships between retailer and manufacturer, there are three possible cases: (I) Revenue-sharing contract, (II) Buy-back contract, (III) Quantity-flexibility contract.

- Case I: Revenue-sharing contract

Under a revenue-sharing contract, a retailer pays a wholesale price for each unit purchased, and a percentage of the revenue the retailer generates. Therefore, the retailer's total profit function is expressed as.

$$
\begin{aligned}
TPB(Q) &= SR - OC - PC - FC - GC - MC - HC \\
&= \rho[p_2 S(Q) - g_r(D - (\beta - \gamma)(p_1 + p_2) - S(Q))] - \frac{A}{n} - w'Q - (F + rQ) \\
&\quad - \frac{1}{2}\frac{Q^2}{D} - c_r Q,
\end{aligned}
\tag{5}
$$

- Case II: Buy-back contract

Under the buy-back contract, the manufacturer will buy back any unsold semi-product from the retailer at the end of period. The retailer's total profit function is expressed as.

$$
\begin{aligned}
TPB(Q) &= SR - OC - PC - GC - FC - MC - HC \\
&= [p_2 S(Q) + \beta(Q - S(Q))] - g_r[D - (\beta - \gamma)(p_1 + p_2) - S(Q)] \\
&\quad - \frac{A}{n} - wQ - (F + rQ) - c_r Q - \frac{1}{2}\frac{Q^2}{D},
\end{aligned}
\tag{6}
$$

- Case III: Quantity-flexibility contract

Under the quantity-flexibility contract, the retailer agrees to acquire $(1 - \delta)Q$ units with the option to restock during the season to optimum quantity points. The agreement constraint is $\delta, 0 \leq \delta \leq 1$. With the different types of demand patterns:

Demand is higher than the quantity agreed upon the retailer.

- Case 1: $D - (\beta - \gamma)(p_1 + p_2) > (1 - \delta)Q$

$$
TPB(Q) = p_2 S(Q) - c_r(1 - \delta)Q - \frac{A}{n} - w(1 - \delta)Q - (F + rQ) - \frac{1}{2}\frac{Q^2}{D},
\tag{7}
$$

Demand is higher that the agreed upon quantity but smaller than the optimal quantity.

- Case 2: $(1 - \delta)Q < D - (\beta - \gamma)(p_1 + p_2) < Q$

$$
TPB(Q) = p_2 S(Q) + w[Q - D - (\beta - \gamma)(p_1 + p_2)] - c_r Q - wQ - \frac{A}{n} - (F + rQ) - \frac{1}{2}\frac{Q^2}{D},
\tag{8}
$$

Demand is higher than the optimum quantity.

- Case 3: $D - (\beta - \gamma)(p_1 + p_2) > Q$

$$
\begin{aligned}
TPB(Q) &= p_2 S(Q) - g_r[D - (\beta - \gamma)(p_1 + p_2) - S(Q)] \\
&\quad - \frac{A}{n} - wQ - (F + rQ) - \frac{1}{2}\frac{Q^2}{D} - c_r Q,
\end{aligned}
\tag{9}
$$

4.2. Manufacturer's Total Profit per Unit Time

The manufacturer's total profit per replenishment cycle is calculated as follows:

1. Sales revenue (SR): The manufacturer's sales revenue per replenishment cycle is expressed as $SR = (1 - \rho)[p_2 S(Q) - g_{ck}(D - (\beta - \gamma)(p_1 + p_2) - S(Q))]$.
2. Wholesale value (WV): $WV = w'nQ$.
3. Setup cost (SC): The manufacturer's setup cost per replenishment cycle is $SC = K\frac{D}{Q}$.
4. Quality degradation function (C_{QD}): Similar to Rong et al. [61], the manufacturer's degradation function per replenishment is $C_{QD} = p_2\left[Q - \frac{D}{b(T)}(1 - e^{-b(T)Q/D})\right]$.

5. Goodwill cost (GC): Goodwill cost of the manufacturer is $GC = g_{ck}[(D - (\beta - \gamma)(p_1 + p_2) - S(Q))]$.

6. Holding cost (HC): Based on Yang et al. [59], the manufacturer's inventory level in a replenishment cycle. The manufacturer's holding cost is calculated as $HC = \frac{h_s n k_T Q^2}{2}\left(\frac{2-n}{P} + \frac{n-1}{D}\right)$.

- CASE I. Revenue-sharing contract

The manufacturer's total profit function at in a replenishment cycle is expressed as

$$
\begin{aligned}
TPF_{ck}(n, T) \quad &= SR + WV - SC - PC - GC - HC - C_{QD}\\
&= -g_{ck}[(D - (\beta - \gamma)(p_1 + p_2) - S(Q))] - K\frac{D}{Q} - c_{ck}nQ + w'nQ\\
&\quad + (1 - \rho)[p_2 S(Q) - g_{ck}(D - (\beta - \gamma)(p_1 + p_2) - S(Q))]\\
&\quad - p_2\left[Q - \frac{D}{b(T)}(1 - e^{-b(T)Q/D})\right] - \frac{h_s n k_T Q^2}{2}\left(\frac{2-n}{P} + \frac{n-1}{D}\right),
\end{aligned}
\tag{10}
$$

- CASE II. Buy-back contract

The manufacturer's total profit function at in a replenishment cycle is expressed as

$$
\begin{aligned}
TPF_{ck}(n, T) \quad &= \beta[Q - S(Q)] - K\frac{D}{Q} - c_{ck}nQ + wnQ - g_{ck}[D - (\beta - \gamma)(p_1 + p_2) - S(Q)]\\
&\quad - p_2\left[Q - \frac{D}{b(T)}(1 - e^{-b(T)Q/D})\right] - \frac{h_s n k_T Q^2}{2}\left(\frac{2-n}{P} + \frac{n-1}{D}\right),
\end{aligned}
\tag{11}
$$

- CASE III. Quantity-flexibility contract

1. Case 1: $D - (\beta - \gamma)(p_1 + p_2) > (1 - \delta)Q$

$$
\begin{aligned}
TPF_{ck}(n, T) \quad &= wn(1 - \delta)Q - K\frac{D}{Q} - c_{ck}n(1 - \delta)Q\\
&\quad - p_2\left[Q - \frac{D}{b(T)}(1 - e^{-b(T)Q/D})\right] - \frac{h_s n k_T Q^2}{2}\left[\frac{2-n}{P} + \frac{n-1}{D}\right],
\end{aligned}
\tag{12}
$$

2. Case 2: $(1 - \delta)Q < D - (\beta - \gamma)(p_1 + p_2) < Q$

$$
\begin{aligned}
TPF_{ck}(n, T) \quad &= wnQ - K\frac{D}{Q} - c_{ck}nQ - p_2\left[Q - \frac{D}{b(T)}(1 - e^{-b(T)Q/D})\right]\\
&\quad - \frac{h_s n k_T Q^2}{2}\left(\frac{2-n}{P} + \frac{n-1}{D}\right),
\end{aligned}
\tag{13}
$$

3. Case 3: $D - (\beta - \gamma)(p_1 + p_2) > Q$

$$
\begin{aligned}
TPF_{ck}(n, T) \quad &= wnQ - K\frac{D}{Q} - c_{ck}nQ - g_{ck}[Q - S(Q)] - p_2\left[Q - \frac{D}{b(T)}(1 - e^{-b(T)Q/D})\right]\\
&\quad - \frac{h_s n k_T Q^2}{2}\left(\frac{2-n}{P} + \frac{n-1}{D}\right),
\end{aligned}
\tag{14}
$$

4.3. The Joint Total Profit per Unit Time

Once the manufacturer and retailer have built a long-term relationship contract, they will jointly determine the best policy for the whole supply chain system. Under this circumstance, for $j = 1, 2, 3$, the joint total profit per unit time for manufacturer and retailer is

$$
\Pi_j(n, Q, T) = \begin{cases} \Pi_1(n, Q, T), & if\ Revenue - sharing\ contract\\ \Pi_2(n, Q, T), & if\ Buy - back\ contract\\ \Pi_3(n, Q, T), \end{cases}
$$

where

$$
\begin{aligned}
\Pi(n, Q, T)_1 \quad &= TPB(Q) + TPF_{ck}(n, T)\\
&= p_2 S(Q) - \frac{A}{n} + (n - 1)w'Q - (F + rQ) - K\frac{D}{Q} - c_{ck}nQ - \frac{1}{2}\frac{Q^2}{D} - c_r Q\\
&\quad - [2g_{ck} + \rho(g_r - g_{ck})][D - (\beta - \gamma)(p_1 + p_2) - S(Q)]\\
&\quad - p_2\left[Q - \frac{D}{b(T)}(1 - e^{-b(T)Q/D})\right] - \frac{h_s n k_T Q^2}{2}\left(\frac{2-n}{P} + \frac{n-1}{D}\right),
\end{aligned}
\tag{15}
$$

$$
\begin{aligned}
\Pi(n,Q,T)_2 \;=\;& TPB(Q) + TPF_{ck}(n,T)\\
=\;& p_2 S(Q) + 2\beta[Q - S(Q)] - \tfrac{A}{n} - (F + rQ) - \tfrac{1}{2}\tfrac{Q^2}{D} - K\tfrac{D}{Q}\\
& -(c_r + c_{ck}n)Q + w(n-1)Q - (g_{ck} + g_r)[D - (\beta - \gamma)(p_1 + p_2) - S(Q)]\\
& -p_2\!\left[Q - \tfrac{D}{b(T)}(1 - e^{-b(T)Q/D})\right] - \tfrac{h_s n k_T Q^2}{2}\!\left(\tfrac{2-n}{P} + \tfrac{n-1}{D}\right),
\end{aligned}
\tag{16}
$$

Nowadays, retailers frequently make challenging decisions such as matching supply with demand. Many retailers are faced with demand uncertainty for product. There are three separate cases (i) $D - (\beta - \gamma)(p_1 + p_2) > (1 - \delta)Q$, (ii) $(1 - \delta)Q < D - (\beta - \gamma)(p_1 + p_2) < Q$ and (iii) $D - (\beta - \gamma)(p_1 + p_2) > Q$.

$$
\Pi_3(n,Q,T) = \begin{cases}
\Pi_{31}(n,Q,T), & \text{if } D - (\beta - \gamma)(p_1 + p_2) > (1 - \delta)Q\\
\Pi_{32}(n,Q,T), & \text{if } (1 - \delta)Q < D - (\beta - \gamma)(p_1 + p_2) < Q\;,\\
\Pi_{33}(n,Q,T), & \text{if } D - (\beta - \gamma)(p_1 + p_2) > Q
\end{cases}
$$

where

$$
\begin{aligned}
\Pi_{31}(n,Q,T) \;=\;& p_2 S(Q) - c_r(1 - \delta)Q - \tfrac{A}{n} - w(1 - \delta)Q - (F + rQ) - \tfrac{1}{2}\tfrac{Q^2}{D}\\
& + wn(1 - \delta)Q - K\tfrac{D}{Q} - c_{ck}n(1 - \delta)Q - p_2\!\left[Q - \tfrac{D}{b(T)}(1 - e^{-b(T)Q/D})\right]\\
& - \tfrac{h_s n k_T Q^2}{2}\!\left(\tfrac{2-n}{P} + \tfrac{n-1}{D}\right),
\end{aligned}
\tag{17}
$$

$$
\begin{aligned}
\Pi_{32}(n,Q,T) \;=\;& p_2 S(Q) - c_r(1 - \delta)Q - \tfrac{A}{n} - w(1 - \delta)Q - (F + rQ) - \tfrac{1}{2}\tfrac{Q^2}{D}\\
& + wn(1 - \delta)Q - K\tfrac{D}{Q} - c_{ck}n(1 - \delta)Q - p_2\!\left[Q - \tfrac{D}{b(T)}(1 - e^{-b(T)Q/D})\right]\\
& - \tfrac{h_s n k_T Q^2}{2}\!\left(\tfrac{2-n}{P} + \tfrac{n-1}{D}\right),
\end{aligned}
\tag{18}
$$

and

$$
\begin{aligned}
\Pi_{33}(n,Q,T) \;=\;& p_2 S(Q) - g_r[D - (\beta - \gamma)(p_1 + p_2) - S(Q)] - \tfrac{A}{n} - wQ - (F + rQ) - \tfrac{1}{2}\tfrac{Q^2}{D} - c_r Q\\
& + wnQ - K\tfrac{D}{Q} - c_{ck}nQ - g_{ck}[Q - S(Q)] - p_2\!\left[Q - \tfrac{D}{b(T)}(1 - e^{-b(T)Q/D})\right]\\
& - \tfrac{h_s n k_T Q^2}{2}\!\left(\tfrac{2-n}{P} + \tfrac{n-1}{D}\right).
\end{aligned}
\tag{19}
$$

5. Solution Procedure

A quantization design was used to find feasible and optimal solutions. The objective is to determine the optimal the number of shipments, lot size per shipment, and temperature at central kitchen that maximizes the joint total profit per unit time of the integrated supply chain. First, the effect of n on the joint total profit per unit time will be examined. Taking the second-order partial derivative of $\Pi_j(n,Q,T)$ with respect to n, we obtain

$$
\frac{\partial^2 \Pi_j(n,Q,T)}{\partial n^2} = -\left[2An^{-3} + h_s k_T Q^2\!\left(\frac{P - D}{PD}\right)\right] < 0,
\tag{20}
$$

This result identifies that $\Pi_j(n,Q,T)$ is a concave function in n for fixed Q and T. Therefore, the search for the optimal shipment number, n^*, is reduced to find a local optimal solution.

5.1. Determination of the Optimal T for Given n and Q

For given n and Q, the first-order necessary condition of $\Pi_j(n,Q,T), j = 1,2,3$, with respect to T. This gives

$$
\frac{\partial \Pi_j(n,Q,T)}{\partial T} = \frac{e^{-b(T)Q/D}}{D[b(T)]^2} b'(T)\left[Qb(T) - D(e^{b(T)Q/D} - 1)\right] = 0, \; j = 1,2,3.
\tag{21}
$$

Next, we start by taking the second derivative of $\Pi_j(n,Q,T)$.

$$
\begin{aligned}
\frac{\partial^2 \Pi_j(n,Q,T)}{\partial T^2} \;=\;& \frac{1}{D^2[b(T)]^3} e^{-b(T)Q/D}\Big\{ D\left[D\left(e^{b(T)Q/D} - 1\right) - Qb(T)\right]\left[2(b'(T))^2 - b(T)b''(T)\right]\\
& - Q^2[b(T)]^2[b'(T)]^2\Big\},
\end{aligned}
\tag{22}
$$

By fixing n and Q, we have the following result.

Proposition 1. *For given feasible n and Q,*

1. *if $2[b'(T)]^2 < b(T)b''(T)$, then the solution T^* which maximizes $\Pi_j(T|n,Q)$ not only exists but also is unique, where $T^* \in [0,\infty)$.*
2. *if $2[b'(T)]^2 > b(T)b''(T)$, then the solution T^* which maximizes $\Pi_1(Q|n,T)$ not only exists but also is unique, where $T^* \in [0,\infty)$.*

5.2. Determination of the Optimal Q for Given n and T

In this section, we first find the optimal lot size per shipment that maximizes $\Pi_j(n,Q,T)$, $j = 1,2,3$, respectively, for a given n and T. Then the optimal number of shipments is derived for a given Q. Taking the first-order partial derivative of $\Pi_j(n,Q,T)$, $j = 1,2,3$, with respect to Q, respectively, we have

$$
\begin{aligned}
\frac{\partial \Pi_1(n,Q,T)}{\partial Q} &= p_2 S'(Q) + (n-1)w' - r + KDQ^{-2} - c_{ck}n - \frac{Q}{D} - c_r \\
&\quad + [2g_{ck} + \rho(g_r - g_{ck})]S'(Q) - p_2\left[1 - e^{-b(T)Q/D}\right] \\
&\quad - h_s n k_T Q\left(\frac{2-n}{P} + \frac{n-1}{D}\right),
\end{aligned}
\tag{23}
$$

$$
\begin{aligned}
\frac{\partial \Pi_2(n,Q,T)}{\partial Q} &= (p_2 + g_{ck} + g_r - 2\beta)S'(Q) + 2\beta - r - \frac{Q}{D} + K\frac{D}{Q^2} \\
&\quad - (c_r + c_{ck}n) + w(n-1) - p_2\left[1 - e^{-b(T)Q/D}\right] \\
&\quad - h_s n k_T Q\left(\frac{2-n}{P} + \frac{n-1}{D}\right),
\end{aligned}
\tag{24}
$$

$$
\begin{aligned}
\frac{\partial \Pi_{31}(n,Q,T)}{\partial Q} &= p_2 S'(Q) - c_r(1-\delta) - w(1-\delta) - r - \frac{Q}{D} + wn(1-\delta) + K\frac{D}{Q^2} \\
&\quad - c_{ck}n(1-\delta) - p_2\left[1 - e^{-b(T)Q/D}\right] - h_s n k_T Q\left(\frac{2-n}{P} + \frac{n-1}{D}\right),
\end{aligned}
\tag{25}
$$

$$
\begin{aligned}
\frac{\partial \Pi_{32}(n,Q,T)}{\partial Q} &= p_2 S'(Q) - c_r - r - \frac{Q}{D} + wn + K\frac{D}{Q^2} - c_{ck}n - p_2\left[1 - e^{-b(T)Q/D}\right] \\
&\quad - h_s n k_T Q\left(\frac{2-n}{P} + \frac{n-1}{D}\right),
\end{aligned}
\tag{26}
$$

$$
\begin{aligned}
\frac{\partial \Pi_{33}(n,Q,T)}{\partial Q} &= (p_2 + g_r)S'(Q) - w - r - \frac{Q}{D} - c_r + wn + KDQ^{-2} - c_{ck}n \\
&\quad - g_{ck}[1 - S'(Q)] - p_2\left[1 - e^{-b(T)Q/D}\right] - h_s n k_T Q\left(\frac{2-n}{P} + \frac{n-1}{D}\right),
\end{aligned}
\tag{27}
$$

Next, we take the second-order partial derivative of $\Pi_j(n,Q,T)$

$$
\begin{aligned}
\frac{\partial^2 \Pi_1(n,Q,T)}{\partial Q^2} &= -2KDQ^{-3}[2g_{ck} - \rho(g_r - g_{ck})]S''(Q) - \frac{1}{D} \\
&\quad - \frac{p_2}{D}\left[b(T)e^{-b(T)Q/D} - 1\right] - h_s n k_T\left(\frac{2-n}{P} + \frac{n-1}{D}\right),
\end{aligned}
\tag{28}
$$

$$
\frac{\partial^2 \Pi_2(n,Q,T)}{\partial Q^2} = (p_2 + g_{ck} + g_r - 2\beta)S''(Q) - 2KDQ^{-3} - p_2\frac{b(T)}{D}e^{-b(T)Q/D},
\tag{29}
$$

and
$$
\frac{\partial^2 \Pi_{31}(n,Q,T)}{\partial Q^2} = p_2\left[S''(Q) - b(T)e^{-b(T)Q/D}\right] - K\frac{D}{Q^3} - 1 - h_s n k_T\left(\frac{2-n}{P} + \frac{n-1}{D}\right),
\tag{30}
$$

$$
\frac{\partial^2 \Pi_{32}(n,Q,T)}{\partial Q^2} = p_2\left[S''(Q) - \frac{b(T)}{D}e^{-b(T)Q/D}\right] - \frac{1}{D} - 2\frac{KD}{Q^3} - h_s n k_T\left(\frac{2-n}{P} + \frac{n-1}{D}\right),
\tag{31}
$$

and

$$
\begin{aligned}
\frac{\partial^2 \Pi_{33}(n,Q,T)}{\partial Q^2} &= (p_2 + g_r + g_{ck})S''(Q) - p_2\left[b(T)e^{-b(T)Q/D} - 1\right]\frac{1}{D} - 2KDQ^{-3} \\
&\quad - h_s n k_T\left(\frac{2-n}{P} + \frac{n-1}{D}\right).
\end{aligned}
\tag{32}
$$

By fixing n and T, we have the following results.

Proposition 2. *For given n and T,*

1. *if $p_2 + g_{ck} + g_r > 2\beta$, then the solution Q^* which maximizes $\Pi_2(Q|n,T)$ not only exists but also is unique, where $Q^* \in [0,\infty)$.*
2. *if $p_2 + g_{ck} + g_r < 2\beta$, then the solution Q^* which maximizes $\Pi_2(Q|n,T)$ not only exists but also is unique, where $Q^* \in [0,\infty)$.*

6. Application Example

The selected bakery is 2017's fastest-growing company (Ding-Dang convenience store) in Huizhou, Guangdong Province. The company's brand is derived from the "Internet +" wave and delivers high-quality services to consumers under the new retailing framework. Considering that it is a small and medium-sized enterprise, its founding team originated from Alibaba's and Sf-express's CEO. This company owns a new retailing framework with online marketing (digital marketing) and offline marketing (convenience store marketing), the number of platform users reached 600,000, and the number of service outlets reached 100 in Huizhou. The case company raised more than CN¥200 million of annual revenue, and the plan is to expand convenience stores in 16 new cities. In this section, we constructed two scenarios that consist of new retailing and food supply chain.

6.1. New Retailing Framework

This section presents an implementation of the proposed model in a real-life bakery logistics network operating in China. Nowadays, using artificial intelligence and big data to change the business model of the bakery industry in China, Hsiung Mao (Panda) Pu Tsou (called "case company") uses social commerce platforms to offer online group buying on a social commerce platform. The main products, five-star cakes, were made of the best raw materials in the medium and high-end market. The case company will be a Personal Service Provider (PSP) via Panda dressed up, Panda distribution. The customers make e-commerce purchases through the Wechat, Meituan, and Elema app ordering systems. Correct handling of cakes, ingredients, and packaging materials from material purchasing occurred through the production process and cold storage in the center kitchen. During Panda distribution to the consumer, it is essential to optimize cake quality. The traditional O2O models are advanced to a new retailing framework by integrating online and offline logistics. The core of the new retailing framework is online mobile payment, product standardization in the central kitchen, distributed personalization, and providing big data services.

- Online mobile payment

The formation of community-owned stores through online–offline channel integration. The case company has opened more physical stores to attract young customers to achieve the new retailing framework. Online Quick Reference Guide, an easy access tool, provides customers to order. Besides that, online customer satisfaction or product categories provide customers more flexible options.

- Product standardization in central kitchen

Four-Tier model of food supply chain consists of central kitchen, cold chain, branch warehouse, Panda distribution, and authorized store (Self-raising). Then, customer orientation plays an important role in market success achievement. Product customization helps brands boost sales on new retailing framework, EX: Sculpture cakes or fictionalized cakes. A rise in cold chain with data warehouse and central kitchen system will enhance the quality of cake, reduce the waiting time. Mercier et al. [12] discussed a significant challenge for an efficient clod chain is the different requirements of perishable food product categories (dairy, eggs, fruits, cakes, and vegetables) to maximize shelf-life and commercial potential, including different optimal temperature ranges (ambient = 15 °C to 20 °C; cool = 2 °C to 15 °C; cold = −9 °C to 2 °C; and frozen ≤ −10 °C).

- Immediate distribution for personal services

On-time 3 h delivery services include online order, materials handling, finished product manufacturing, branch's self-storage, and surprise services. The case company provides customers with activity planning, including surprise events. Traditional Hakka culture is incorporated into a personal-service activity ex: Traditional Hakka cake, the unique Hakka cultural experience.

- Big data as a service (BDaaS)

In order to find the target customers through big data analytics, online platform could not only help case company to better discriminate the most profitable customers, but also encourage more customer to return. Offline services can attract customers online and offer precision marketing based on a customer's exact age and birth date. Since marketing is all about reaching the right customers at the right time, big data can be used to understand the behavior of bakery customers.

- Customer image capture

For Big data as a Service (BDaaS) and Internet Presence Provider (IPP), customer data analytics refer to the analysis of customer characteristics and behaviors so as to support the organization's customer management strategies. Customer image, online marketing, and site selection are used to revolutionize offline shopping experiences. Data collected from e-payment system can develop the marketing strategy to attract customers to return.

In view of the preceding decision variables, there were few empirical studies of new retailing framework. Figure 1 shows the cake supply chain summary. The food supply chain-case study is then presented, with a thorough description of cold chain Hub.

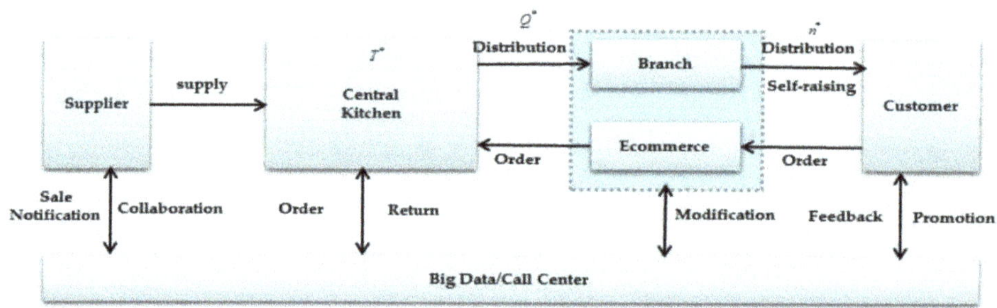

Figure 1. New retailing business model.

6.2. Food Supply Chain-Case Study

The food supply chain (FSC) contains several members, including one central kitchen and numerous branch warehouses. The FSC consists of a four-tier structure such as a central kitchen, cold distribution services, branch warehouse, food-panda delivery, and customer. For quick service retailers, the distribution of cakes from the central kitchen to branch warehouses is throughout the cold chain distribution process, which is normally at a constant temperature between 0 °C and 5 °C. To improve distribution efficiency, the case company owns three reefer vehicles to transport cake and cake products from central kitchen to branch warehouse (72 cakes per vehicle; 8 rounds per day; transport unit at 1700 units per day). Moreover, Panda distribution hit a 90% completion rate from branch warehouse to customer and 10% completion rate from branch warehouse to an authorized store (self-raising). Four branches can own 30 Panda vehicles and 8 electric vehicles. Maximum quantity is 800 units per delivery, 24 units per delivery for one Panda vehicle,

12 units per delivery for one electric vehicle. The simple Figure 2 seeks to capture the cake supply chain summary.

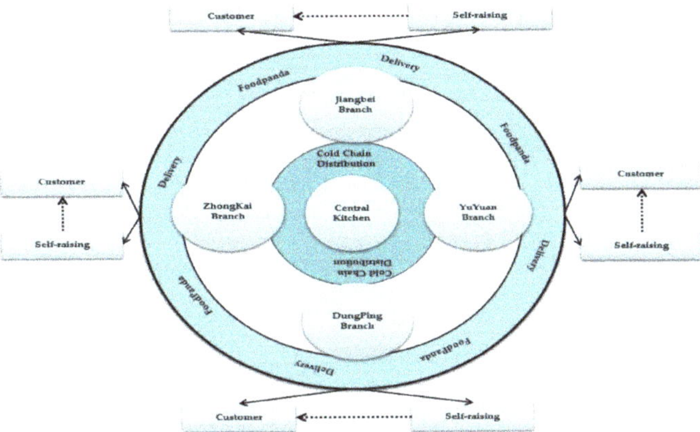

Figure 2. Cake supply chain summary.

Chen et al. [62] demonstrated the reliability changes of performance parameters with the importance measures in inventory systems. Few empirical studies have been done on this issue. The examination to carry out this study was using an algorithm, which solve optimal solutions between a single manufacturer and single retailer.

6.3. Alogorithm

- Step 1. Choose the initial value of $n = 1$.
- Step 2. Evaluate the solution of Q according to Equations (23)–(25).
- Step3. Use Propositions 1 to determine $\max\{\Pi_1(n, Q, T), \Pi_2(n, Q, T), \Pi_3(n, Q, T)\}$ and the corresponding value of Q.
- Step 4. Let $n = n + 1$ and repeat Steps 2–3.
- Step 5. If $\Pi_j(n, Q^*_{(n)}, T^*) \geq \Pi_j(n - 1, Q^*_{(n-1)}, T^*)$, then return to Step 4; otherwise, execute Step 6.
- Step 6. Let $(n, Q^*_{(n)}, T^*) = (n - 1, Q^*_{(n-1)}, T^*)$; therefore, (n^*, Q^*, T^*) is the optimal solution and the maximum total profit per unit time is $\Pi_j(n^*, Q^*, T^*)$.

7. Numerical Example

In this section, we consider the numerical example with different situations to verify the obtained analytical results. The base settings of parameters for the example are listed in Table 2. Then the sensitivity analysis is also tabulated for exploring the variety trend of optimal polies.

Table 2. The value of parameters for Example 1. with new retailing problem.

$D = 1000$	$P = 2000$	$w = 9$	$p_1 = 170$
$p_2 = 160$	$\rho = 0.1$	$A = 80$	$F = \$10$
$r = \$5$	$c_r = \$8$	$k_T = \$0.1$	$c_{ck} = \$5$
$g_r = \$15$	$g_{ck} = \$1$	$\beta = 0.55$	$\gamma = 0.4$
$m = 6$	$T_c = 5$	$h_s = 40$	$K = 100$
$\delta = 0.5$	$s_{ck} = \$4$		
$S(Q) = a - bQ = 100 - 0.2 \times Q$			

7.1. Comparison among Decisions in Online and Offline Stratedgy

The decisions made in the two O2O strategies are compared in this section. A coordination mechanism was designed to clarify the relative contribution on the supply chain, and the three known contracts were compared to explore the application of O2O strategy. Assessment of performance measures occurred under mixture policy (online and offline) and single policy (offline). The results represented in Table 3 present the result of comparison of Revenue-sharing agreement (Case I), Buy-back contract agreement (Case II), and Quantity flexibility contract (Case III).

Table 3. Computation results of Example 1. for online and offline.

Online + Offline	Case I	Case II	Case III
Q^*	46.3743	45.3534	47.8674
T^*	2.4070	2.29413	2.30921
n^*	32	32	10
TPB	13,108.8	10,937.5	13,754.8
TPF_{ck}	18,921.1	6683.78	2019.6
TP	29,692.4	17,621.3	15,774.4
Offline	**Case I**	**Case II**	**Case III**
Q^*	42.1495	43.634	46.8146
T^*	2.29675	2.34995	2.38194
n^*	30	30	12
TPB	1338.26	10,672.1	14,880.1
TPF_{ck}	909.386	561.133	1214.06
TP	2247.646	11233.3	16,094.2

The retailer, the manufacturer, and total supply chain profit in the three O2O policies are presented in Figures 3 and 4. Figure 3 presents revenue-sharing agreement larger than other polies. It implied that revenue-sharing contracts in a general supply chain model with revenues are determined by each retailer's purchase quantity and price.

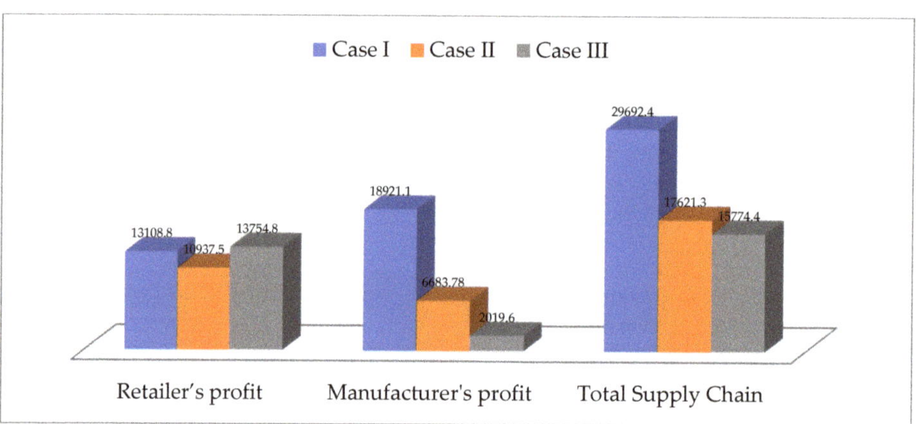

Figure 3. Mixture situation measures under three policies.

Obviously, Figure 4 presents the performance of offline policy less than mixture policy. It implied that the channel members would like to adopt new online technologies in a smaller, safe environment with the retailer and manufacturer before they scaled out to retail partners.

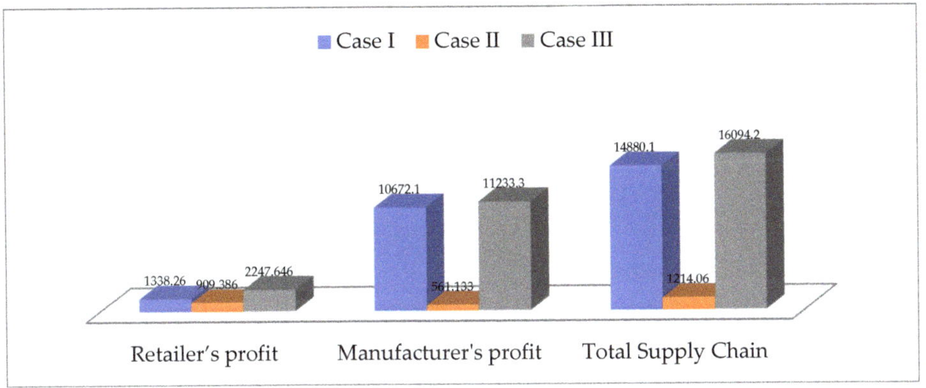

Figure 4. Offline situation measures under three policies.

7.2. Sensitivity Analysis

To perform data analysis, the relative contributions of these parameters (g_{ck}, g_r, F, r, c_r, ρ, T_c, m) were calculated on the values of Q^*, T^*, n^*, $TP_B(Q^*, T^*, n^*)$ and $TPF_{ck}(Q^*, T^*, n^*)$. Each parameter was adjusted separately by +50%, +25%, −25%, or −50%. Our analytical results are in Table 4.

Table 4. Effect of changes in various parameters of the model.

Para.	%	CASE I		CASE II		CASE III	
		(Q^*, T^*, n^*)	(TP_B, TPF_{ck})	(Q^*, T^*, n^*)	(TP_B, TPF_{ck})	(Q^*, T^*, n^*)	(TP_B, TPF_{ck})
g_{ck}	−50%	(46.54, 2.61, 34)	(13,005.1, 20, 060.3)	(45.89, 2.21, 32)	(10, 955.2, 7439.55)	(48.14, 2.43, 10)	(11,692.5, 2123.1)
	−25%	(46.42, 2.59, 33)	(13,098.2, 19, 661.3)	(45.86, 2.22, 32)	(10, 945.1, 7223.54)	(48.02, 2.36, 10)	(12,991.3, 2117.9)
	0	(46.37, 2.40, 32)	(13,108.8, 18, 921.1)	(45.35, 2.29, 32)	(10, 937.5, 6683.78)	(47.86, 2.30, 10)	(13,754.8, 2019.6)
	25%	(46.32, 2.39, 32)	(14,098.2, 18, 842.2)	(43.83, 2.36, 32)	(10, 627.5, 6791.53)	(42.78, 2.26, 10)	(17,128.4, 2007.1)
	50%	(46.28, 2.38, 32)	(14,099.3, 18, 432.7)	(43.80, 2.36, 32)	(10, 628.5, 6975.53)	(42.66, 2.21, 10)	(17,195.0, 2001.6)
g_r	−50%	(46.44, 2.56, 32)	(14,740.1, 19, 253.6)	(46.49, 2.27, 32)	(17, 035.2, 7043.44)	(48.50, 2.29, 10)	(17,115.6, 1921.5)
	−25%	(46.41, 2.49, 32)	(14,418.5, 19, 252.7)	(46.16, 2.27, 32)	(13, 830.6, 7025.41)	(48.20, 2.29, 10)	(13,910.8, 1940.4)
	0	(46.37, 2.40, 32)	(13,108.8, 18, 921.1)	(45.35, 2.29, 32)	(10, 937.5, 6683.78)	(47.86, 2.30, 10)	(13,754.8, 2019.6)
	25%	(46.33, 2.38, 32)	(9775.51, 18,250.9)	(43.53, 2.43, 32)	(7421.73, 6989.81)	(-,-,-)	(-,-)
	50%	(46.29, 2.31, 32)	(9454.01, 18,250.1)	(43.22, 2.44, 32)	(4217.41, 6972.22)	(-,-,-)	(-,-)
F	−50%	(46.37, 2.40, 32)	(14,102.1, 18,921.1)	(45.35, 2.29, 32)	(10,851.1, 6683.78)	(-,-,-)	(-,-)
	−25%	(46.37, 2.40, 32)	(13,099.5, 18,921.1)	(45.35, 2.29, 32)	(10,854.5, 6683.78)	(-,-,-)	(-,-)
	0	(46.37, 2.40, 32)	(13,108.8, 18,921.1)	(45.35, 2.29, 32)	(10,937.5, 6683.78)	(-,-,-)	(-,-)
	25%	(46.37, 2.40, 32)	(10,094.5, 18,921.1)	(45.35, 2.29, 32)	(11,149.5, 6683.78)	(-,-,-)	(-,-)
	50%	(46.37, 2.40, 32)	(10,092.1, 18,921.1)	(45.35, 2.29, 32)	(11,172.1, 6683.78)	(-,-,-)	(-,-)
r	−50%	(47.67, 2.43, 32)	(13,184.3, 19,282.1)	(45.93, 2.40, 32)	(10,976.9, 7067.72)	(-,-,-)	(-,-)
	−25%	(47.00, 2.41, 32)	(13,140.2, 19,267.1)	(45.38, 2.40, 32)	(10,951.4, 7037.41)	(-,-,-)	(-,-)
	0	(46.37, 2.40, 32)	(13,108.8, 18,921.1)	(45.35, 2.29, 32)	(10,937.5, 6683.78)	(-,-,-)	(-,-)
	25%	(45.76, 2.09, 33)	(10,054.8, 18,236.4)	(43.33, 2.29, 33)	(10,901.1, 6678.07)	(-,-,-)	(-,-)
	50%	(45.17, 1.93, 33)	(10,013.4, 18,220.9)	(40.38, 2.29, 33)	(10,820.5, 6596.76)	(-,-,-)	(-,-)
c_r	−50%	(48.50, 2.33, 31)	(13,238.7, 19,299.7)	(45.83, 2.46, 31)	(10,977.9, 6604.68)	(-,-,-)	(-,-)
	−25%	(47.40, 2.38, 32)	(13,166.6, 19,276.1)	(45.71, 2.45, 31)	(10,966.7, 6655.55)	(-,-,-)	(-,-)
	0	(46.37, 2.40, 32)	(13,108.8, 18,921.1)	(45.35, 2.29, 32)	(10,937.5, 6683.78)	(-,-,-)	(-,-)
	25%	(45.41, 2.45, 33)	(10,029.9, 18,227.2)	(43.03, 2.26, 32)	(10,586.2, 6960.57)	(-,-,-)	(-,-)
	50%	(44.50, 2.47, 34)	(9964.94, 18,202.2)	(42.25, 2.21, 32)	(10,546.9, 6914.61)	(-,-,-)	(-,-)
ρ	−50%	(46.56, 2.30, 32)	(10,081.7, 19,590.1)	(45.35, 2.29, 32)	(10,937.5, 6683.78)	(-,-,-)	(-,-)
	−25%	(46.44, 2.31, 32)	(10,064.3, 19,580.8)	(45.35, 2.29, 32)	(10,937.5, 6683.78)	(-,-,-)	(-,-)
	0	(46.37, 2.40, 32)	(13,108.8, 18,921.1)	(45.35, 2.29, 32)	(10,937.5, 6683.78)	(-,-,-)	(-,-)
	25%	(46.28, 2.42, 32)	(13,139.9, 18,912.4)	(45.35, 2.29, 32)	(10,937.5, 6683.78)	(-,-,-)	(-,-)
	50%	(46.19, 2.43, 32)	(13,182.9, 18,872.9)	(45.35, 2.29, 32)	(10,937.5, 6683.78)	(-,-,-)	(-,-)

Table 4. *Cont.*

Para.	%	CASE I		CASE II		CASE III	
		(Q^*, T^*, n^*)	(TP_B, TPF_{ck})	(Q^*, T^*, n^*)	(TP_B, TPF_{ck})	(Q^*, T^*, n^*)	(TP_B, TPF_{ck})
T_c	−50%	(43.85, 0.07, 34)	(13,626.1, 17,007.4)	(45.35, 0.06,34)	(10,626.1, 7007.54)	(-,-,-)	(-,-)
	−25%	(43.85, 1.07, 33)	(13,626.1, 17,007.4)	(45.35, 1.06,34)	(10,626.1, 7007.54)	(-,-,-)	(-,-)
	0	(46.37, 2.41, 32)	(13,108.8, 18,921.1)	(45.35, 2.29,32)	(10,937.5, 6683.78)	(-,-,-)	(-,-)
	25%	(-,-,-)	(-,-)	(-,-,-)	(-,-)	(-,-,-)	(-,-)
	50%	(-,-,-)	(-,-)	(-,-,-)	(-,-)	(-,-,-)	(-,-)
m	−50%	26.3743	(10,069.1, 6335.23)	(26.37, 0.67, 32)	(10,069.1, 7335.23)	(-,-,-)	(-,-)
	−25%	43.8481	(10,626.1, 7007.54)	(43.84, 1.53, 32)	(10,626.1, 7007.54)	(-,-,-)	(-,-)
	0	46.3743	(13,108.8, 18,921.1)	(45.35, 2.29, 32)	(10,937.5, 6683.78)	(-,-,-)	(-,-)
	25%	(-,-,-)	(-,-)	(-,-,-)	(-,-)	(-,-,-)	(-,-)
	50%	(-,-,-)	(-,-)	(-,-,-)	(-,-)	(-,-,-)	(-,-)

Note: Para. = Parameter; % = Change %.

7.3. The Discussion of Results

This section will compare and analyze the three contracts. The data analysis has highlighted a most interesting possibility: Reputation management and transportation performance. Several management implications can be drawn from Table 5. Marketing policies are changes in sales or other results that can be expected from a particular strategy. For example, ensuring food quality and security during the transportation and postponing strategy.

Table 5. Effect of changes in various parameters of managerial insights.

Increasing Parameter(s)	Case I	Case II	Case III
g_{ck}	Company's goodwill is an intangible asset owned by and associated with operation of a company. Thus, a trademark is often an important investment in protecting the intellectual property of a manufacturer.	More specifically, in a high critical ratio environment revenue sharing contracts are more profitable for the manufacturer than buyback contracts.	With goodwill cost exists, the channel coordination provides allocation of the supply chain's profit if $g_{ck} < \rho g_r$.
g_r	The contract offers accepted by the retailer, two-part tariff and quantity discount increases the efficiency of the channel in the terms of channel profit.	The contract provides an enormous motivation for retailer to give orders more than usual without concerning stock out in inventory.	With goodwill cost exists, the channel coordination provides allocation of the supply chain's profit if $g_{ck} < \rho g_r$.
F	The online retailer provides the freight subsidy to the manufacturer increase the profit of the total profit.	Allows a retailer to return unsold inventory up to a specified amount at an agreed upon price/freight fee, resulting in higher product availability and higher profits for both the retailer and the manufacturer.	
r	A wholesale price is plausibly below transportation fee, the members may have adopted a coordinating contract.	Focusing on the transportation costs. The retailer and the manufacturer will be covering the transportation cost.	

Table 5. *Cont.*

Increasing Parameter(s)	Case I	Case II	Case III
c_r	If $w' > c_r$, which a wholesale price is greater than marginal cost, which is in sharp contrast to the optimal wholesale price under a revenue-sharing contract.	The manufacturer would like to increase his own profit by increasing the wholesale price and buy-back price, where $\beta' > 0, c_r > 0$.	
ρ	With revenue-sharing fraction, the manufacturer willingly reduces its wholesale price and makes money by sharing the retailer's revenue.	If effort is significant (i.e., $Q > 1900$), the effect dominates the quantity effect, and the wholesale price falls.	
T_c	To monitor the food quality in the distribution, the absolute temperature decreased by a coefficient of $b(T)$, the manufacturer will increase the channel profit.	To monitor the food quality in the distribution, the absolute temperature decreased by a coefficient of $b(T)$, the manufacturer will increase the channel profit.	
m	To monitor the food quality in the distribution, the absolute temperature decreased by a coefficient of $b(T)$, the manufacturer will increase the channel profit.	To monitor the food quality in the distribution, the absolute temperature decreased by a coefficient of $b(T)$, the manufacturer will increase the channel profit.	

8. Conclusions

Due to the evolution of O2O technologies, the coordination mechanism will be incorporated into the classic inventory model. The authors develop a food supply chain among one manufacturer and one retailer under the O2O model in this research. The study also compared the three coordination mechanisms to fill the gap knowledge between theoretical and empirical in terms of inventory management. Relying on the existing theoretical results, this study has attempted to point out the optimal solution through examining decision-makers with an algorithm. Therefore, the authors can assert based on having understood the shortcomings of O2O technology that our theoretical claim about how a manufacturer and its compromising retailer evaluate a suitable ordering process through a comparison of three commitment contracts. Meanwhile, there are some potential bottlenecks in the development of the O2O model, including (1) users and offline sellers lack rational knowledge of mobile e-commerce; (2) the security of e-payments; and (3) the lack of innovations. Moreover, the findings have been proved by several examples to clarify the proposed model and solution. A total of the proof examples highlighted that (1) Mobile payment lays the foundation for online/offline integration strategy. With advancements in e-payment systems and face recognition/fingerprint recognition, both companies and consumers discover new technology to pay online quickly; (2) the IoT-Driven cold chain tracking is reducing logistics cost and improving business effectiveness, (3) Instant distribution impact on customer online satisfaction. Among the many topics to be explored in future research, some important ones are as follows: Estimating an effective after-seller's services mechanism, strengthening consumers' loyalty to offline sellers, and improving the innovative capacity of the O2O platform.

Author Contributions: Conceptualization, Y.-F.H. and N.X.; methodology, M.-W.W.; software, M.-W.W.; validation, Y.-F.H., N.X., M.-W.W., and M.-H.D.; formal analysis, N.X.; investigation, N.X.; resources, N.X.; data curation, N.X.; writing—original draft preparation, N.X., M.-W.W., and M.-H.D.; writing—review and editing, N.X., M.-W.W., and M.-H.D.; visualization, Y.-F.H.; supervision, Y.-F.H.;

project administration, Y.-F.H. All authors have read and agreed to the published version of the manuscript.

Funding: This research received no external funding.

Institutional Review Board Statement: Not applicable.

Informed Consent Statement: Not applicable.

Data Availability Statement: Due to its proprietary nature <or ethical concerns>, supporting data cannot be made openly available. Further information about the data and conditions for access are available at http://www.xiongmaodangao.com/.

Acknowledgments: The authors would like to thank the editor and anonymous reviewers for their valuable and constructive comments, which have led to a significant improvement in the manuscript. The author's research was partially supported by the Ministry of Science and Technology of the Republic of China under Grant MOST 109-2622-H-324-001.

Conflicts of Interest: The authors declare no conflict of interest.

Abbreviations

In addition: the following notations and assumptions are used throughout this article.

System Parameters

D	Retailer's base demand.
P	Manufacturer's production rate, where $P > D$.
A	Manufacturer's ordering cost per order.
K	Manufacturer's the setup cost per order.
α	Self-price sensitivity.
β	Cross-price sensitivity.
p_1	Retailer's price per unit offline.
p_2	Retailer's price per unit online.
g_{ck}	Goodwill cost of the manufacturer.
g_r	Goodwill cost of the retailer, $g_{ck} < \rho g$.
F	Fixed costs per shipment.
r	All-unit freight charged per unit.
w	Retailer's unit wholesale price, where $w > c$.
c_{ck}	Manufacturer's unit production cost.
c_r	Retailer's the marginal cost.
h_s	Manufacturer's holding cost per unit per unit time.
Q	Order quantity.
$S(Q)$	Expected sales.
β'	Buy-back price of unsold items (paid by the manufacturer toward the retailer).
w'	New wholesale price.
w	Wholesale price.
ρ	Revenue-sharing fraction, $0 \leq \rho \leq 1$.
δ	Quantity-flexibility fraction, $0 \leq \delta \leq 1$.
k_T	The ratio between the COP at the cooling temperature, set by the manufacturer for treating the product (i.e.T).
TPB	Total profit of the retailer during cycle time.
TPF_{ck}	Total profit of the manufacturer during production cycle.

Decision Variables

Q	Optimal order quantity for supply chain coordination, $t_1 \geq 0$.
n	Number of shipments from the supplier to the retailer per production.
T	Optimal temperature [°C].

References

1. Ma, J. 2017 Letter to Shareholders. 2017. Available online: https://medium.com/swlh/5-retail-insights-from-alibabas-new-retail-revolution-1855bc39bc64 (accessed on 7 December 2020).
2. CNNIC. China's Online Shopping Market Research Report in 2012. 2013. Available online: http://www.cnnic.cn/hlwfzyj/hlwxzbg/dzswbg/201304/t20130417_39290.htm (accessed on 7 December 2020).
3. Du, R.; Jiang, J. New retailing: Connotation, development motivation and key issues. *Price Theory Pract* **2017**, *2*, 139–141.
4. Maimaiti, M.; Zhao, X.; Jia, M.; Ru, Y.; Zhu, S. How we eat determines what we become: Opportunities and challenges brought by food delivery industry in a changing world in China. *Eur. J. Clin. Nutr.* **2018**, *72*, 1282–1286. [CrossRef] [PubMed]
5. Chen, X.; Wang, X.; Jiang, X. The impact of power structure on the retail service supply chain with an O2O mixed channel. *J. Oper. Res. Soc.* **2016**, *67*, 294–301. [CrossRef]
6. He, Z.; Cheng, T.C.E.; Dong, J.; Wang, S. Evolutionary location and pricing strategies for service merchants in competitive O2O markets. *Eur. J. Oper. Res.* **2016**, *254*, 595–609. [CrossRef]
7. Smith, D.; Sparks, L. Temperature controlled supply chains. In *Food Supply Chain Management*; Blackwell Publishing: Oxford, UK, 2004; pp. 179–198.
8. Van der Vorst, J.; Van Kooten, O.; Marcelis, W.J.; Luning, P.A.; Beulens, A.J.M. Quality controlled logistics in food supply chain networks: Integrated decision-making on quality and logistics to meet advanced customer demands. In Proceedings of the 14th International EurOMA Conference, Ankara, Turkey, 17–20 June 2007.
9. Trienekens, J.; Zuurbier, P. Quality and safety standards in the food industry, developments and challenges. *Int. J. Prod. Econ.* **2008**, *113*, 107–122. [CrossRef]
10. Zanoni, S.; Zavanella, L. Chilled or frozen? Decision strategies for sustainable food supply chains. *Int. J. Prod. Econ.* **2012**, *140*, 731–736. [CrossRef]
11. Ahuja, R.K.; Huang, W.; Romeijn, H.E.; Morales, D.R. A heuristic approach to the multi-period single-sourcing problem with production and inventory capacities and perishability constraints. *INFORMS J. Comput.* **2007**, *19*, 14–26. [CrossRef]
12. Mercier, S.; Villeneuve, S.; Mondor, M.; Uysal, I. Time-Temperature Management Along the Food Cold Chain: A Review of Recent Developments. *Compr. Rev. Food Sci. Food Saf.* **2017**, *16*, 647–667. [CrossRef]
13. Lu, C.J.; Lee, T.S.; Gu, M.; Yang, C.T. A multistage sustainable production-inventory model with carbon emission reduction and price-dependent demand under stackelberg game. *Appl. Sci.* **2020**, *10*, 4878. [CrossRef]
14. Kelepouris, T.; Pramatari, K.; Doukidis, G. RFID-enabled traceability in the food supply chain. *Ind. Manag. Data Syst.* **2007**, *107*, 183–200. [CrossRef]
15. Manzini, R.; Accorsi, R. The new conceptual framework for food supply chain assessment. *J. Food Eng.* **2013**, *115*, 251–263. [CrossRef]
16. Yu, M.; Nagurney, A. Competitive food supply chain networks with application to fresh produce. *Eur. J. Oper. Res.* **2013**, *224*, 273–282. [CrossRef]
17. Gouda, S.K.; Saranga, H. Sustainable supply chains for supply chain sustainability: Impact of sustainability efforts on supply chain risk. *Int. J. Prod. Res.* **2018**, *56*, 5820–5835. [CrossRef]
18. Goyal, S.K. An integrated inventory model for a single supplier-single customer problem. *Int. J. Prod. Res.* **1977**, *15*, 107–111. [CrossRef]
19. Darwish, M.; Odah, O.M. Vendor managed inventory model for single-vendor multi-retailer supply chains. *Eur. J. Oper. Res.* **2010**, *204*, 473–484. [CrossRef]
20. Lin, Y.J.; Ouyang, L.Y.; Dang, Y.F. A joint optimal ordering and delivery policy for an integrated supplier–retailer inventory model with trade credit and defective items. *Appl. Math. Comput.* **2012**, *218*, 7498–7514. [CrossRef]
21. Ouyang, L.Y.; Ho, C.H.; Su, C.H.; Yang, C.T. An integrated inventory model with capacity constraint and order-size dependent trade credit. *Comput. Ind. Eng.* **2015**, *84*, 133–143. [CrossRef]
22. Vijayashree, M.; Uthayakumar, R. A single-vendor and a single-buyer integrated inventory model with ordering cost reduction dependent on lead time. *J. Ind. Eng. Int.* **2017**, *13*, 393–416. [CrossRef]
23. Pang, Q.; Chen, Y.; Hu, Y. Coordinating three-level supply chain by revenue-sharing contract with sales effort dependent demand. *Discrete. Dyn. Nat. Soc.* **2014**, *2014*, 561081. [CrossRef]
24. Tsao, Y.C.; Lee, P.L. Employing revenue sharing strategies when confronted with uncertain and promotion-sensitive demand. *Comput. Ind. Eng.* **2020**, *139*, 106200. [CrossRef]
25. Zhao, X.; Wu, F. Coordination of agri-food chain with revenue-sharing contract under stochastic output and stochastic demand. *Asia-Pac. J. Oper.* **2011**, *28*, 487–510. [CrossRef]
26. Cachon, G.P.; Lariviere, M.A. Supply chain coordination with revenue-sharing contracts: Strengths and limitations. *Manag. Sci.* **2005**, *51*, 30–44. [CrossRef]
27. Sang, S. Revenue sharing contract in a multi-echelon supply chain with fuzzy demand and asymmetric information. *Int. J. Comput. Int. Sys.* **2016**, *9*, 1028–1040. [CrossRef]
28. Yao, Z.; Leung, S.C.H.; Lai, K.K. The effectiveness of revenue-sharing contract to coordinate the price-setting newsvendor products' supply chain. *Supply Chain. Manag.* **2008**, *13*, 263–271. [CrossRef]
29. Govindan, K.; Malomfalean, A. A framework for evaluation of supply chain coordination by contracts under O2O environment. *Int. J. Prod. Econ.* **2019**, *215*, 11–23. [CrossRef]

30. Gong, Q. Optimal buy-back contracts with asymmetric information. *Int. J. Manag. Mark. Res.* **2008**, *1*, 23–47.
31. Wang, Y.; Zipkin, P. Agents' incentives under buy-back contracts in a two-stage supply chain. *Int. J. Prod. Econ.* **2009**, *120*, 525–539. [CrossRef]
32. Hou, J.; Zeng, A.Z.; Zhao, L. Coordination with a backup supplier through buy-back contract under supply disruption. *Transp. Res. E Logist. Transp. Rev.* **2010**, *46*, 881–895. [CrossRef]
33. Ding, X.S.; Zhang, J.H.; Chen, X. A joint pricing and inventory control problem under an energy buy-back program. *Oper. Res. Lett.* **2012**, *40*, 516–520. [CrossRef]
34. Tibrewala, R.; Tibrewala, R.; Meena, P.L. Buy-back policy for supply chain coordination: A simple rule. *Int. J. Oper. Res.* **2012**, *31*, 1–28. [CrossRef]
35. Zhang, L.H.; Li, T.; Fan, T.J. Inventory misplacement and demand forecast error in the supply chain: Profitable RFID strategies under wholesale and buy-back contracts. *Int. J. Prod. Res.* **2018**, *56*, 5188–5205. [CrossRef]
36. Dai, T.; Li, Z.; Sun, D. Equity-based incentives and supply chain buy-back contracts. *Decis. Sci. J.* **2012**, *43*, 661–686. [CrossRef]
37. Sainathan, A.; Groenevelt, H. Vendor managed inventory contracts-coordinating the supply chain while looking from the vendor's perspective. *Eur. J. Oper. Res.* **2019**, *272*, 249–260. [CrossRef]
38. Tsay, A.; Lovejoy, W.S. Quantity flexibility contracts and supply chain performance. *Manuf. Serv. Oper. Manag.* **1999**, *1*, 89–111. [CrossRef]
39. Li, C.L.; Kouvelis, P. Flexible and risk-sharing supply contracts under price uncertainly. *Manag. Sci.* **1999**, *45*, 1378–1398. [CrossRef]
40. Sethi, S.; Yan, H.; Zhang, H. Quantity flexibility contracts: Optimal decisions with information updates. *Decis. Sci.* **2004**, *35*, 691–712. [CrossRef]
41. Wu, J. Quantity flexibility contracts under Bayesian updating. *Comput. Oper. Res.* **2005**, *32*, 1267–1288. [CrossRef]
42. Milner, J.M.; Kouvelis, P. Order quantity and timing flexibility in supply chains: The role of demand characteristics. *Manag. Sci.* **2005**, *51*, 970–985. [CrossRef]
43. Bag, S.; Chakraborty, D.; Roy, A.R. A production inventory model with fuzzy random demand and with flexibility and reliability considerations. *Comput. Ind. Eng.* **2009**, *56*, 411–416. [CrossRef]
44. Li, X.; Lian, Z.; Choong, K.K.; Liu, X. A quantity-flexibility contract with coordination. *Int. J. Prod. Econ.* **2016**, *179*, 273–284. [CrossRef]
45. Lian, Z.; Deshmukh, A. Analysis of supply contracts with quantity flexibility. *Eur. J. Oper. Res.* **2009**, *196*, 526–533. [CrossRef]
46. Aung, M.M.; Chang, Y.S. Temperature management for the quality assurance of a perishable food supply chain. *Food. Control.* **2014**, *40*, 198–207. [CrossRef]
47. Lorite, G.S.; Selkälä, T.; Siploa, T.; Palenzuela, J.; Jubete, E.; Viñuales, A.; Cabañero, G.; Grande, H.J.; Tuominen, J.; Uusitalo, S.; et al. Novel, smart and RFID assisted critical temperature indicator for supply chain monitoring. *J. Food Eng.* **2017**, *193*, 20–28. [CrossRef]
48. Kuo, J.C.; Chen, M.C. Developing an advanced multi-temperature joint distribution system for the food cold chain. *Food. Control.* **2010**, *21*, 559–566. [CrossRef]
49. Lawrence, J.J.; Pasternack, B.A. *Applied Management Science: Modeling, Spreadsheet Analysis, and Communication for Decision Making*, 2nd ed.; Wiley: New York, NY, USA, 2002.
50. Aksen, D. Loss of customer goodwill in the uncapacitated lot-sizing problem. *Comput. Oper. Res.* **2007**, *34*, 2805–2823. [CrossRef]
51. Finn, Z.S.; Anderson, T.; Lund, D. Spoiled brands: Protecting your company's goodwill and assets from food contamination claims. *CPCU eJournal* **2008**, *61*, 1–8.
52. Mitreva, E.; Gjurevska, S. The need for a quality system in a company for bread and bakery production in Macedonia. *Quality* **2018**, *19*, 118–124.
53. Hobbs, J.E. Food supply chains during the COVID-19 pandemic. *Can. J. Agric. Econ.* **2020**, *68*, 171–176. [CrossRef]
54. Singh, S.; Kumar, R.; Panchal, R.; Tiwari, M.K. Impact of COVID-19 on logistic systems and disruptions in food supply chain. *Int. J. Prod. Res.* **2020**, 1–16. [CrossRef]
55. Rizou, M.; Galanakis, J.M.; Aldawoud, T.M.S.; Galanakis, C.M. Safety of foods, food supply chain and environment within the COVID-19 pandemic. *Trends Food Sci. Technol.* **2020**, *102*, 293–299. [CrossRef] [PubMed]
56. He, Z.; Han, G.; Cheng, T.C.E.; Fan, B.; Dong, J. Evolutionary food quality and location strategies for restaurants in competitive online-to-offline food ordering and delivery markets: An agent-based approach. *Int. J. Prod. Econ.* **2019**, *215*, 61–72. [CrossRef]
57. Wang, O.; Somogyi, S.; Charlebois, S. Food choice in the e-commerce era a comparison between business-to-consumer (B2C), online-to-offline (O2O) and new retail. *Br. Food J.* **2020**, *122*, 1215–1237. [CrossRef]
58. Li, C.; Mirosa, M.; Bremer, P. Review of online food delivery platforms and their impacts on sustainability. *Sustainability* **2020**, *12*, 5528. [CrossRef]
59. Yang, C.T.; Ho, C.H.; Lee, H.M.; Ouyang, L.Y. Supplier-retailer production and inventory models with defective items and inspection errors in non-cooperative and cooperative environments. *RAIRO-Oper. Res.* **2018**, *52*, 453–471. [CrossRef]
60. Peleg, M.; Engel, R.; Gonzalez-Martinez, C.; Corradini, M.G. Non-Arrhenius and non-WLF kinetics in food systems. *J. Sci. Food Agric.* **2002**, *82*, 1346–1355. [CrossRef]

61. Rong, A.; Akkerman, R.; Grunow, M. An optimization approach for managing fresh food quality throughout the supply chain. *Int. J. Prod. Econ.* **2011**, *131*, 421–429. [CrossRef]
62. Chen, L.; Kou, M.; Wang, S. On the use of importance measures in the reliability of inventory systems, considering the cost. *Appl. Sci.* **2020**, *10*, 7942. [CrossRef]

applied sciences

MDPI

Article

Preparation of the Orange Flavoured "Boba" Ball in Milk Tea and Its Shelf-Life

Ying Liu [1,2], Huan Cheng [2] and Dan Wu [2,*]

[1] Department of Food Nutrition and Detection, Hangzhou Vocational & Technical College, Hangzhou 310018, China; 2016010012@hzvtc.edu.cn
[2] College of Biosystems and Food Science, Zhejiang University, Hangzhou 310058, China; huancheng@zju.edu.cn
* Correspondence: wudan2008@zju.edu.cn; Tel./Fax: +86-0571-8898-2156

Featured Application: The preparation technology of fruit-flavoured "boba" balls can be applied by food ingredient companies. Through the further optimization of the fruit ball core formula, healthy substances (such as flavones, active proteins, sterols, enzymes, probiotics, and vitamins) can be embed-ded in fruit balls and applied to drinks, overcoming the traditional technical bot-tleneck, en-hancing the stability and compatibility of the substances, and effectively extending the shelf life of the products.

Abstract: Boba milk tea is very popular around the world. The "boba" balls in milk tea are usually made of tapioca. Reports on calcium alginate ball encapsulation in fruit-flavoured drinks have rarely been seen. The preparation method for this kind ball was studied. The "boba" balls were obtained by membrane formation on the interface through the addition of calcium chloride fluids into a sodium alginate solution. The operation conditions were studied, including drop height, flow velocity, sodium alginate and calcium chloride solution concentration. The diameter, mechanical strength, loading ratio and encapsulation rate of the "boba" balls are discussed. The optimized preparation conditions were as follows: the diameter of adding tube was 8 mm, the drop height was 25 cm, the drop flow rate was 60 mL/min, 1.0% sodium alginate, 1.0% calcium chloride. The prepared "boba" balls were stored at different temperatures. No microorganisms were detected in 90 days, and the sensory quality decreased with storage time. Shelf life was predicted using the Arrhenius equation; when the storage temperature was less than 10 °C, it could be stored for more than 1 year. This preparation technology of "boba" balls has potential for application by milk tea ingredient companies or relevant beverage manufacturing factories.

Keywords: boba milk tea; calcium alginate ball; preparation method; shelf life

Citation: Liu, Y.; Cheng, H.; Wu, D. Preparation of the Orange Flavoured "Boba" Ball in Milk Tea and Its Shelf-Life. *Appl. Sci.* **2021**, *11*, 200. https://doi.org/10.3390/app11010200

Received: 10 December 2020
Accepted: 24 December 2020
Published: 28 December 2020

1. Introduction

Boba milk tea originated in Taiwan in the 1980s, became very popular in Asia in the 1990s, and has been increasingly popular around the world since 2000 [1,2]. The "boba" balls, which are also called "pearl" balls, are the central ingredient of the milk tea. It is usually made of tapioca, or tapioca mixed with other ingredients, such as konjac powder, sweet potato, chamomile, and so on [3–5]. There have been some studies on the calory content and obesity risk of milk tea [6], but there have been few reports related to the improvement of milk tea technology, including that related to the "boba" balls.

Orange is rich in nutrition. It not only contains nutrient elements needed by the human body, it also contains plant phenols and flavonoids, containing an especially high content of Vc [7,8]. It is one of the most popular fruits in the world. Neves, Trombin, and Marques et al. analysed the global consumption of frozen concentrate orange juice from 2013 to 2018. They summarized the consumer behaviour of three kinds of orange juice in the main market worldwide [9]. A large number of studies have shown that eating a large amount of citrus fruits is beneficial to anti-cancer functions and prevention of cardiovascular diseases [10–13].

In this project, orange juice was wrapped in calcium alginate film to obtain fruit balls, which were used in milk tea, forming a new type of fruit ball drink. This technique was reported in cell immobilization during the last century. For example, Rickert et al. used calcium alginate gel to embed bacteria for the fermentation of organic acids [14]. By dropping Calcium ions (Ca^{2+}) into sodium alginate solution, Ca^{2+} diffused to the periphery and reacted with sodium alginate to form a gel film to wrap the liquid core, and hollow micro-capsules were prepared. Some scholars have performed studies on the influence of processing conditions of calcium alginate gel (CAG). The physical properties of CAG beads change with heat treatment, including rupture strength, size, sphericity and porous structures [15,16].

In general, calcium alginate gel is a solid gel; the CAG studied in this paper has an orange-flavoured drink core. When you bite it, the juice flows out of the inner core, like jam, giving people a sweet and sour, delicious, unprecedented eating experience. It is also a calcium alginate immobilized embedding technology, which has the potential to maintain the stability of sensitive food ingredients, especially some natural health ingredients, such as flavones, active proteins, sterols, enzymes, probiotics and vitamins, etc. By applying calcium alginate ball embedding methods, important breakthroughs could result in some key food industry technologies, thus achieving ideal sensory, physical and chemical indicators, and shelf life, while also achieving functional food design. For example, Lin et al. wrapped astaxanthin in the CAG. The results showed that after storage at 25 °C for 21 days, the contents of astaxanthin in the CAG both remained above 90% of their original amount [17]. This provides an effective way of developing the stability of sensitive compounds. Jiang et al. fixed bifidobacteria in carboxymethyl chitin–calcium alginate gel, the bifidobacteria could be delivered to the gut without being dissolved in the gastric juice [18]. Here, "boba" balls made of calcium alginate for use in milk tea were studied. Operational conditions such as drop height, flow velocity, and sodium alginate and calcium chloride solution concentration were investigated. The diameter, mechanical strength, loading ratio and encapsulation rate of the "boba" balls are discussed. The sensory and microorganisms changes of the "boba" balls at different temperatures were studied, and their shelf life is predicted by using the Arrhenius equation.

2. Materials and Methods

2.1. Materials

Orange juice (sugar: 66°Bx, acidity: 4.16%), modified starch (hydroxypropyl distarch phosphate from potato, stable to temperature, acidity and shear force), calcium chloride anhydrous and sodium alginate were provided by Hangzhou Bodo Industry and Trade Co., LTD., Hangzhou, China. Sucrose and table salt were bought from a market; citric acid anhydrous was purchased from Shandong Zhongchuang Lemon Biochemical Co., LTD., Anqiu, Shandong, China. Sunset yellow, lemon yellow, and orange essence were purchased from Hangzhou Lvjing Flavor Co., LTD., Shandong, China, Xanthan gum and Guar bean gum were purchased from Hebei Jinfeng Chemical Co., LTD., Hengshui, Hebei, China. The companies above were located in China. All of the above materials were of food grade.

2.2. The "boba" Ball Preparation

The composition of the orange-flavoured drink, which was the core of the "boba" ball, was carried out with reference to Wu et al. [19] and was improved for processing. It included sucrose (22.80%), citric acid anhydrous (0.50%), salt (0.04%), sunset yellow (0.0012%), citric yellow (0.006%), orange essence (0.07%), Xanthan (0.1%), and guar gum (0.25%). All the ingredients were put into pure boiling water (100 °C) to be fully dissolved, after which the mixture was cooled to 50–60 °C, and concentrated orange juice (5%), modified starch (3.75%), calcium chloride anhydrous (1%) were added and blended together. Homogenizer (Shanghai Zollo Instrument Co., Ltd., Shanghai, China) was used to homogenize the orange flavoured drink at the pressure of 18–22 mpa for 30 min.

The preparation process of the "boba" ball is shown in Figure 1. The orange fruit-flavoured drink was added to the sodium alginate solution drop by drop using the Automatic liquid filler (Newking Pump Co., Ltd., Shangdong, China). After 1 min of reaction, calcium alginate balls were filtered out with a stainless-steel filter spoon, and were washed with distilled water and then transferred into 5% calcium chloride solution to react for 15 min. Finally, the residual calcium chloride outside of the "boba" ball was washed off with distilled water again, and the preparation was completed. All the above operations were carried out at room temperature. The prepared "boba" balls were stored in 0.9% sodium chloride solution for property detection.

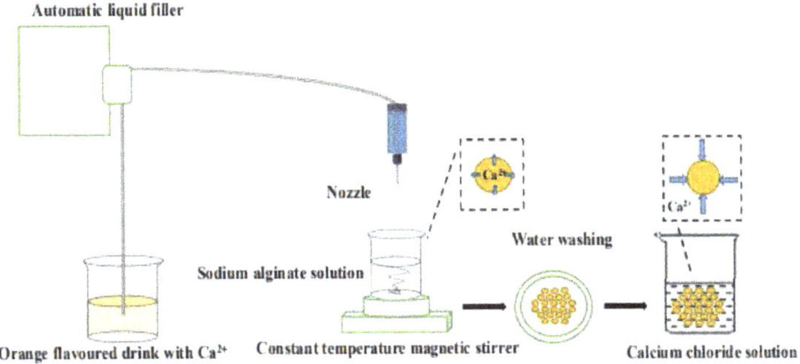

Figure 1. Schematic diagram of "boba" ball preparation.

2.3. Preparation of "boba" Ball Storage Solution

CAG showed a three-dimensional network structure, known as an "egg box" structure, that enables small molecules of water-soluble substances to pass through freely [20]. The swelling property of CAG was affected by pH value and stabilised at solution pH < 6. This indicates that calcium alginate would gradually decompose in salt solution with pH greater than 5.5 [21]. Therefore, "boba" balls must be stored in environments with pH values less than 5.5. There were some sugar, acid, pigment, essence and other small molecule water-soluble substances in the "boba" balls. To prevent the loss of solute and maintain the pH of the environment, the formulation of the "boba" ball storage solution was prepared as follows: sucrose 22.80%, citric acid 0.50%, salt 0.04%, sunset yellow 0.0012%, citric yellow 0.006%, orange essence 0.07%. The pH value of the storage solution was measured to be about 2.05 ± 0.05 using a pH meter (INESA Scientific Instrument Co., Ltd., Shanghai, China).

2.4. Determination of "boba" Ball Properties

Measurement of diameter: 20 capsules were randomly selected, the diameters were measured one by one using Vernier calipers (Shanghai Meinite Industrial Co., Ltd., Shanghai, China), and the average value was taken [22].

Determination of mechanical strength: The single-axis pressing method was used for measurement. A "boba" ball was placed on the tray of the analytical balance (Mettler-Toledo, LLC, Shanghai, China), and positive pressure was applied to the "boba" ball until it burst. The balance showed a sudden change of pressure from low to high, and then to low; the maximum positive pressure was recorded, namely the mechanical strength of the "boba" ball. Twenty "boba" balls were randomly selected for each detection, and the mechanical strength was represented by the average value [20].

The orange fruit-flavoured drink loading ratio and encapsulation rate:

$$\text{Loading ratio} = \text{Total weight of orange fruit-flavoured drink in the"boba" ball/Total weight of the "boba" ball} \tag{1}$$

$$\text{Encapsulation rate} = \text{Total weight of orange fruit-flavoured drink in the ``boba'' ball/Total weight of orange fruit-flavoured drink} \tag{2}$$

The encapsulation rate is an important parameter in the preparation of "boba" balls. The higher, the better.

2.5. Shelf Life Determination and Prediction

The "boba" balls were placed in a sealed glass jar by adding the storage solution, pasteurized at 70–80 °C for 30 min, and were then stored at 20 °C, 25 °C, 30 °C, 35 °C and 40 °C, respectively.

Total bacterial colony determination was conducted with reference to GB 4789.2— National Standard Food Microbiology Examination for Food Safety [23].

The sensory evaluation indexes included colour, aroma, taste and appearance, and the standards are shown in Table 1. The panel of 12 assessors was aged from 20 to 36 years and possessed expertise in the sensory evaluation of food. Approximately 20 "boba" balls stored at a constant temperature were taken out, kept in a plate at room temperature for 30 min, and served to each panellist along with the questionnaire. The samples were presented one at a time, with about a 5 min wait between samples. The intervals for sensory evaluation of the "boba" balls varied at different storage temperatures: balls stored at 20–30 °C were evaluated every 5–9 days, and balls stored at 35 °C and 40 °C were evaluated every 1–2 days. The intensities of the sensory attributes are presented as the mean of the scores provided by the 12 panel assessors. If the sensory score was below 80, the product was considered to be unable to meet consumers' expectations and reach the shelf-life limit.

Table 1. Standard of sensory evaluation.

Sensory Index	Aroma			Appearance			Taste		
characteristic description	pure fruit aroma with long duration	light fruit aroma, short retention time	No obvious aroma or odor	normal colour, no stratification Pale, no broken	not clear Colour difference, Capsule breakage <5%	solution stratified or turbidness, Capsule breakage >5%	sweet and sour, mellow taste	Partial sour or partial sweet	taste relatively light Sour, difficult to accept
score	20–30	10–20	0–10	20–30	10–20	0–10	30–40	15–30	0–15

According to [24,25], the sensory evaluation response function F(X) at a certain temperature can be expressed as follows:

$$F(X) = k\, dt \tag{3}$$

where k is the sensory evaluation change reaction rate, and t is the period for which the food was stored. Here, the k value at a certain temperature can be derived from the slope of the regression equation between the response value F(X) and time. The effect of temperature on the change reaction of sensory evaluation can described by the Arrhenius relationship.

Taking the logarithm on both sides of the Arrhenius function:

$$\ln k = -\,Ea/RT + \ln k_0 \tag{4}$$

where Ea is the activation energy in J/mol, R is the gas constant in J/(K mol) and is equal to 8.314, k_0 is a pre-exponential factor, and T is absolute temperature in K (273 °C).

Here, the Ea value and k_0 can be derived from the slope of the regression equation between ln k and 1/T.

2.6. Statistical Analysis

The Data Processing System (DPS) software v13.5 was applied to fix the experiment data and establish the mathematical model [26]. The Origin software (OriginLab Corporation, Northampton, MA, USA) was applied to paint the response surface.

3. Results and Discussion

3.1. Selection of Hose Diameter

The diameter of the straws used for "boba" milk tea is generally about 10.00–12.00 mm. To allow the "boba" ball pass through the straw smoothly, the diameter of the prepared balls generally needs to be 1.00–2.00 mm lower than this. The objective of this project was to prepare "boba" balls with a diameter of about 8–10 mm; therefore, it was advisable to use a hose with a diameter of 8 mm.

3.2. Influence of Drop Height

Figure 2 shows the influence of drop height on the "boba" balls. The velocity of the automatic liquid filler was 60 mL/min, the sodium alginate solution was 0.8%. It was found that if the drop height wasn't high enough, calcium alginate bal would easily float on the surface, and the final ball was often elliptical (see the drop height of 5 cm in Figure 2); with increasing drop height, the probability of elliptical balls gradually decreased (see the drop height 5–20 cm in Figure 2). When the drop height was greater than 20 cm, the "boba" balls were basically spherical. Therefore, it is advisable to choose a drop height greater than 20 cm. The drop height selected here was 25 cm.

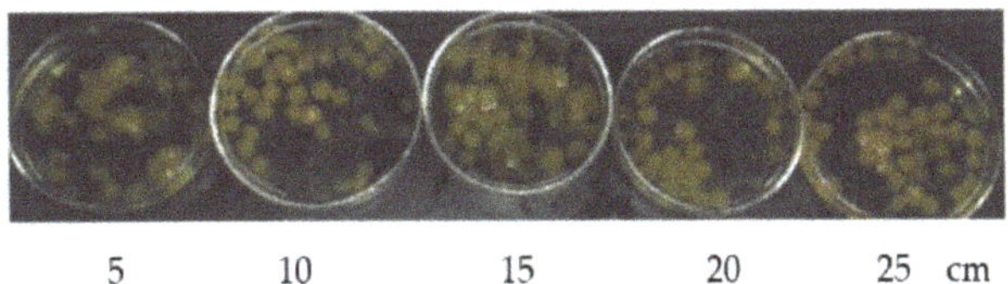

Figure 2. Effects of drop height on the "boba" ball.

3.3. Influence of Flow Velocity

Table 2 and Figure 3a present the effect of flow velocity on the "boba" ball properties. The diameter of the hose was 8 cm, the drop height was 25 cm, and the concentration of sodium alginate solution was 0.8%. It was found that the flow velocity had a positive effect on the diameter of the "boba" balls (Table 2). When the flow rate was less than 40 mL/min or greater than 100 mL/min, the produced ball diameters were sometimes smaller than 8 cm or larger than 10 cm. With the increasing flow velocity, there was a directly proportional increase in particle size distribution. This was in good agreement with the report on alginate micro spheres [27]. These balls wouldn't be applied to the milk tea, as they would result in a bad consumption experience. Therefore, in order to obtain the required "boba" with suitable diameters of between 8–10 mm, the flow velocity should be controlled at 60–80 mL/min. Figure 3a shows that the loading ratio reached its optimal value (76.15 ± 1.51) when the flow velocity was 60 mL/min. Under this preparation condition, the flow velocity had little effect on encapsulation efficiency and mechanical strength. The encapsulation efficiencies of the "boba" balls were all 100%, and the mechanical strength maintained stability.

Table 2. Influence of flow velocity on the "boba" ball diameter.

The Velocity (mL/min)	The "boba" Ball Diameter (cm)							The Ratio of "boba" Ball Diameter 8–10 mm (%)
	6–7	7–8	8	9	10	11	>11	
	Ratio (%)							
40	0	40	55	5	0	0	0	60
60	0	0	25	35	40	0	0	100
80	0	0	30	40	30	0	0	100
100	5	5	30	35	25	0	0	90
120	5	5	25	30	35	5	0	90
140	5	10	15	10	35	10	15	85

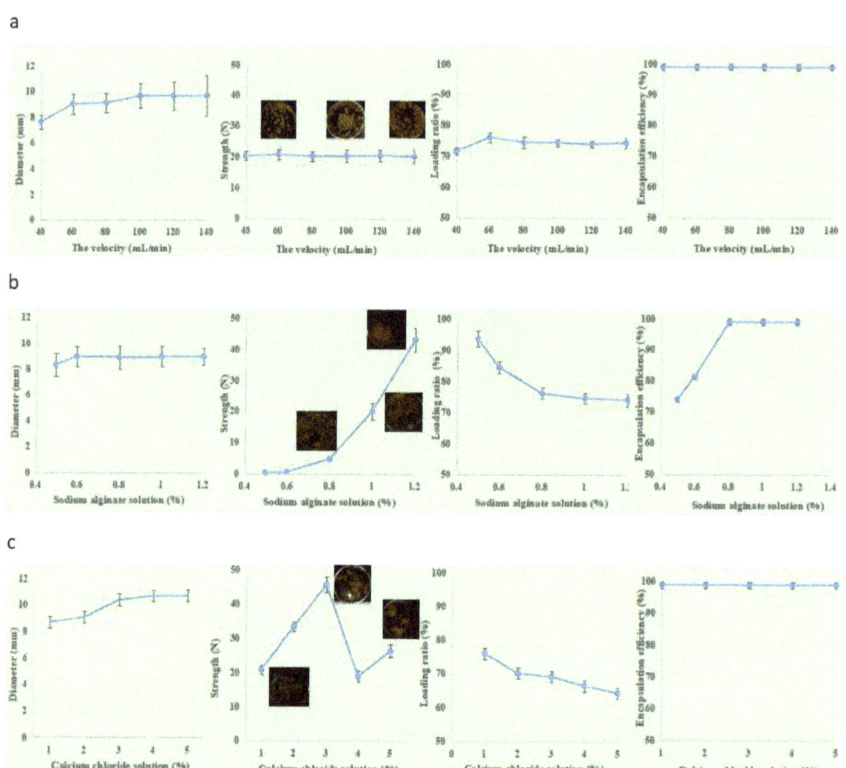

Figure 3. Effects of the preparation situation on the characteristics of the "boba" balls. (**a**) Flow velocity; (**b**) Sodium alginate solution concentration; (**c**) Calcium chloride solution concentration.

3.4. Influence of Sodium Alginate Solution Concentration

Figure 3b shows the influence of different concentrations of sodium alginate on the "boba" ball properties. The diameter of the hose was 8 cm, the drop height was 25 cm, the flow rate was 60 mL/min, and the concentration of calcium chloride solution was 1.0%. It was found that the sodium alginate solution concentration had positive effects on the diameter, mechanical strength, loading ratio and encapsulation efficiency of the "boba" balls. when the sodium alginate solution concentration was greater than 0.6%, the diameter maintained stability. However, if its concentration was below 0.5%, there were some "boba" balls with diameters <8.00 mm. The hinge structure formed by sodium alginate and calcium ions changed with the concentration of sodium alginate, and affected the diameter, mechanical strength, loading ratio and encapsulation efficiency of the "boba" ball [27–29]. The encapsulation efficiency of the "boba" ball was 74.07 ± 0.87% (0.5% sodium alginate) and 81.48 ± 0.75% (0.6% sodium alginate). When the concentration of sodium alginate was higher than 0.8%, the encapsulation reached 100%. However, when the concentration of sodium alginate reached' 1.2%, the "boba" balls tended to stick together and were not conducive to being cleaned separately; this situation would increase their preparation time. The mechanical strength increased exponentially with increasing concentration of sodium alginate. The relationship between the concentration of sodium alginate and the strength of the "boba" balls was as follows, and correlation coefficient was 0.9786:

$$y = 0.0132e^{7.0139x} \tag{5}$$

When the concentration of sodium alginate increased from 0.8% to 1.0%, the strength increased from 4.83 ± 0.50 to 20.14 ± 2.60, and the crushing resistance increased by more than four times. To give the "boba" balls a certain crushing resistance and 100% encapsulation efficiency in production, 1% sodium alginate was deemed to be the best.

3.5. Influence of Calcium Chloride Solution Concentration

Figure 3c shows the influence of different concentrations of calcium chloride on the "boba" ball properties. The diameter of the hose was 8 cm, the drop height was 25 cm, the flow rate was 60 mL/min, and the concentration of sodium alginate solution concentration was 1.0%. It was found that the calcium chloride concentration had positive effects on the diameter, mechanical strength, and the loading ratio of the "boba" balls. The encapsulation efficiency was 100% under the condition of 1–5%. With the increase of calcium chloride concentration, the loading ratio decreased, and the mechanical strength first went up, then went down, and then went up again. When the calcium chloride solution was 3%, the strength reached its maximum. The calcium ion/alginate ratio here was found to present an optimized ratio. The increase of calcium content could increase the crosslinking density and shrinkage of matrix [30], but also produced more carbon dioxide, leading to a more porous and weak matrix [31]. The "boba" balls' average diameter gradually increased from 8.70 ± 0.70 to 10.73 ± 1.30, and was then maintained at a certain level. When the calcium chloride concentration reached or exceeded 2%, the "boba" balls stuck together easily, showed the astringency of calcium chloride and presented a poor sensory evaluation experience. Taking the above factors into consideration, the 1% calcium chloride concentration was determined to be the best.

3.6. Microbial Colonies/Sensory Changes and Prediction of Shelf Life for the "boba" Ball

The "boba" balls were prepared according to the optimized experimental conditions described above, and stored at different temperatures. During the 90 days of storage, no microorganisms were detected. This was due to pasteurization and sealed filling. Therefore, the microbiological safety of the product could be guaranteed. The product's shelf life is dependent on changes in sensory properties. Colour, aroma, stratification and capsule breakage rate were the main factors affecting sensory indicators.

Figure 4 shows the response of the sensory evaluation of "boba" balls stored at different temperatures. The sensory evaluation values of the product were linearly correlated with storage time, the correlation coefficient R^2 of the fitting regression curve was above 0.94 (Table 3), and the k value was derived, which exhibited a constant value at a constant temperature. The linear equation of ln k against 1/T was obtained according to Equation (4) (Figure 5). The activation energy Ea (93.45 kJ/mol) and the pre-finger factor k_0 (7.46 × 10^{15}) were calculated. Then, the mathematical model equation reflecting the change of the quality of the "boba" balls was established, as follows:

$$F(x) = 100 - 7.46 \times 10^{15} \times exp\,(-93.45 \times 10^{3}/RT) \times t \tag{6}$$

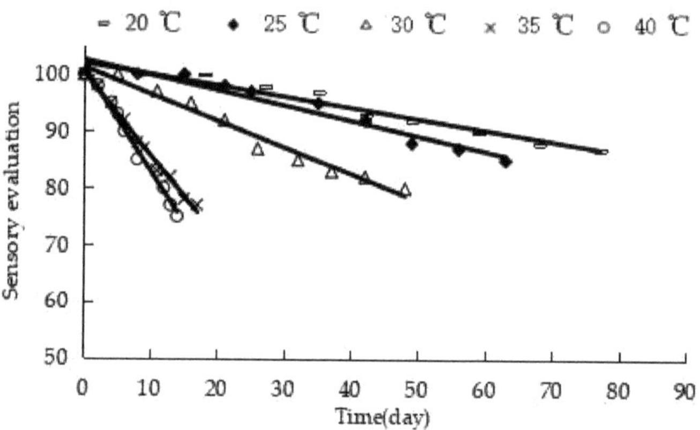

Figure 4. Response of the sensory evaluation of "boba" balls stored under different conditions.

Table 3. The Arrhenius kinetic parameters of "boba" ball quality.

Temp. (°C)	k	R^{2} [a]	Ea ± 95%CI [b] (kJ/mol)	R^{2} [c]	Predicted Time to End Point (Day/°C)
20	0.1961	0.9454			1991 day/0 °C
25	0.2676	0.9412			465 day/10 °C
30	0.4698	0.9735	93.45 ± 4.67	0.9504	120 day/20 °C
35	1.4407	0.9905			34 day/30 °C
40	1.8118	0.9847			10 day/40 °C

[a] Correlation coefficients of regression lines of the sensory evaluation value against time. [b] Ea, activation energy; CI, confidence intervals. [c] Correlation coefficients of regression lines of ln k against 1/T.

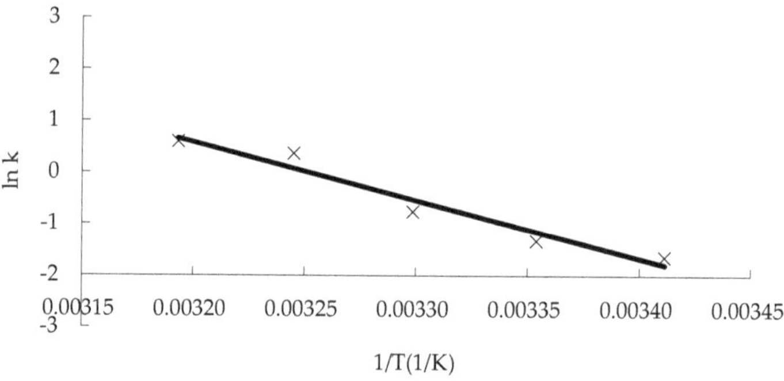

Figure 5. Arrhenius behaviour curve of "boba" ball quality.

The response surface of the sensory evaluation changed with time and temperature was determined (Figure 6). When the sensory score was below 80, the product was considered to be unable to meet consumers' exp ectations and reach the shelf-life limit. Therefore, product shelf life could be predicted for different storage temperatures (Table 3). When the storage temperature was lower than 10 °C, the shelf life of the "boba" balls was 465 days, exceeding one year.

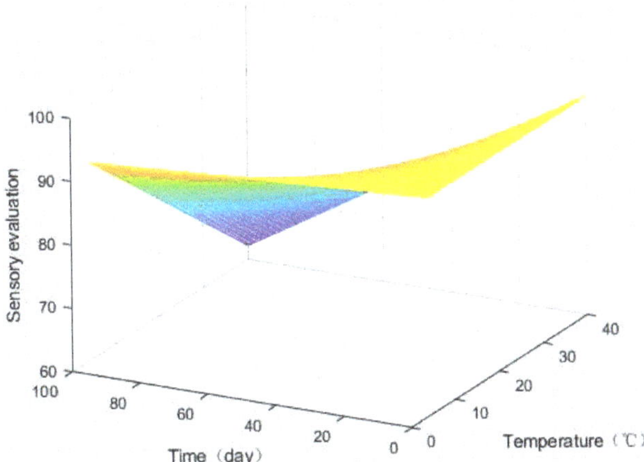

Figure 6. The response surface of the sensory evaluation changes with time and temperature of the "boba" balls.

4. Conclusions

The "boba" balls used in milk tea, which are made of calcium alginate, were studied here. The optimal technological parameters for the preparation of "boba" balls were discussed. The diameter of the liquid adding tube was 8 mm, the drop height was 25 cm, the sodium alginate solution concentration was 1.0%, the flow velocity was 60 mL/min, and the concentration of calcium chloride concentration was 1%. The prepared "boba" balls were stored under different temperature conditions. No microbes were detected after storage for 90 days. The sensory evaluation value of the "boba" ball decreased gradually with storage time at different temperatures. The mathematical model for predicting shelf life was as follows Equation (6).

According to the above mathematical model, the shelf life of the "boba" ball developed could reach more than 1 year in a refrigerated environment at temperatures lower than 10 °C.

Author Contributions: Carrying out the experiment and writing the manuscript, Y.L. and H.C.; Designing the experiment and assisting in editing the manuscript, D.W. All authors have read and agreed to the published version of the manuscript.

Funding: This research was supported by Hangzhou Science and Technology Development Project (Grant Number: 20180432B34), Provincial Key R&D project (Grant Number: 2020C02033) and the national research program of China (Grant Numbers: 2017YFD0400704 and 2014BAD04B01).

Data Availability Statement: Data sharing not applicable.

Acknowledgments: The authors thank the Teaching Reform Project of Zhejiang University for the research funding. Thanks to the teachers Li Yang, Pro. Jianchu Chen and Xingqian Ye for offering technical assistance and the experimental facilities for our experiment.

Conflicts of Interest: The authors declare no conflict of interest.

References

1. CNN. Bubble Tea. *CNN News*. 13 November 2015. Available online: http://edition.cnn.com/2015/11/13/asia/gallery/bubble-tea/index.html (accessed on 29 January 2016).
2. Lin, J.H.; Chang, C.W.; Chi, H.C. Transnational Bubbles: Making Taiwanese Bubble Tea in the UK (in Mandarin). *J. Geogr. Sci.* **2017**, *87*, 1–20.
3. Min, J.E.; Green, D.B.; Kim, L. Calories and sugars in boba milk tea: Implications for obesity risk in Asian Pacific Islanders. *Food Sci. Nutr.* **2016**, *5*, 38–45. [CrossRef] [PubMed]
4. Sun, Y.Y. Research on Processing Technology of Konjac Pearl Milk Tea. *Food Nutr. China* **2020**, *26*, 33–37.

5. Fan, N.; Wang, Z.Y. The development of konjac bubble milk tea. *Food Sci. Technol.* **2015**, *10*, 92–96.

6. Lin, P.Y.; Chen, T.C.; Lin, F.Y.; Dong, J.Y.; Chen, W.L.; Sumiko, K.; Indri, K.S.; Hitomi, T.; Shigeru, Y. The Effect of Limiting Tapioca Milk Tea on Added Sugar Consumption In Taiwanese Young Male and Female Subjects. *J. Med. Investig.* **2018**, *65*, 43–49.

7. Ding, X.B.; Zhang, H.; Liu, S.Y.; Liao, Y.J.; Zhou, Z.Q. Current status of the study in citrusnutriology. *Acta Hortic. Sin.* **2012**, *39*, 1687–1702.

8. Perez-Cacho, P.R.; Rouseff, R.L. Fresh Squeezed Orange Juice Odor: A Review. *Crit. Rev. Food Sci. Nutr.* **2008**, *48*, 681–695. [CrossRef]

9. Neves, M.F.; Trombin, V.G.; Marques, V.N.; Martinez, L.F. Global orange juice market: A 16-year summary and opportunities for creating value. *Trop. Plant Pathol.* **2020**, *45*, 166–174. [CrossRef]

10. Miyagi, Y.; Om, A.S.; Chee, K.M.; Bennink, M.R. Inhibition of azoxymethane-induced colon cancer by orange juice. *Nutr. Cancer* **2000**, *36*, 224–229. [CrossRef]

11. Poulose, S.M.; Harris, E.D.; Patil, B.S. Citrus limonoids induce apoptosis in human neuroblastoma cells and have radical scavenging activity. *J. Nutr.* **2005**, *135*, 870–877. [CrossRef]

12. Cesar, T.; Benassi, R.; Ponce, O.; Nasser, M. Orange juice combined to a healthy-eating pattern improved endothelial function and reduced global risk of CHD in metabolic syndrome patients. *Proc. Nutr. Soc.* **2020**, *79*, OCE2. [CrossRef]

13. Franke, S.I.R.; Guecheva, T.N.; João Antonio, P.H.; Daniel, P. Orange Juice and Cancer Chemoprevention. *Nutr. Cancer* **2013**, *65*, 943–953. [CrossRef] [PubMed]

14. David, A.R.A.; Charles, E.G.B.; Bonita, A.G.A. Improved organic acid production by calcium alginate-immobilized propionibacteria. *Enzym. Microb. Technol.* **1998**, *22*, 409–414.

15. Kim, S.; Jeong, C.; Cho, S.; Kim, S.B. Effects of thermal treatment on the physical properties of edible calcium alginate gel beads: Response surface methodological approach. *Foods* **2019**, *8*, 578. [CrossRef] [PubMed]

16. Jeong, C.; Kim, S.; Lee, C.; Cho, S.; Kim, S.B. Changes in the physical properties of calcium alginate gel beads under a wide range of gelation temperature conditions. *Foods* **2020**, *9*, 180. [CrossRef]

17. Lin, S.; Chen, Y.; Chen, R.; Chen, L.; Ho, H.; Tsung, Y.; Shen, M.; Liu, D. Improving the stability of astaxanthin by microencapsulation in calcium alginate beads. *PLoS ONE* **2016**, *11*, 61–70. [CrossRef]

18. Jiang, Y.; Wei, H.; Hua, P.; Cai, X. Research and Application of Immobilized Bifidobacterium Carboxymethyl Chitin/Calcium Alginate. *Sci. Technol. Eng.* **2018**, *18*, 37–43.

19. Wu, D.; Zhang, Z.; Zhang, A.; Ye, X.; Chen, J. Application of mixture design to optimizing orange and strawberry flavored beverages. *Sci. Technol. Food Ind.* **2014**, *35*, 202–206.

20. Amsden, B.; Turner, N. Diffusion characteristics of calcium alginate gels. *Biotechnol. Bioeng.* **2015**, *65*, 605–610. [CrossRef]

21. Dainty, A.L.; Goulding, K.H.; Robinson, P.K.; Simpkins, I.; Trevan, M.D. Stability of alginate-immobilized algal cells. *Biotechnol. Bioeng.* **2010**, *28*, 210–216. [CrossRef]

22. Chen, G.; Huang, S.F. A new method for the preparation of alginate-Ca capsule with liquid core. *J. Shanxi Univ. (Nat. Sci. Ed.)* **2008**, *31*, 119–123.

23. Ministry of Health of the People's Republic of China. *Microbiological Examination of Food Hygiene (GB/T4789-2008)*; China National Standard: Beijing, China, 2008; (In Chinese).

24. Wu, D.; Wang, Y.; Chen, J.; Ye, X.; Wu, Q.; Liu, D. Preliminary study on time–temperature indicator (tti) system based on urease. *Food Control.* **2013**, *34*, 230–234. [CrossRef]

25. Cheng, H.; Liu, Y.; Wu, D.; Zhao, X.; Mao, C.; Wang, S.; Chen, J. Black Tea Extract on the Preservation of Cantonese Sausage. *Am. J. Biochem. Biotechnol.* **2020**, *16*, 417–423. [CrossRef]

26. Tang, Q.Y.; Zhang, C.X. Data Processing System (DPS) software with experimental design, statistical analysis and data mining developed for use in entomological research. *Insect Sci.* **2013**, *20*, 254–260. [CrossRef] [PubMed]

27. Davoud, S.; Atefeh, S.; Ali, S.; Alexander, M. Preparation of internally-crosslinked alginate microspheres: Optimization of process parameters and study of pH-responsive behaviors. *Carbohydr. Polym.* **2020**. [CrossRef]

28. Zou, X.; Zhao, X.; Ye, L.; Wang, Q.; Li, H. Preparation and drug release behavior of pH-responsive bovine serum albumin-loaded chitosan microspheres. *J. Ind. Eng. Chem.* **2015**, *21*, 1389–1397. [CrossRef]

29. Blandino, A.; Maeias, M.; Cantero, D. Formation of calcium alginate gel capsules: Influence of sodium alginate and CaCl$_2$ concentration on gelation kinetics. *J. Biosci. Bioeng.* **1999**, *88*, 686–689. [CrossRef]

30. Silva, C.M.; Ribeiro, A.J.; Figueiredo, I.V.; Gonçalves, A.R.; Veiga, F. Alginate microspheres prepared by internal gelation: Development and effect on insulin stability. *Int. J. Pharm.* **2006**, *311*, 1–10. [CrossRef]

31. Chan, L.; Lee, H.; Heng, P. Production of alginate microspheres by internal gelation using an emulsification method. *Int. J. Pharm.* **2002**, *242*, 259–262. [CrossRef]

applied sciences

Article

Application of a Multiscale Approach in the Substitution and Reduction of NaCl in Costeño-Type Artisan Cheese

Martha L. Diaz-Bustamante *, Luis H. Reyes  and Oscar Alberto Alvarez Solano

Grupo de Diseño de Productos y Procesos (GDPP), Department of Chemical and Food Engineering, Universidad de Los Andes, Bogotá 111711, Colombia; lh.reyes@uniandes.edu.co (L.H.R.); oalvarez@uniandes.edu.co (O.A.A.S.)
* Correspondence: ml.diaz@uniandes.edu.co; Tel.: +57-1-339-4949 (ext. 1779)

Received: 6 November 2020; Accepted: 19 November 2020; Published: 17 December 2020

Abstract: The effects on the texture, rheology, and microstructure of costeño-type artisan cheese caused by the substitution and reduction of NaCl and the increase in cooking temperature during cheese production were studied using a multiscale approach that correlates responses at the macroscopic and microscopic levels. The decrease in the NaCl content, the partial substitution by KCl, and the increase in the cooking temperature before the serum drainage showed physicochemical, textural, and rheological differences between the cheeses. The microstructure was not affected by the reduction in salt or by modifications in the cheese making. The cheeses with an increase in the cooking temperature before the whey drainage stage and reduced NaCl by 5% and 7.5% (Q_2 and Q_3, respectively) showed similarity with the physicochemical composition and textural attributes of the control cheese (Q_C). Overall, this study contributes to the design of cheeses with specific functionalities through multiscale modeling.

Keywords: costeño-type cheese; sodium chloride; texture; rheology; microstructure

1. Introduction

Salt (NaCl) is the primary source of sodium in the human diet. Its high consumption has been linked to hypertension and, consequently, an increased risk of stroke and premature death due to cardiovascular diseases [1,2]. Reducing sodium chloride intake represents one of the most important goals for advancing public health worldwide [2]. The maximum recommended daily intake of salt is 5 g per day, equivalent to 2 g of sodium per day [3]. However, approximately 80% of the ingested salt is added to food during manufacturing [4,5]. Cheese is perceived as a highly concentrated food source of sodium [6]. Therefore, importance should be given to reducing salt intake in this dairy product.

Sodium reduction in cheese is complex and challenging due to salt's multifunctional behavior in the product [5]. It could be achieved by reducing NaCl's amount in the product or using substitutes for mineral salts such as potassium chloride (KCl) [7–9]. The addition of KCl in cheeses has given good results concerning the rheological, textural, sensory, and stability properties of the product, and it has been considered a salt compound chemically similar to NaCl when compared to other substitute salts, such as $CaCl_2$ and $MgCl_2$ [6–10]. In addition, the dietary intake of KCl can decrease the effect of sodium-induced hypertension and can reduce calcium excretion in the urine [6]. However, completely replacing NaCl with KCl is not recommended because the latter gives food a bitter taste [11,12].

Reducing the salt content in cheeses accelerates protein hydration, significantly influencing the cheese's physical properties and quality [5,10,13]. Lucey et al. (2003) [14] indicated that the cheese's texture would generally remain smooth even when the moisture content is adjusted to the typical sodium content. The texture is also influenced by the chemical composition of the cheeses and the manufacturing process. Some studies have even attempted to manufacture reduced-salt cheese by standard methods and have confirmed that moisture retention increases with salt reduction in cheese [5,9].

In Colombia, the costeño-type cheese is one of the fresh artisan cheeses made on the Colombian Caribbean coast, characterized by its high salt content from approximately 10% (*w/w*) [15–17]. Consequently, it leads to increased hardness and dryness [13]. Costeño-type cheese is considered crumbly, allowing it to be crushed and sprinkled on food [17].

Considering that the potential reduction of NaCl depends on many factors associated with the nature of the product—its composition, the type of processing, and the manufacturing conditions [6]—it is crucial when varying salt concentrations to understand the interactions existing between the production process, the cheese properties, and the product. Studies have been carried out on the rheological, textural, and microstructural properties of cheeses, but many were limited to the analysis of product–property relationships [18] and product–process [5] without considering the relationship as a whole (product–process–property).

Multiscale modeling visualizes food structure at various scales, creating a geometric model that predicts macroscale behavior consistent with the structure of matter at underlying scales without requiring excessive computing resources [19]. This methodology allows an understanding of the link between the macroscopic and microscopic properties of cheese. This method has also been used in other studies and has produced promising results [20,21].

Reducing the NaCl content in the costeño-type cheese may reduce salt consumption in the country. However, there is no information on the existing phenomenology when making a costeño-type cheese with low salt concentrations or a NaCl substitute. We hypothesized that by increasing the cooking temperature during the cheese-making procedure, one could reduce the curd's moisture content. This is because higher temperatures increase hydrophobic interactions, causing more syneresis and contraction of the curd. This decreases the amount of moisture retained in the matrix of cheese [14,22,23], which would create a costeño-type cheese with characteristic hardness with lower salt content.

The increasing attention that consumers pay to healthy foods is considered an essential factor influencing products' development of health benefits. Therefore, by its very nature, a multiscale model can potentially provide a more accurate description of how foods change during processing operations so that new products can be designed [24].

Based on the preceding, this research's objective was to study the effect of sodium reduction and partial substitution by KCl and to increase the cooking temperature in the production of costeño-type cheese through a multiscale approach. This approach studied the relationships between composition, cooking temperature, and properties (microscopic and macroscopic) when reducing NaCl concentration.

2. Materials and Methods

2.1. Costeño-Type Cheese-Making

Experimental cheeses were prepared according to Ballesta (2014) [25], with some modifications. Each trial was made using 2 L of raw milk, pasteurized at 63 °C for 30 min. The milk was cooled to 32 °C, and calcium chloride was added at a rate of 0.02% (*w/w*) with stirring for 2 min. The coagulant (Microclerici, 1 g per 150 L) was added to the milk with slow stirring. After 60 min, the curd was cut into cubes of 1 cm^3 and held for another 5 min, followed by stirring. Before the whey drainage, the curd was cooked to a standard cooking temperature of 45 °C for 15 min. An increase in temperature was made for some experimental tests at this stage, as indicated in Table 1.

The curd was transferred to cheesecloth to drain the whey and obtain a percentage of lactic acid of 0.1% (*v/v*). The salt was added depending on the amount of milk used during the cheese making. It was made with dry salt crystals, either single NaCl or a combination of salt (NaCl and KCl), according to the weight (Table 1). The percentages of NaCl substitution by KCl were chosen based on the maximum decrease of sodium in the cheeses and its possible effects on the protein structure.

Table 1. Stage drainage of the whey and salting modified during the production of the costeño-type cheese.

Experimental Cheese	Cook Temperature (°C)	Added Salt (%)	
		NaCl (*w/w*)	KCl (*w/w*)
Q_c (Control)	45	10.00	0.00
Q_1	45	7.50	0.00
Q_2	55	5.00	0.00
Q_3	55	2.50	0.00
Q_4	45	5.00	5.00
Q_5	45	2.50	7.50

All the cheeses were molded and pressed at 1.5 bar in a pneumatic press, making an initial unmold after 4 h. Subsequently, the pressure was increased to 3 bars, keeping these conditions constant for 3 h. Finally, the cheeses were stored at a temperature of 4 °C. For each formulation, the production and analysis of cheeses were done in duplicate and carried out after processing over the course of one day.

2.2. Physicochemical Analysis

The physicochemical characterization of the milk (pH, titratable acidity, and fat analysis) was carried out according to AOAC (1995) [26]. The density was determined using a lactodensimeter and the moisture content using a digital thermobalance XM 60-HR (Precisa Gravimetrics AG, Dietikon, Switzerland). For the characterization of costeño-type cheese, moisture content was determined by oven drying [27]. Fat content was analyzed by the Gerber method [27]. Fat-in-dry-matter (FDM, %) was calculated as fat%/(100 − moisture%) × 100, and moisture-in-nonfat-substance (MNFS, %) was calculated as moisture%/(100 − fat%) × 100 [28]. All physicochemical determinations were made in duplicate.

2.3. Texture Profile Analysis

The texture profile was determined using a TAHD plus texturometer (Stable Micro Systems, Godalming, UK). The analysis was conducted on cylindrical pieces of cheese (14 mm height, 13 mm diameter). A 35 mm diameter stainless steel probe was fitted to the texture analyzer, calibrated using a 5 kg load cell. The samples were analyzed in duplicate at 18 ± 2 °C and a compression of 70% using two compression cycles [29,30]. This was used to determine texture attributes such as hardness (force necessary to achieve a given deformation) and cohesiveness (resistance of the internal ties that make up the body of a product, $area_2/area_1$) [31,32].

2.4. Rheological Analysis

The rheological measurements were made on samples of 1 mm in height and 20 mm in diameter with an ARG2 rheometer (TA Instruments, New Castle, DE, USA). A corrugated surface was placed on the upper and lower plate to eliminate the slippage of the sample. The lower plate temperature of the measuring system was maintained by circulating water at 25 °C. The analysis in each experimental test was carried out at 1 Hz frequency and at a deformation of 0.1%, under which the properties of the cheeses remained within the linear viscoelastic region, where the product can still be recovered and where there is a linear relationship between the stress applied and the strain obtained [31–33]. The dynamic rheological data collected included the two components of the complex shear modulus: the storage module or elastic component (G′) and the loss module or viscous component (G″) [31,32]. Results are presented as the average of two sweeps. The modules were plotted against frequency (Hz) for comparative purposes.

2.5. Scanning Electron Microscopy

The cheese samples (approximately 1 mm × 20 mm) were stored at a temperature of −80 °C. Then, they underwent a drying process using a lyophilizer. Later they were coated in gold with a DV-TSC metallizer (Denton Vacuum LLC, Moorestown, NJ, USA) to be observed in a Phenom Pro X Scanning Electron Microscope (Phenom-World, Thermo Fisher Scientific, Waltham, MA, USA) operated at 10 kV. The fields were randomly selected over the sample area. The images were recorded at 4500× magnification and were used to determine the average pore size using ImageJ v.1.43s software (National Institute of Health, Bethesda, MD, USA).

2.6. Statistical Analysis

In order to determine whether the samples differed in their physicochemical composition and textural attributes, the data were analyzed using the Student's t test for two samples, assuming equal variances. It was determined whether the samples' specific differences were significant, with a confidence interval (CI) of at least 95%. The software used for analysis was Microsoft Office Excel 2019.

3. Results and Discussion

3.1. Physicochemical Analysis of Milk

Table 2 shows the average physicochemical composition of the raw milk used in the production of costeño-type cheese.

Table 2. Physicochemical composition of milk used in the production of costeño-type cheese.

Composition	Raw Bovine Milk
Density (g/cm^3)	1.027 ± 0.010
Titratable acidity (°D)	14.20 ± 1.17
pH	6.70 ± 0.12
Fat (%, w/w)	3.20 ± 0.32
Total Solids (%, w/w)	10.70 ± 0.44

Mean values ± standard deviation.

3.2. Physicochemical Analysis in Costeño-Type Cheese

The reduction in NaCl content and the increase in the cooking temperature did not significantly affect the physicochemical composition of the costeño-type cheese (Figure 1). Reduction in the salt content of cheese generally causes an increase in the moisture due to the higher capacity of water retention in the protein matrix [5,7,13]. Grummer and T. C. Schoenfuss (2011) [34] attempted to produce reduced-salt cheese by standard processes and confirmed that moisture retention increases with salt reduction in cheese; this effect was observed in cheese Q_1.

Conversely, the experimental cheeses Q_2 and Q_3 with NaCl addition of 5.0% and 2.5%, respectively, showed higher similarity with the physicochemical composition of the control cheese (Q_C). This could be due to the increase in cooking temperature before draining the whey, which could increase hydrophobic interactions. This would cause more syneresis and contraction of the curd, decreasing the amount of moisture retained in the matrix of cheese [14,22,23].

The effect of temperature on moisture retention has been previously reported by Ganesan et al. [5]. They achieved the same moisture content in mozzarella production by decreasing the temperature during stretching with reduced salt.

The fat content tends to decrease in Q_1 cheese (*p*-value < 0.01) made through the standard process and reduced NaCl content. This tendency contrasts with the cheeses made with increased cooking temperature and salt mixtures. Salt reduction increases the moisture content, thus decreasing the fat content [31]. Moreover, within the standard cheese-making process, fat reduction also increases moisture content [32].

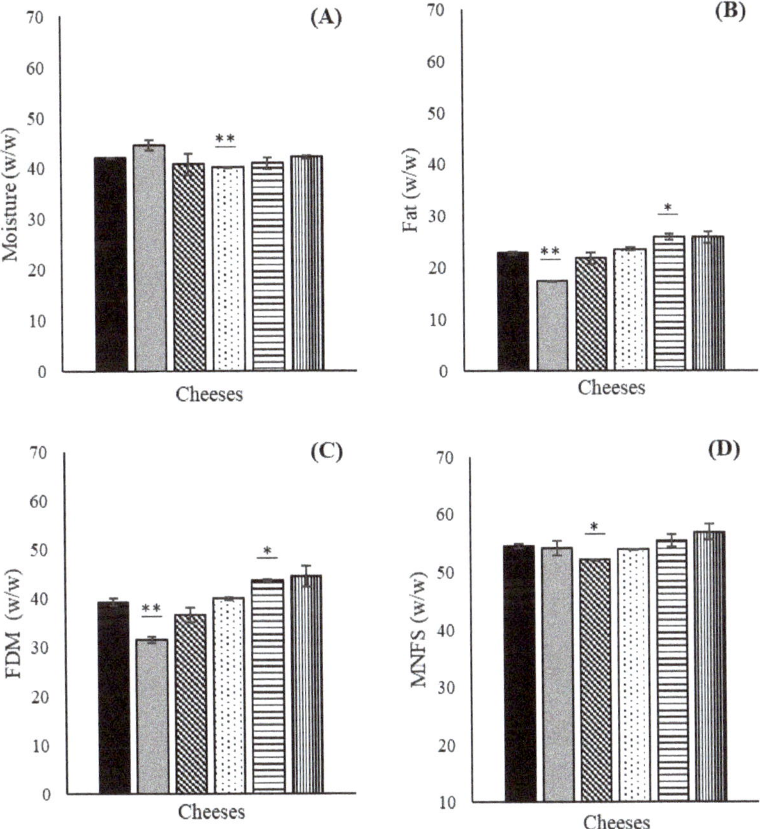

Figure 1. Physicochemical composition in percent of the experimental costeño-type cheese: (**A**) Moisture; (**B**) Fat; (**C**) Fat-in-dry-matter (FDM); (**D**) Moisture-in-non-fat-substances (MNFS) in Q_C (black bar), Q_1 (gray bar), Q_2 (diagonal striped bar), Q_3 (point bar), Q_4 (horizontal striped bar) and Q_5 (vertical striped bar). Error bars represent one standard deviation. Student's t-tests were performed to assess significance compared to the control cheese (Q_C), where (*) corresponds to a p-value < 0.05, and (**) to a p-value < 0.01.

Cheeses with mixtures of NaCl and KCl showed an increase in fat content compared to cheeses with only NaCl. This behavior indicates a possible interaction between the added KCl and the amount of fat trapped in the Q_4 and Q_5 cheeses' protein networks. This phenomenon requires further research outside the scope of this study because several studies indicate that the fat content is not affected by the different salt mixtures in the cheeses compared to the control cheese and the cheeses made by standard process [7,13,35].

All the experimental cheeses could be classified as semi-fat, because their FDM content <25%, and hard cheese, because the MNFS content was between 49 and 56% [36]. In hard cheeses like cheddar, FDM content decreases when the moisture increases; this behavior is similar to the one observed in Q_1 cheese [26,36].

Overall, slight differences were observed in the cheeses' physicochemical properties, with changes in the salt content and the cheese-making process.

3.3. Texture Profile Analysis

Figure 2 shows the texture attributes evaluated instrumentally in the experimental costeño-type cheeses. Control cheese (Q_C) had the highest hardness and cohesiveness compared with reduced-salt cheese.

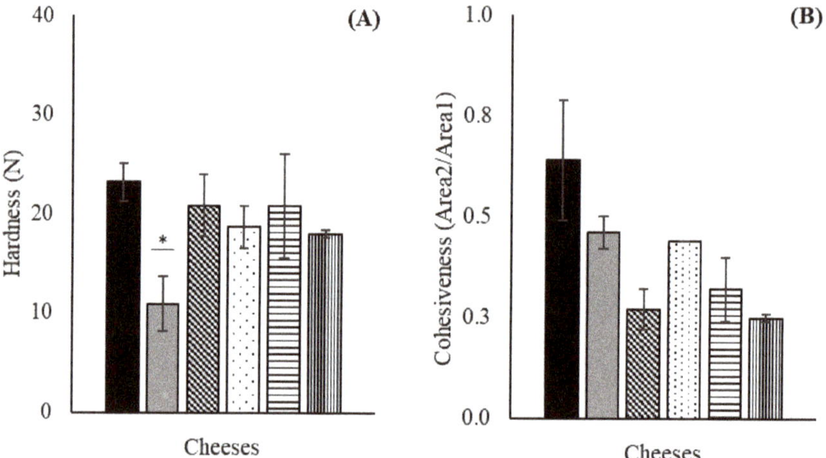

Figure 2. Texture attributes of costeño-type cheese: (**A**) hardness; (**B**) cohesiveness. Q_C (black bar), Q_1 (gray bar), Q_2 (diagonal striped bar), Q_3 (point bar), Q_4 (horizontal striped bar), and Q_5 (vertical striped bar). Error bars represent standard deviation. Student's t-tests were performed to assess significance compared to the control cheese (Q_C), where (*) corresponds to a *p*-value < 0.05.

Pastorino et al. (2003) [37] reported that the decrease in the salt content of muenster cheese decreased its hardness. This effect was seen in the hardness results reported in this study, except for Q_1 (*p*-value < 0.05). Studies report that the firmness of the cheese increases as the level of fat decreases. This phenomenon indicates that the moisture content present in Q_1 (Figure 1) and the adjustment of the fat globules within the protein matrix may have affected this cheese [18,37]. The hardness also significantly increases as the moisture content decreases, and the FDM content increases [14]. This phenomenon was observed in this study (Table 1).

On the other hand, Q_2 cheese with lower addition of NaCl compared to Q_1 had a hardness with values similar to those reported in the control cheese (Q_C) and the cheeses with partial substitution of NaCl for KCl (Q_4 and Q_5). These similarities could be due to the increase in cooking temperature applied during the making of these cheeses (Table 1) [10,14,24].

McSweeney (2007) [38] indicated that the cooking process increases the hydrophobic junction's relative strength, resulting in particles' aggregation into more extensive and easily processed curds with higher density and strength. Consequently, there is an increase in the degree of serum separation.

Concerning cohesiveness, Gunasekaran and Ak (2003) [32] reported that an increase in hardness creates more brittle and less cohesive cheese texture. Nevertheless, costeño-type cheese with maximum salt content (Q_C) had the highest cohesion than reduced-salt cheese and cheeses with NaCl and KCl. However, it is impossible to establish statistical differences in this texture variable due to the sample size.

Partial substitution of NaCl with KCl resulted in a difference in cohesiveness compared to control cheese (Q_C). Various studies have demonstrated that mixtures of NaCl and KCl affect the cheese's texture [7,8,13]. KCl has a decreased ionic strength towards NaCl, which results in a decrease in the "salinity" or solubility of proteins, with a direct effect on the cheese matrix [8].

3.4. Rheological Analysis

An increase in the storage and loss modulus was evident and proportional to the increase in frequency, demonstrating dominant viscoelastic properties (Figure 3A,B). These observations agree with previous rheological research in cheeses [31,39]. Figure 3 also shows a correlation between rheological behavior and NaCl's addition in the cheese—except for Q_1, which showed low storage modulus compared to Q_2, which had a smaller NaCl addition.

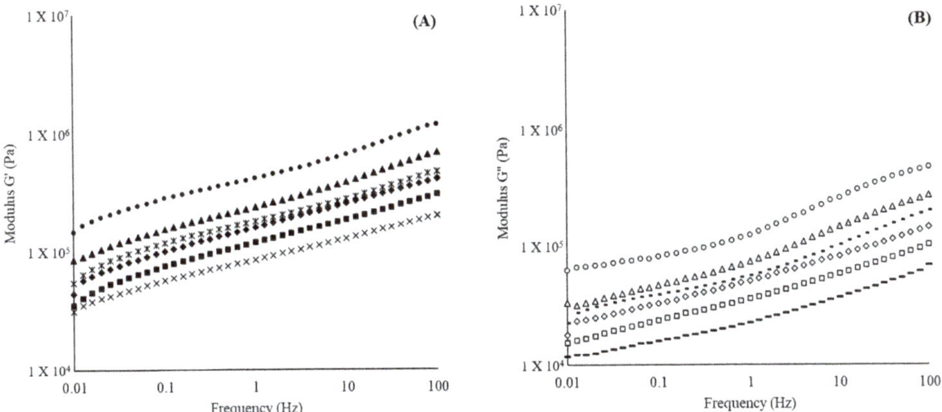

Figure 3. Frequency sweep in experimental costeño-type cheese. (**A**) The storage modulus (G′) for the cheeses Q_C (black circles), Q_1 (black squares), Q_2 (black diamonds), Q_3 (crosses), Q_4 (black diamonds) and Q_5 (asterrisks), and (**B**) loss modulus (G″) for the cheeses Q_C (white circles), Q_1 (white squares), Q_2 (white diamonds), Q_3 (short dashes), Q_4 (white triangles) and Q_5 (long dashes).

The effect of salt on cheese protein hydration affects the cheese's viscoelastic character [18]. Some studies have held compositional parameters constant and shown a role for salt in cheese rheology primarily because of its effect on protein hydration [13].

The temperature at which the curd is heated affects the rheological properties to a certain extent [32]. Therefore, increasing the cooking temperature before draining the whey could influence the rheological behavior of the cheeses Q_2 and Q_3. Lucey et al. (2003) [14] stated a more rapid relaxation of protein bonds at high temperatures and a change to a more liquid character, indicating a decrease in the modules and a greater probability that the chemical bonds will break.

As part of the cheese-making process, the cooking temperature correlates with the rheological behavior and the previously mentioned texture attributes.

3.5. Scanning Electron Microscopy

Figure 4A–F shows the microstructure of the experimental costeño-type cheeses. The cheese microstructure can be viewed as a continuous protein gel network disrupted with interspersed fat globules [37,40]. The scanning electron micrographs showed that all the experimental costeño-type cheese had a discontinuous structure with numerous irregular cavities. Similar observations have previously been reported by Tunick and Van Hekken [28] in fresh Mexican cheese.

From a rheological viewpoint, the structural discontinuities may result in the lack of tensile strength in many kinds of cheese, which in practical terms may be reflected as crumbliness, shortness, and fracturability, as occurs in cheeses such as Feta, Stilton, and Cheshire [41].

The average pore size of the cheese microstructures evidences a variability in the pore distribution of the microstructures (Figure 4G). The cheeses Q_C, Q_4, and Q_5 had more homogeneous pore sizes (Figure 4A,E,F) compared to cheeses with less NaCl addition (Q_1, Q_2, and Q_3) (Figure 4B–D). This pore variability can be attributed to changes in the texture of the cheeses.

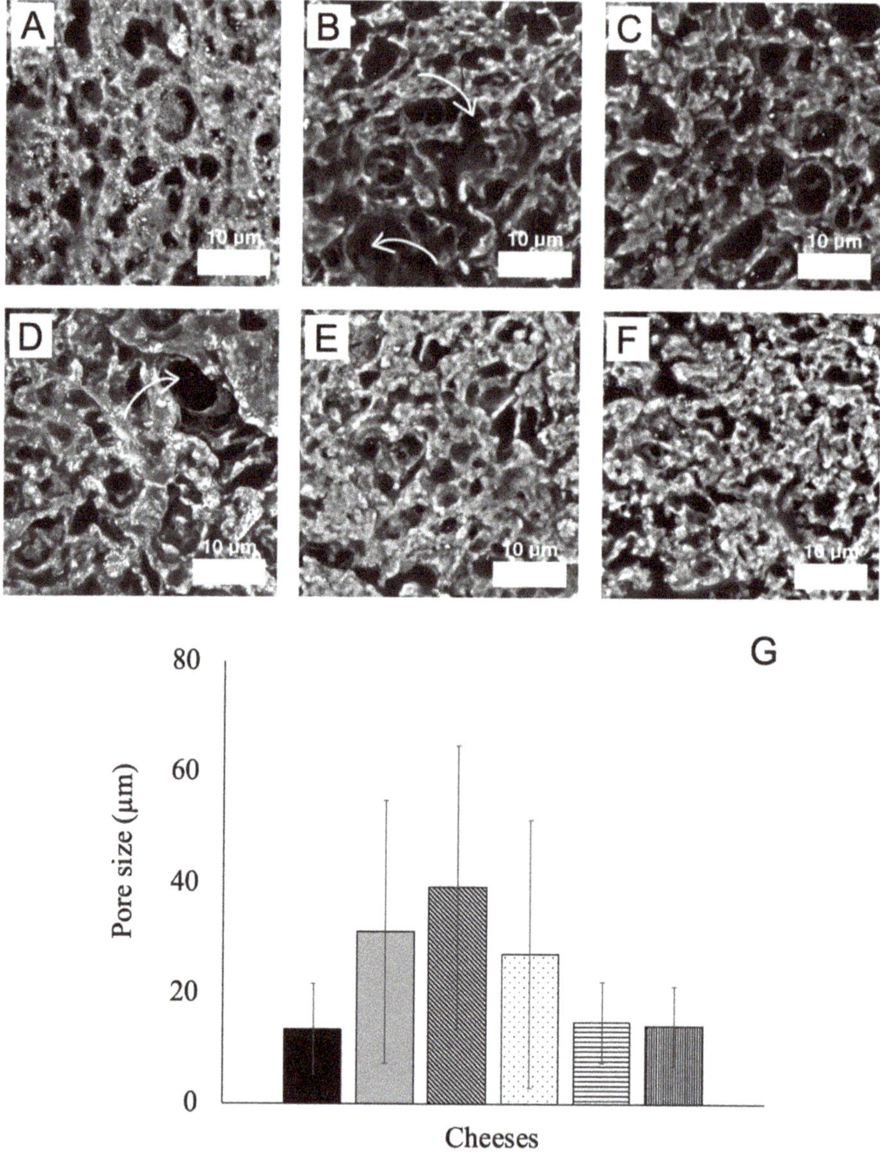

Figure 4. Scanning electron micrograph of costeño-type cheese at 10 μm scale at 4500× . Qc (**A**), Q_1 (**B**), Q_2 (**C**), Q_3 (**D**), Q_4 (**E**), Q_5 (**F**), and average pore size (**G**), where error bars represent one standard deviation. The light areas in the micrograph are the protein phase, and the dark areas are the fat or serum phase. Arrows indicate coalesced fat globules.

Lamichhane et al. (2018) [22] indicated that temperature influences the cheese structure through its effect on the components of cheese and their interactions, including changes in the physical state of fat. The molecular interactions between the casein show a relationship between the influence of the increase in cooking temperature on the microstructure of these cheeses. However, Q_2 and Q_3 cheeses' microstructures do not establish a relationship between increasing the cooking temperature on the cheese's microstructure.

Cheese Q_1 showed a microstructure with coalescences of fat globules observed as larger and irregularly shaped openings and discontinuities in the para-casein matrix (Figure 4B) [41]. Coalescence of fat globules occurs during manufacturing because of the combined effects of shear stress on the fat globule membrane and shrinkage of the surrounding para-casein matrix, which forces the occluded globules into close contact [41].

4. Conclusions

Cheeses with salt reduction generally show alterations in their structure. Increasing the cooking temperature during the Q_2 and Q_3 cheese production influenced their physicochemical composition and textural and rheological properties. These cheeses, with a reduction in NaCl of 5% and 2.5%, respectively, showed similarities with the physicochemical composition and textural properties of control cheese (Q_C). This could be possible without partial substitution of NaCl with KCl. This study established that the increase in cooking temperature before the whey drainage stage during the cheese making can also influence macrostructural properties (rheology and texture).

Furthermore, applying a multiscale approach allowed the macro and microstructural properties' responses to be correlated with the process, the product, and the properties. The results evidenced existing changes between the texture variables and the rheological behavior when making variations in the NaCl addition and the manufacturing process of the costeño-type cheese.

Author Contributions: Conceptualization, M.L.D.-B. and O.A.A.S.; methodology, M.L.D.-B. and O.A.A.S.; validation, L.H.R. and O.A.A.S.; formal analysis, M.L.D.-B., L.H.R. and O.A.A.S.; investigation, M.L.D.-B.; writing—original draft preparation, M.L.D.-B.; writing—review and editing, M.L.D.-B., L.H.R. and O.A.A.S.; visualization, M.L.D.-B.; supervision, O.A.A.S. All authors have read and agreed to the published version of the manuscript.

Funding: This research received no external funding.

Acknowledgments: The authors would like to thank the Department of Chemical and Food Engineering at Universidad de Los Andes for providing funding and access to laboratory infrastructure. All authors acknowledge financial provided by the Vice Presidency for Research & Creation publication fund at the Universidad de los Andes.

Conflicts of Interest: The authors declare no conflict of interest.

References

1. Doyle, M.E.; Glass, K.A. Sodium Reduction and Its Effect on Food Safety, Food Quality, and Human Health. *Compr. Rev. Food Sci. Food Saf.* **2010**, *9*, 44–56. [CrossRef]
2. He, F.; Brown, M.; Tan, M.; MacGregor, G.A. Reducing population salt intake-An update on latest evidence and global action. *J. Clin. Hypertens.* **2019**, *21*, 1596–1601. [CrossRef]
3. World Health Organization. Guideline: Sodium Intake for Adults and Children. Available online: https://www.who.int/nutrition/publications/guidelines/sodium_intake_printversion.pdf (accessed on 20 August 2020).
4. Arboatti, A.S.; Olivares, M.; Sabbag, N.G.; Costa, S.C.; Zorrilla, S.E.; Sihufe, G.A. The influence of sodium chloride reduction on physicochemical, biochemical, rheological and sensory characteristics of Mozzarella cheese. *Dairy Sci. Technol.* **2014**, *94*, 373–386. [CrossRef]
5. Ganesan, B.; Brown, K.; Irish, D.A.; Brothersen, C.; McMahon, N.J. Manufacture and sensory analysis of reduced and low-sodium Cheddar and Mozzarella cheeses. *J. Dairy Sci.* **2014**, *97*, 1970–1982. [CrossRef] [PubMed]
6. Cruz, A.G.; Faria, J.A.; Pollonio, M.A.; Bolini, H.M.; Celeghini, R.M.; Granato, D.; Shah, N.P. Cheeses with reduced sodium content: Effects on functionality, public health benefits and sensory properties. *Trends Food Sci. Technol.* **2011**, *22*, 276–291. [CrossRef]
7. Carmi, I.K.; Benjamin, O. Reduction in sodium content of fresh, semihard Tzfat cheese using salt replacer mixtures: Taste, texture and shelf-life evaluation. *Int. J. Dairy Technol.* **2016**, *70*, 354–364. [CrossRef]
8. Gomes, A.P.; Cruz, A.G.; Cadena, R.S.; Celeghini, R.M.S.; Faria, J.A.F.; Bolini, H.M.A.; Pollonio, M.A.R.; Granato, D. Manufacture of low-sodium Minas fresh cheese: Effect of the partial replacement of sodium chloride with potassium chloride. *J. Dairy Sci.* **2011**, *94*, 2701–2706. [CrossRef]

9. Grummer, J.; Bobowski, N.; Karalus, M.; Vickers, Z.; Schoenfuss, T. Use of potassium chloride and flavor enhancers in low sodium Cheddar cheese. *J. Dairy Sci.* **2013**, *96*, 1401–1418. [CrossRef]

10. Ayyash, M.; Sherkat, F.; Shah, N. The effect of NaCl substitution with KCl on Akawi cheese: Chemical composition, proteolysis, angiotensin-converting enzyme-inhibitory activity, probiotic survival, texture profile, and sensory properties. *J. Dairy Sci.* **2012**, *95*, 4747–4759. [CrossRef]

11. Horita, C.; Morgano, M.; Celeghini, R.; Pollonio, M. Physico-chemical and sensory properties of reduced-fat mortadella prepared with blends of calcium, magnesium and potassium chloride as partial substitutes for sodium chloride. *Meat Sci.* **2011**, *89*, 426–433. [CrossRef]

12. Grummer, J.; Karalus, M.; Zhang, K.; Vickers, Z.; Schoenfuss, T. Manufacture of reduced-sodium Cheddar-style cheese with mineral salt replacers. *J. Dairy Sci.* **2012**, *95*, 2830–2839. [CrossRef] [PubMed]

13. Guinee, T.P.; Fox, P.F. *Salt in Cheese: Physical, Chemical and Biological Aspects BT-Starter Cultures: General Aspects*, 4th ed.; Elsevier Ltd.: Amsterdam, The Netherlands, 2004; Volume 1.

14. Lucey, J.; Johnson, M.; Horne, D. Invited Review: Perspectives on the Basis of the Rheology and Texture Properties of Cheese. *J. Dairy Sci.* **2003**, *86*, 2725–2743. [CrossRef]

15. Acevedo, D.; Jaimes, J.D.; Espitia, C.R. Efecto de la Adición de Lactosuero al Queso Costeño Amasado. *Inf. Tecnológica* **2015**, *26*, 11–16. [CrossRef]

16. Serpa, J.G.; Pérez, T.I.; Hernández, E.J. Effect of pasteurization and starter cultures on physicochemical and microbiological properties of costeño cheese. *Rev. Fac. Nac. Agron. Medellín* **2016**, *69*, 8007–8014. [CrossRef]

17. Tirado, D.F.; Acevedo, D.; Torres-Gallo, R. Effect of deformation history on the stress relaxation behaviour of Colombian Caribbean coastal cheese from goat milk. *Food Sci. Technol. Int.* **2018**, *24*, 487–496. [CrossRef] [PubMed]

18. Floury, J.; Camier, B.; Rousseau, F.; Lopez, C.; Tissier, J.-P.; Famelart, M.-H. Reducing salt level in food: Part 1. Factors affecting the manufacture of model cheese systems and their structure–texture relationships. *LWT* **2009**, *42*, 1611–1620. [CrossRef]

19. Cussler, E.L.; Moggridge, G.D.; Cussler, E.L.; Moggridge, G.D. An Introduction to Chemical Product Design. In *Chemical Product Design*; Cambridge University Press: Cambridge, UK, 2012; pp. 1–16.

20. Mora, H.B.; Piñeros, M.; Moreno, D.E.; Restrepo, S.R.; Jaramillo, J.C.; Solano, Ó.A. Álvarez; Fernandez-Niño, M.; Barrios, A.G. Multiscale design of a dairy beverage model composed of Candida utilis single cell protein supplemented with oleic acid. *J. Dairy Sci.* **2019**, *102*, 9749–9762. [CrossRef]

21. Pradilla, D.; Vargas, W.L.; Alvarez, O.A. The application of a multi-scale approach to the manufacture of concentrated and highly concentrated emulsions. *Chem. Eng. Res. Des.* **2015**, *95*, 162–172. [CrossRef]

22. Lamichhane, P.; Kelly, A.L.; Sheehan, J.J. Symposium review: Structure-function relationships in cheese. *J. Dairy Sci.* **2018**, *101*, 2692–2709. [CrossRef]

23. McSweeney, P.L.H.; Fox, P.F. *Advanced Dairy Chemistry*; Springer: New York, NY, USA, 2009; ISBN 9780387848648.

24. Ho, Q.T.; Carmeliet, J.; Datta, A.K.; Defraeye, T.; Delele, M.A.; Herremans, E.; Opara, L.; Ramon, H.; Tijskens, E.; Van Der Sman, R.; et al. Multiscale modeling in food engineering. *J. Food Eng.* **2013**, *114*, 279–291. [CrossRef]

25. Ballesta Rodriguez, I. Evaluación de la Calidad del Queso Costeño Elaborado con Diferentes Tipos de Cuajo (Animal y Microbiano) y la Adición o no de Cultivos Lácticos (Lactococcus Lactis Subps. Lactis y Lactococcus Lactis Subps. Cremoris). Master's Thesis, Universidad Nacional De Colombia Sede Medellin Facultad, Medellín, Columbia, 2014.

26. *AOAC Official Methods of Analysis of AOAC International*, 16th ed.; AOAC: Rockville, MD, USA, 1995; Volume 2.

27. *AOAC Official Methods of Analysis of AOAC International*; AOAC: Rockville, MD, USA, 2005; ISBN 0935584544.

28. Tunick, M.H.; Van Hekken, D.L. Rheology and Texture of Commercial Queso Fresco Cheeses Made from Raw and Pasteurized milk. *J. Food Qual.* **2010**, *33*, 204–215. [CrossRef]

29. Bähler, B.; Back, R.; Hinrichs, J. Evaluation of oscillatory and shear strain behaviour for thermo-rheological plasticisation of non-ripened cheese curd: Effect of water, protein, and fat. *Int. Dairy J.* **2015**, *46*, 63–70. [CrossRef]

30. Katsiari, M.; Voutsinas, L.; Alichanidis, E.; Roussis, I. Manufacture of Kefalograviera cheese with less sodium by partial replacement of NaCl with KCl. *Food Chem.* **1998**, *61*, 63–70. [CrossRef]

31. Foegeding, E.; Drake, M. Invited Review: Sensory and Mechanical Properties of Cheese Texture. *J. Dairy Sci.* **2007**, *90*, 1611–1624. [CrossRef]

32. Gunasekaran, S.; Ak, M.M. *Cheese Rheology and Texture*; CRC Press: Boca Raton, FL, USA, 2003; ISBN 1587160218.

33. Castro, A.C.; Novoa, C.F.; Algecira, N.; Buitrago, G. *Reología y Textura de Quesos Bajos en Grasa*; Revista de Ciencia y Tecnología: Bogotá, Colombia, 2014; Volume 16, pp. 58–66.

34. Grummer, J.; Schoenfuss, T. Determining salt concentrations for equivalent water activity in reduced-sodium cheese by use of a model system. *J. Dairy Sci.* **2011**, *94*, 4360–4365. [CrossRef] [PubMed]

35. Kamleh, R.; Olabi, A.; Toufeili, I.; Daroub, H.; Younis, T.; Ajib, R. The effect of partial substitution of NaCl with KCl on the physicochemical, microbiological and sensory properties of Akkawi cheese. *J. Sci. Food Agric.* **2015**, *95*, 1940–1948. [CrossRef]

36. Codex Alimentarius. Codex Standard for Milk and Milk Products. Group standard for cheese including fresh cheese CODEX STAN 221-2001. General standard for cheese CODEX STAN 283-1978. 2011. Available online: http://www.fao.org/home/en/ (accessed on 23 November 2020).

37. Pastorino, A.; Hansen, C.; McMahon, D.J. Effect of Salt on Structure-Function Relationships of Cheese. *J. Dairy Sci.* **2003**, *86*, 60–69. [CrossRef]

38. McSweeney, P.L.H. *Cheese Problems Solved*; Elsevier: Amsterdam, The Netherlands, 2007; ISBN 9781845690601.

39. Brown, J.; Foegeding, E.A.; Daubert, C.; Drake, M.; Gumpertz, M. Relationships Among Rheological and Sensorial Properties of Young Cheeses. *J. Dairy Sci.* **2003**, *86*, 3054–3067. [CrossRef]

40. Rogers, N.; McMahon, D.; Daubert, C.; Berry, T.; Foegeding, E.A. Rheological properties and microstructure of Cheddar cheese made with different fat contents. *J. Dairy Sci.* **2010**, *93*, 4565–4576. [CrossRef]

41. O'Callaghan, D.J.; Guinee, T.P. Rheology and Texture of Cheese BT-Starter cultures: General aspects. In *Cheese: Chemistry, Physics and Microbiology*, 3rd ed.; Fox, F.P., McSweeney, P.L.H., Cogan, T.M., Guinee, T.P., Eds.; Elsevier: Amsterdam, The Netherlands, 2004; Volume 1, pp. 511–540.

Publisher's Note: MDPI stays neutral with regard to jurisdictional claims in published maps and institutional affiliations.

applied sciences

Article

Effectiveness of Bacteriocin-Producing Lactic Acid Bacteria and *Bifidobacterium* Isolated from Honeycombs against Spoilage Microorganisms and Pathogens Isolated from Fruits and Vegetables

Chrysa Voidarou [1], Athanasios Alexopoulos [2], Anastasios Tsinas [1], Georgios Rozos [2], Athina Tzora [1], Ioannis Skoufos [1], Theodoros Varzakas [3,*] and Eugenia Bezirtzoglou [4]

1   School of Agriculture, University of Ioannina, 47100 Arta, Greece; xvoidarou@uoi.gr (C.V.); actsinas@uoi.gr (A.T.); tzora@uoi.gr (A.T.); jskoufos@uoi.gr (I.S.)
2   Department of Agricultural Development, Laboratory of Microbiology, Biotechnology and Hygiene, Democritus University of Thrace, 67100 Xanthi, Orestiada, Greece; alexopo@agro.duth.gr (A.A.); clevervet@hotmail.com (G.R.)
3   Department of Food Science and Technology, University of the Peloponnese, Antikalamos, 24100 Kalamata, Greece
4   Medical School, Laboratory of Hygiene and Environmental Protection, Democritus University of Thrace, 67100 Alexandroupolis, Greece; empezirt@yahoo.gr
*   Correspondence: theovarzakas@yahoo.gr or t.varzakas@us.uop.gr; Tel.: +30-2721045279

Received: 29 August 2020; Accepted: 15 October 2020; Published: 19 October 2020

Abstract: Screening natural products for bacteriocin-producing bacteria may be the equilibrium point between the consumer demand for mild processing and the industry's need for hazard control. Raw unprocessed honeycombs filled with oregano honey from the alpine mountainous territory of Epirus, Greece were screened for bacteriocinogenic lactic acid bacteria and *Bifidobacterium* spp., with inhibitory action towards some pathogens and spoilage microorganisms isolated from fresh fruits and vegetables (number and type of strains: three *E. coli*, two *L. monocytogenes*, two *Salmonella* spp., two *B.cereus*, two *Erwinia* spp., one *Xanthomonas* spp., *L. innocua* (ATCC 33090TM) and *E. coli* 0157:H7 (ATCC 69373)). Among the 101 collected isolates (73 *Lactobacillus*, 8 *Lactococcus*, 8 *Leuconostoc* and 12 *Bifidobacterium* species) from the oregano honeycombs (an original finding since there are no other reports on the microbial biodiversity of the flora of the oregano honey), 49 strains of lactic acid bacteria (LAB) and *Bifidobacterium* spp. were selected and tested for their bacteriocin-producing capacity (34 *Lactobacillus*, 6 *Lactococcus*, 5 *Leuconostoc* and 4 *Bifidobacterium*). The antibacterial activity exerted by the tested LAB and *Bifidobacterium* strains was not of the same potency. Our results suggest that the main molecules involved in the antimicrobial activity are probably bacteriocin-like substances (a conclusion based on reduced antibacterial activity after the proteolytic treatment of the cell-free supernatant of the cultures) and this antimicrobial activity is specific for the producing strains as well as for the target strains. The spoilage bacteria as well as the reference microorganisms showed increased resistance to the bacteriocin-like substances in comparison to the wild-type pathogens.

Keywords: LAB; *Bifidobacterium*; BLS; fruits; vegetables; Oregano honey

---

1. Introduction

According to EUROSTAT (ec.europa.eu) surveys, half of the EU population eats at least one portion of fruits and vegetables on a daily basis and this trend is increasing (the southern states are in the top three EU Member States in daily intake of fruit: Italy (85%), Portugal (81%) and Spain (77%)) [1]. Unsurprisingly, the WHO/FAO Joint Expert Consultation Report on Diet, Nutrition and the Prevention of Chronic Diseases states that at least 400 g of fruits and vegetables per day are needed to prevent heart disease, certain types of cancer, diabetes and obesity [2]. Given these statistics, important food quality and food safety issues emerge. Fruits and vegetables are rich in humidity and carbohydrate content and hence characterized as ideal habitats for bacterial and fungal growth [3]. Among others, factors such as the presence of various parasites and the quality and origin of manure, harvest and postharvest treatment and skin lesions are of utmost importance for their contamination with pathogens and spoilage microorganisms. Various studies have shown the presence of *Salmonella* spp., *E. coli* O157:H7, *Listeria monocytogenes*, *Bacillus cereus*, *Campylobacter* spp., *Yersinia enterocolitica*, *Clostridium botulinum* and other pathogens, in addition to some viruses and parasites in fruits and vegetables [4,5]. Spoilage bacteria such as *Erwinia carotovora* and *Xanthomonas campestris* cause various types of lesions regarding texture and color thus reducing their quality and commercial value [6].

There are modern trends in food processing concerning food safety and prolonging shelf life. Consumers prefer minimally processed foods that are free of chemical preservatives. Given these demands, and the increasing resistance of pathogens and spoilage bacteria to antibiotics and other chemicals, the food industry is seeking alternative means of food preservation [7,8].Consequently, there is an increasing interest in so-called "green technologies," including novel approaches to the minimal processing of food as well as the use of microbial metabolites such as bacteriocins on an industrial scale for "biopreservation" [9].

In order to control, or even prevent, the growth of such microorganisms, the application of bacteriocins-producing lactic acid bacteria (LAB) has been proposed [10]. LAB can be isolated from many raw wild fruits, vegetables and flowers [11,12]. Moreover, lactic and acetic acid bacteria can promote the spoilage of fruits, vegetables, fruit juices and beverages as residents at the outer layer of the skin of fruits and vegetables [6]. These bacteria are known for the fermentation of carbohydrates and the production of various organic acids such as lactic acid, which significantly lower the pH of fermented foods [12]. They also produce other compounds with antimicrobial action such as hydrogen peroxide, acetaldehyde and bacteriocins [13–15].

Bacteriocins are peptides with natural antibacterial activity, and there is a reason for their comparison to antibiotics. Many researchers propose the term "biological preservatives of foods" and stress the fact that bacteriocins are not used for clinical therapeutic purposes as antibiotics are [16]. They are synthesized in the ribosomes of the bacterial cell and secreted extracellularly. Bacteriocins are most effective against Gram-positive bacteria. The spectrum of their activity varies from narrow (against one species) to broad (against several species) [17,18]. The main advantage of bacteriocins is that their presence does not change the sensorial characteristics of foods. Their usage enables the reduction of the intensity of other means of preservation such as heat. These two characteristics make bacteriocins' application compatible with modern consumer demands for the minimal and more natural treatment of foods [19]. Bacteriocins can be preferably added to foods as compounds rather than cultures of bacteriocinogenic LAB because, in the latter case, LAB can ferment the carbohydrates of foods [20].

Nisin, pediocin, enterocin AS-48, bovicin, enterocin 416K1 and bificin C6165 are some bacteriocins already tested against spoilage bacteria and pathogens but only the first two were granted approval as food additives. However, they are mostly used in other foods and their usage in the fruit and vegetable industry is still limited [19].

LAB have been established as "generally regarded as safe" (GRAS), a fact that makes them attractive candidates for industrial utilization [21,22]. There is ongoing research for new strains with potentially superior properties, such as being probiotic, and the production of active bacteriocins. Sources of such LAB are various natural products, which could be an ideal ecological niche for these microorganisms. Oregano honey is a very special and extremely rare natural product. It is produced by honeybees grazing on wild oregano plants in the alpine mountainous territory of Epirus, Greece. It is difficult to find it since there are very few producers. Unlike other types of honey, oregano honey has a bitter taste. It is consumed by rural populations of the area and data from local people suggest they believe oregano honey possesses therapeutic properties against various infections and gastrointestinal disorders.

The aim of this study is to screen the diverse autochthonous microbiota isolated from the honeycombs of oregano-grazing bees for LAB and *Bifidobacterium* strains and determine the antibacterial activity of these bacteriocin-producing isolates against some pathogens and spoilage microorganisms isolated from fresh fruits and vegetables.

2. Materials and Methods

2.1. Sample Collection

2.1.1. Fruits and Vegetables

All samples were purchased from open fruit markets in different areas of Epirus, Greece. In total, 20 each of pears, apples, peaches, tomatoes, cucumbers and red peppers, 40 white cabbages and 60 carrots were collected in sterile plastic bags and brought within an hour to the laboratory for analysis. The samples were of reasonably good quality and chosen for their slight lesions or minor skin ruptures (Figure 1). Fifty commercially sealed bags of minimally processed fresh-cut products ready-to-use (RTU) were sampled from various supermarkets and other retail stores.

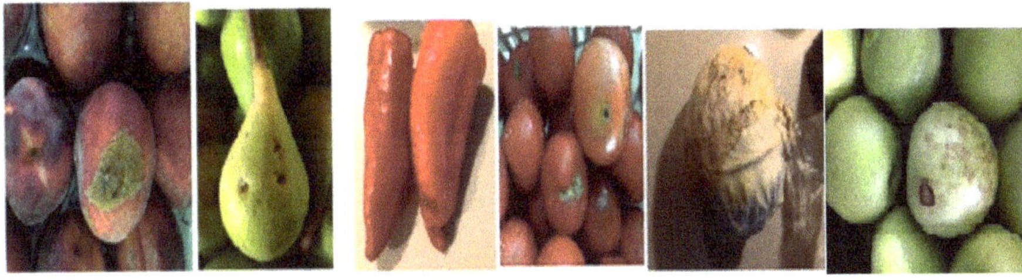

Figure 1. Photos of the samples. Their quality was reasonably good and they were chosen for their slight lesions and minor skin ruptures in order to isolate pathogens and spoilage bacteria.

2.1.2. Honeycomb Filled with Oregano Honey

A total of 30 raw unprocessed honeycombs filled with oregano honey, were received from local producers. A portion of 1000 g samples (from each local producer) were weighed and placed in a dark, sterile glass container under sterile conditions and homogenized. The samples did not contain any additives or diluents and had not been previously heated. They were evaluated for their microbiological quality by inoculation into blood agar (Columbia agar base with 5% sheep blood, Becton Dickinson), egg yolk agar, mannitol-egg yolk-polymyxin (MYP) agar and incubated aerobically at 30 °C and 37 °C for 48 h. Samples that showed growth of bacteria or growth of more than 4–5 colonies of yeasts were excluded from the study; only 4 samples were excluded from our study. The samples were stored at 5 °C in the dark in the Laboratory of School of Agricultural Science (University of Ioannina) to prevent photodegradation until being used.

2.2. Isolation of Spoilage and Pathogens Organisms

The bacterial flora of fruits or vegetables was collected by the following method. The entire product was placed in a sterile sample bag and a sufficient volume of enrichment broth (sterile 0.1% buffered peptone water (BPW, Oxoid)) was added in order to totally submerge the fruit or vegetable. The samples were then agitated and rubbed separately in the nylon sterile bag for 10 min to suspend surface microbes [23]. At the same time, the areas and tissues with lesions were removed using a sterile knife and placed into the same sterile bag as above. Then, various dilutions under aseptic conditions were made and different selective media and incubation conditions were used to isolate specific bacterial species, as described in the following sections. After incubation, the numbers of CFU were counted and different types of colonies were isolated. The distinct colonies were screened and selected on the basis of morphology and cultural characteristics and identified on the basis of standard tests.

2.2.1. Detection of *E.coli*

To enumerate the *E. coli*, a technique employing two media(eosin methylene blue (EMB) agar (Merck) at 37 °C for 24 h and chromogenic medium tryptone bile agar containing 5-bromo-4-chloro-3-indolyl-B D-glucuronic acid (BCIG) (tryptone bile × glucuronide agar) (Merck)) was used. The second medium relies on the use of bile salts and an elevated incubation at 44 °C to suppress competitor organisms. The chromogenic TBX was used to indicate the presence of β-glucuronidase activity, which is common in 95% of *E. coli* strains [24].

2.2.2. Detection of *Salmonella* Spp.

A method based on primary enrichment in Erlenmeyer flasks containing 225 mL of lactose broth was used (the procedure consisted of very thorough stirring at first then leaving the flasks at room temperature for 60 min. pH was measured and adjusted to 6.8 ± 0.2 with 1N NaOH or 1N HCl followed by incubation at 35 ± 2 °C for 24 ± 2 h). Subsequently, a secondary selective enrichment took place with the use of two media (Rappaport–Vassiliadis (RV) medium and tetrathionate (TT) broth) and incubation at 42 ± 0.2 °C for 24 ± 2 h (circulating, thermostatically controlled water bath), incubated at 35 ± 2.0 °C for 24 ± 2 h, respectively. Finally, 10 μL from each incubated TT broth and RV medium were streaked on bismuth sulfite (BS) agar, xylose lysine deoxycholate (XLD) agar and Hektoen enteric (HE) agar. Then, using a range of biochemical tests, suspect *Salmonella* spp. isolates were identified using the VITEK system [23,25–27].

2.2.3. Detection of *Shigella* spp.

The enrichment *Shigella* broth containing 3.0 μg/mL of novobiocin was used and the plates were incubated anaerobically (using the Anaerocult® A system (Merck) in an anaerobic jar) at 44 °C for 20 h. Next, surface spreading took place (spread plate method) on MacConkey agar followed by anaerobic incubation at 42 °C for 24 h. All the isolated colonies were characterized biochemically [26].

2.2.4. Detection of *L. monocytogenes*

Detection began with the enrichment of *Listeria* spp. using the Fraser broth (Oxoid), which contains selective agents including nalidixic acid, acriflavine and lithium chloride, to suppress most of the competitors. Afterwards, the broth was streaked onto Listeria Palcam agar (Merck) and chromogenic medium (agar Listeria according to Ottaviani and Agosti (ALOA)) incubated at 37 °C for 24 h [28]. Characteristic colonies were Gram-stained and tested for motility, oxidase and catalase activity, followed by identification with the API Listeria system (bioMérieux®, Marcyl'Etoile, France) [24].

2.2.5. Detection of *P. fluorescens*

For *P. fluorescens*, the plate count agar (PCA) and cetrimide agar (bioMérieux®) were used. The incubation conditions were 35 °C and 48 h and 30 °C and 48 h, respectively. The isolates were stored at −20 °C for further tests, such as Gram-staining, the catalase, oxidase and motility tests, starch hydrolysis, fluorescent pigment and gelatin liquefaction [29].

2.2.6. Detection of *B. cereus*

For the qualitative detection of *B. cereus*, a procedure for strengthening the strains of this species took place with the use of trypticase-soy-polymyxin (TSB) broth (Oxoid) at 30 °C for 48 h. Then, cultures were streaked on the mannitol-egg yolk-polymyxin (MYP) agar plate (Oxoid) and chromogenic plate (BACARA™, bioMérieux) and incubated at 30 °C for 24 h. Colonies with a pink sparkle in blue or blue-green precipitation on the chromogenic plate were picked for further testing to identify the presence of *B. cereus* in the examined samples of fruits and vegetables (parasporal crystal observation root growth observation was conducted, along with hemolysis, catalase, motility, nitrate reduction, casein decomposition, lysozyme tolerance, glucose utilization and acetyl methyl alcohol tests) [30].

2.2.7. Detection of Other Spoilage Bacteria, Yeasts and Molds

For the estimation of other spoilage bacteria, yeasts and molds, the nutrient agar, oxytetracycline glucose yeast extract (OGYE) (Sigma-Aldrich, St. Louis, MO, USA), dichloran Rose Bengal chloramphenicol (DRBC) agar (Fluka, Germany) and a sterile fresh medium of potato dextrose agar (PDA) (Merck) were used and incubated at 28 °C for at least 5 days or until fungal proliferation occurred on the medium surface [31–33].

In the present study, crystal violet pectate (CVP) medium [34] was used. Surface spreading took place from a dilution series prepared in saline supplemented with 0.05% *w/v* ascorbic acid, followed by plating on the CVP medium [35].

2.2.8. Detection of *C. perfringens*

For the detection of *C. perfringens* (vegetative and spore forms), a pre-enrichment technique was performed in PBS for 2 h, followed by decimal dilutions in lactose sulfite broth (LS), which was the growth medium used. The composition of the broth as per the Council of Europe [36] is as follows: 5 g tryptic digest of casein, 2.5 g yeast extract, 2.5 g sodium chloride, 2.5 g lactose, 0.3 g L-cysteine hydrochloride and 1 L distilled water. The broth (9 mL) was distributed into screw cap tubes, which contained inverted Durham tubes, and sterilized by autoclaving at 115 °C for 20 min. After cooling to 45–50 °C, a filter-sterilized solution of 1.2% sodium metabisulfite (0.5 mL) and 1.0% ferric ammonium citrate (0.5 mL) was added to each tube. Reduced conditions in the medium were accomplished via boiling and due to the presence of cysteine. To examine the samples for vegetative cells, filter membranes were placed into tubes and incubation was performed aerobically in a water bath at 46 °C for 24 h. Additionally, a second aliquot (100 mL) of each sample was heated for 15 min at 80 °C for spore germination (and vegetative cell destruction), cooled under running water and filtered. The filter was then placed into the LS broth as described above. The interpretation of results (positive samples) was based on the resulting turbidity from lactose fermentation, the presence of iron sulfide (black precipitate) and the presence of gas (H_2S) visible in the inverted Durham tube within 24 h at 46 °C [37,38].

2.3. Isolation of LAB and Bifidobacterium

In order to investigate the LAB in the stored samples of honeycombs filled with oregano honey, 100 g of each honeycomb was weighed, placed in a sterile Stomacher blender bag under completely sterile conditions and homogenized for 2 min using a Stomacher 400 Circulator. Then, 900 mL of peptone saline diluent were added to give a total volume of 1 L (0.9% *w/v* NaCl, 0.1% *w/v* Tween 80 and 0.1% *w/v* peptone). The above homogenization technique was repeated for 2 min [39] and 10-fold serial dilutions were prepared in normal saline diluents. Aliquots of 0.1 mL of each 10-fold serial dilution were spread on modified MRS agar (Oxoid) adjusted to an agar content of 1.5% *w/v* (Agar, Merck) and supplemented with 0.3% *w/v* $CaCO_3$ (Merck) [40]. The plates for the recovery of facultative and strict anaerobe LAB and *Bifidobacterium* were incubated for 3–4 days at 37 °C under anaerobic conditions using anaerobic jars with Anaerocult® A gas packs (Merck) [41]. Following incubation, colonies from the modified MRS agar of a different morphology with a surrounding clear zone in the agar, indicating acid production, were transferred to the MRS agar without $CaCO_3$ to grow pure cultures under conditions as described above. In a general approach, cultures were identified as LAB species or *Bifidobacterium* by evaluating their morphological, cultural, physiological and biochemical characteristics by the procedures described in Bergey's Manual of Systematic Bacteriology. At least

10 selected isolates from each plate were initially identified based on the following: microscopic examination of Gram-stained cells, motility, catalase and oxidase reactions, gas production from glucose and growth in anaerobic conditions. Finally, phenotypic observations and biochemical tests were performed using the VITEK2 Compact System [42,43]. The presumptive LAB isolates were further characterized according to their registered protein and peptide information analysis results with the help of the Microflex LT MALDI-TOF MS (matrix-assisted laser desorption ionization time-of-flight mass spectrometry) (Bruker Daltonics, Bremen, Germany), which is a new technology for identifying LAB cultures. Mass spectra were processed using BioTyper software (version 3.0; Bruker Daltonics) running with the BioTyper database (version DB-5989). Matching between experimental MALDI-TOF MS profiles obtained from bacteria isolates and the reference MALDI-TOF MS profiles was expressed by BioTyper according to a Log(Score). Analyses were conducted in accordance with the manufacturer's instructions and performed twice for each isolate. Results (BioTyper score value, created by an automatic mass spectra comparison with the Bruker library) with a score value >2.3 were interpreted as identified at the species level with a high probability. A score value between 1.7 and 2.0 implies probability only at the genus level [44,45].

2.4. Determination of the Antibacterial Activity of LAB and Bifidobacterium against the Isolated Pathogens and Spoilage Bacteria

Preparation of the Cell-Free Supernatant (CFS)

In short, MRS broth (20 mL) was inoculated separately with the LAB isolates previously characterized and incubated anaerobically at 37 °C for 72 h. After incubation, a cell-free supernatant was obtained by centrifuging the bacterial culture at 10,000× g for 10 min at 4 °C, followed by filtration of the supernatant through a 0.22 µm syringe filter with a low protein binding capacity membrane (cellulose acetate, Rotilabo; syringe filter, Carl Roth).

For *Bifidobacterium* isolates, the procedure was carried out as follows. The strains were inoculated into the MRS broth at 1% (*v/v*). After 16 h of incubation at 37 °C, each culture supernatant was collected by centrifugation at 22,100× g for 30 min. The supernatant was sterilized by filtration through 0.45 and 0.22 µm poresize filters sequentially and precipitated with 70% ammonium sulfate. The precipitate was stored at 4 °C for 18 h. Subsequently, the precipitate was collected after centrifugation at 10,000× g for 30 min and dissolved in a minimum amount of deionized water.

The agar diffusion bioassay was used to screen for bacteriocin-producing LAB or *Bifidobacterium* strains among the 49 isolates from honeycomb filled with oregano honey, against selected (the most typical) bacterial pathogens and spoilage, isolated from fresh fruits and vegetables purchased in open markets and retail stores. These comprised: 3 *E. coli* strains, 2 *Salmonella* spp. strains, 2 *L. monocytogenes* strains, 2 *B. cereus* strains, 2 *Erwinia* spp. strains and 1 *Xanthomonas* spp. strain. Targeted indicator organisms were used as reference: *L. innocua* (ATCC 33090TM), *E. coli* 0157:H7 (ATCC 69373) [17,46].

One milliliter of each pathogen or spoilage fruits or vegetables organism (approximately 7×10^5 cfu/L) was inoculated into 15 mL of semisolid BHI agar (BHI broth powder), supplemented with 0.7% agar (Merck) maintained at 50 °C and then poured into a Petri dish. After solidification, three wells of 5 mm diameter and 2 mm depth were made in the agar. (a) In the first well, 50 µL of the untreated CFS aliquot from each LAB or *Bifidobacterium* isolate were added. (b) The CFS was treated further and was adjusted to pH 6.0 with 1 mol/L NaOH in order to rule out possible observed antibacterial activity caused by acid formation. Then, 50 µL of the pH-adjusted CFS were filtered and added to the second well. (c) The neutralized CFS described above was then treated with a final concentration of 1 mg mL^{-1} of catalase (Merck) at 20–25 °C for 30 min to eliminate the possible inhibitory action of H_2O_2. It was filtered and immediately was placed in the third well. In summary, if inhibition zones were present in the third well, the LAB or *Bifidobacterium* isolates were considered to be able to produce bacteriocin-like inhibitory substances (BLIS). The BHI plates were incubated aerobically at 37 °C for 24 h or anaerobically at 37 °C for 24 h (in order to obtain the ideal growth requirements for each strain). The inhibition zone was measured using an electronic caliper with a digital display [15,41]. Each screening test for the bacteriocin was performed in triplicate and the results are reported as mean values.

The isolated crude CFS was evaluated to discover whether inhibition was due to BLIS. For that purpose, the selected enzymes were tested on the cell-free supernatant. The CFS displaying antimicrobial potential after acid neutralization and H_2O_2 elimination was treated with proteolytic enzymes, among others, including trypsin, pepsin, lipase and α-amylase. The proteolytic enzymes were dissolved in 40 mM Tris-HCl (pH, 8.2), 0.002 M HCl (pH, 7) and 0.05 M sodium phosphate (pH, 7.0), respectively, to a concentration of 0.1 mg mL^{-1}, while lipase and α-amylase were dissolved in 0.1 M potassium phosphate (pH, 6.0) and 0.1 M potassium phosphate (pH, 7.0), respectively, to a final concentration of 0.1 mg/mL. Equal aliquots of both filter-sterilized CFS of each test strain and enzyme solution were mixed. Each enzyme was incubated at 37 °C for 2 h and heated to 100 °C for 5 min to inactivate the enzymes. These sample mixtures and controls (without enzyme treatment) were inoculated with the indicator strains *L. innocua* (ATCC 33090TM), *E. coli* 0157:H7 (ATCC 69373). The treated inoculated CFS aliquots were then incubated in a solution at 37 °C for 2 h, after which the retention of bacteriocin-like substances (BLS) in the treated samples was determined by the agar diffusion method as described above. If the antibacterial activity was negative after the enzymatic treatment, the isolate was regarded as positive for producing a bacteriocin [47,48].

2.5. Statistical Analysis

Inhibition zones from the susceptibility experiments were estimated in millimeters and categorized into four classes according to their diameter, i.e., Class I: 0–10 mm, Class II: 10.1–12 mm, Class III: 12.1–14 mm and Class IV: 14.1–18 mm. A Pearson's chi-squared test was used to screen for independence between categories and treatments of CFS (A–C, as described above) and classes of inhibition zones (I–IV) [49]. Analysis was performed using SPSS v. 19 (IBM Corp, Armonk, NY, USA) at a significance (alpha (α)) level of 0.05.

3. Results

Table 1 shows the isolated LAB and *Bifidobacterium* strains from the oregano honeycombs. In total, 101 strains were isolated belonging to 16 species, a novel finding regarding the biodiversity of the natural microflora of the oregano honeycombs. The majority of these strains were *Lactobacillus* ($n = 73$, 72.27%), followed by *Bifidobacterium* ($n = 12$, 11.88%), *Lactococcus* ($n = 8$, 7.92%) and *Leuconostoc* ($n = 8$, 7.92%).

For the purposes of the present study, 49 strains of LAB and *Bifidobacterium* were selected and tested for their bacteriocin-producing capacity. The selected strains had a higher BioTyper score value (not including *L. insectis*) than the others and thus were considered the most typical of their species. Of these strains, 34 belonged to the *Lactobacillus* genus, 6 to the *Lactococcus* genus, 5 to the *Leuconostoc* genus and 4 to the *Bifidobacterium* genus. The 34 *Lactobacillus* strains were classified into 12 genera as follows: three *L. insectis* strains, two *L. kunkeei* strains, three *L. casei* strains, three *L. fermentum* strains, two *L. curvatus* strains, three *L. paracasei* subsp. *paracasei* strains, two *L. plantarum* strains, three *L. reuteri* strains, three *L. sakei* strains, two *L. salivarius* strains, three *L. pentosus* strains and five *L. kefiri* strains. The six *Lactococcus* strains were classified into two species: three *L. lactis* strains and three *L. lactis* subsp. *cremoris* strains.

Table 2 shows the pathogen and spoilage bacteria isolated from the fruit and vegetable samples ($n = 311$) purchased from local open markets and retail stores. It is obvious that apart from the spore formation of *C. perfringens* ($n = 117$, 37.62%), *Salmonella* spp. is the most frequently isolated microorganism ($n = 27$, 8.68%), its isolates originating from the RTU salad (12 strains out of 27, 44.44%), tomatoes (6 out of 27 strains), white cabbage (5 out of 27 strains) and apples (4 out of 27 strains). *L. monocytogenes* ($n = 3$), *Listeria* spp. ($n = 10$) and *E. coli* ($n = 20$) were also isolated.

In general, inhibition zones from 0 to 18 mm were observed. All pathogens and spoilage bacteria ($n = 14$) showed susceptibility to the bacteriocins produced by the LAB and *Bifidobacterium* strains.

However, the degree of this susceptibility varied significantly depending on the bacteriocin-like substances (BLS)-producing strain as well as the pathogen target cell (Table 3). In order to assess these discrepancies, the inhibition zones were classified into four classes: Class I, 0–10 mm, characterized a weak inhibitory result; Class II, 10.1–12 mm, characterized by a moderate inhibitory result; Class III, 12.1–14 mm, characterized by a moderate to strong inhibitory result; and Class IV, 14.1–18 mm, characterized by a strong inhibitory result. All LAB and *Bifidobacterium* strains were classified with respect to the inhibition zone they produced under each of the three treatments and for each of the 14 pathogens and spoilage bacteria (Table 3). The difference in the distribution (frequencies) of the inhibitory result of the LAB and *Bifidobacterium* strains was statistically assessed (chi-squared test, significance level of $p < 0.05$).

The CFS was left untreated or neutralized (pH = 6.0). In the case of neutralization, it was treated with catalase to eliminate hydrogen peroxide. For the sake of convenience, we named the three types of CFS as follows: (a) untreated CFS, (b) neutralized CFS and (c) neutralized CFS treated with a 1 mg mL^{-1} concentration of catalase. In order to assess the effect of the different treatments (Treatments A–C) to the distribution of the inhibitory potency of the LAB and *Bifidobacterium* strains, the comparison concerned the different categories/treatments (A–C) for every pathogen (same pathogen but different treatment). For example, in the case of *E. coli* strain No.1 (Table 3), there is no significant difference in the distribution of the inhibitory potency of the LAB and *Bifidobacterium* strain for the three treatments. In other words, the different treatments did not significantly affect the number of LAB and *Bifidobacterium* strains in the different potency classes (I–IV). The same conclusion was derived for the other two *E. coli* strains, the *E. coli* O157:H7 strain, the two *L. monocytogenes* strains, the two *Salmonella* spp. strains and the two *B. cereus* strains (Table 3). In the cases of the above-mentioned pathogens, the different categories/treatments (A–C) did not affect the distribution of the inhibitory potency of the LAB and *Bifidobacterium*. However, the picture is totally different in the cases of the plant pathogens (two *E. carotovora* strains and one *Xanthomonas* spp. strain) as well as the *L. innocua*strain (ATCC 33090TM). The inhibitory potency of the LAB and *Bifidobacterium* strains was reduced from Category A to Category/Treatment B and from Category/Treatment B to Category/Treatment C when the target cell was *L. innocua* ($\chi^2 = 13.7806$, $p = 0{,}008029$, df = 6).In all plant pathogens, Categories/Treatments B and C significantly reduced the inhibitory potency of the LAB and *Bifidobacterium* strains ($\chi^2 = 18.2566$, $p = 0.005622$, df = 6 for *E. carotovora* strain No. 1, $\chi^2 = 16.6212$, $p = 0.01078$, df = 6 for *E. carotovora* strain No. 2 and $\chi^2 = 39.404$, $p < 0.0001$, df = 6 for *Xanthomonas* spp.).

In order to compare the difference in susceptibility of the different pathogens and spoilage bacteria (target cells of the BLS) to the BLS of the LAB and *Bifidobacterium* strains, a statistical comparison was made between pathogens of the same species and for the same category/treatment (Category A, or the same treatment as either B or C, but different target cell pathogens of the same species). For example, the three *E. coli* strains showed no significant difference in the distribution of inhibitory potency to the various classes of BLS-producing LAB and *Bifidobacterium* strains under the same category/treatment (A (untreated), B or C). However, when the *E. coli* O157:H7 strain was introduced in the comparison scheme, the perspective changed dramatically. It appears that the latter strain is far more resistant (less susceptible) to the BLS than the other three *E. coli* strains, no matter the treatment ($\chi^2 = 29.0012$, $p < 0.0001$, df = 9 for Treatment A, $\chi^2 = 33.7242$, $p < 0.0001$, df = 9 for Treatment B and $\chi^2 = 30.4993$, $p < 0.0001$, df = 9 for Treatment C). The different treatments did not significantly affect the two *Salmonella* strains with respect to the distribution of the inhibitory potency to the various classes of BLS-producing LAB and Bifidobacteria strains. The same conclusion is also valid for the two strains of *L. monocytogenes*. However, when the two *L. monocytogenes* strains were compared to the *L. innocua* (ATOC 33090TM) strain belonging to the same genus, the differences became significant, showing that the *L. innocua* (ATOC 33090TM) strain is more resistant (less susceptible) to the BLS for Treatments B and C but not Treatment A ($\chi^2 = 10.0342$, $p = 0.039856$, df = 4 for Treatment B and $\chi^2 = 14.8155$, $p = 0.0051$, df = 4 for Treatment C). *B. cereus* strains showed significant differences under all treatments with strain No. 1 being more vulnerable than strain No. 2 ($\chi^2 = 4.3457$, $p = 0.037102$, df = 1 for Treatment A, $\chi^2 = 5.0178$,

$p = 0.025069$, df = 1 for Treatment B and $\chi^2 = 7.1273$, $p = 0.007592$, df = 1 for Treatment C). Finally, the two *E. carotovora* strains showed no significant differences for Treatments A and B but showed significant differences for Treatment C ($\chi^2 = 9.7695$, $p = 0.020631$, df = 3), meaning that when the CFS of LAB and *Bifidobacterium* strains was treated under Treatment C, *E. carotovora* strain No. 1 was more vulnerable than strain No. 2.

The CFS from the isolated LAB and *Bifidobacterium* strains from oregano honey were treated with different enzymes to verify the proteinaceous nature of the inhibitory substances (Table 4). For that purpose, proteolytic enzymes, among others such as trypsin, pepsin, lipase and α-amylase, were used. The data show that almost all LAB and *Bifidobacterium* strains were susceptible and no inhibition zones were observed after the enzymatic treatment of the CFS, a finding suggesting that the causative agent of the inhibition zone observed prior to enzymatic treatment was due to substances of protein origin like BLS.

Table 1. Isolated species of lactic acid bacteria (LAB) and *Bifidobacterium* spp. From the oregano honey honeycomb.

Species	*n*
Lactobacillus insectis	10
Lactobacillus kunkeei	10
Lactobacillus casei	7
Lactobacillus fermentum	6
Lactobacillus curvatus	4
Lactobacillus paracasei subsp. *paracasei*	5
Lactobacillus plantarum	7
Lactobacillus reuteri	4
Lactobacillus sakei	7
Lactobacillus salivarius	3
Lactobacillus pentosus	5
Lactobacillus kefiri	5
Lactococcus lactis	4
Lactococcus lactis subsp. *cremoris*	4
Leuconostoc spp.	8
Bifidobacterium spp.	12
Total	101

Table 2. Number of isolated microorganisms from the surface of various fruits or vegetables and the entire vegetable content in ready-to-eat (RTU) commercially available salads.

Isolated Species	Apples ($n = 20$) *	Pears ($n = 20$)	Peaches ($n = 20$)	Tomatoes ($n = 20$)	Cucumbers ($n = 20$)	Red Pepper ($n = 20$)	White Cabbage ($n = 40$)	Carrots ($n = 40$)	RTU ($n = 50$)
E. coli	1	-	-	6	3	4	3	-	3
Salmonella spp.	4	-	-	6	-	-	5	-	12
Shigella spp.	-	-	-	1	-	-	-	3	-
Listeria spp.	-	-	-	2	-	-	-	6	2
L. monocytogenes	-	-	-	-	-	-	2	-	1
B. cereus	-	-	1	2	-	4	2	3	2
C. perfringens (vegetative forms)	-	1	5	-	2	-	4	-	9
C. perfringens (spore forms)	11	12	13	12	7	6	20	25	11
Clostridium spp.	2	3	-	3	-	-	3	4	3
Erwinia spp.	-	1	6	3	-	-	-	4	-
Xanthomonas spp.	4	3	2	2	3	-	4	3	-
P. fluorescens	2	1	5	2	2	3	3	-	2
Aspergillus spp.	4	-	3	-	4	-	1	7	-
Penicillium spp.	3	4	-	-	1	2	-	3	-

* Number of samples examined; RTU: packages of ready-to-eat commercially available salads.

Table 3. Distribution of bacteriocin-producing LAB and *Bifidobacterium*, with respect to their class of inhibition zone against various pathogens.

Class of Inhibition Zone	*E. coli* (1)			*E. coli* (2)			*E. coli* (3)			*E. coli* O157:H7 (ATCC 69373)		
	A	B	C	A	B	C	A	B	C	A	B	C
I (0–10 mm)	10	10	10	11	11	11	10	12	11	25	26	34
II (10.1–12 mm)	4	13	15	7	11	12	3	8	13	10	16	9
III (12.1–14 mm)	17	11	13	17	19	16	18	13	14	11	5	6
IV (14.1–18 mm)	18	15	11	14	8	10	18	16	11	3	2	0

Class of Inhibition Zone	*L. monocytogenes* (1)			*L. monocytogenes* (2)			*L. innocua* (ATCC 33090TM)		
I (0–10 mm)	28	28	29	30	30	30	31	28	34
II (10.1–12 mm)	5	6	6	2	6	2	4	15	11
III (12.1–14 mm)	3	4	7	6	6	12	10	5	4
IV (14.1–18 mm)	13	11	7	11	7	5	4	1	0

Class of Inhibition Zone	*Salmonella* spp. (1)			*Salmonella* spp. (2)		
I (0–10 mm)	9	10	10	8	9	10
II (10.1–12 mm)	2	11	17	4	5	7
III (12.1–14 mm)	10	11	11	8	16	13
IV (14.1–18 mm)	18	17	21	20	19	19

Class of Inhibition Zone	*B. cereus* (1)			*B. cereus* (2)		
I (0–10 mm)	39	40	40	46	47	48
II (10.1–12 mm)	8	7	6	3	2	1
III (12.1–14 mm)	1	2	3	0	0	0
IV (14.1–18 mm)	1	0	0	0	0	0

Class of Inhibition Zone	*Erwinia* spp. (1)			*Erwinia* spp. (2)			*Xanthomonas* spp.		
I (0–10 mm)	7	13	18	3	4	5	5	7	13
II (10.1–12 mm)	12	19	17	8	20	22	4	21	23
III (12.1–14 mm)	17	13	12	24	19	19	26	18	12
IV (14.1–18 mm)	13	4	2	14	6	3	14	3	1

A: cell-free supernatant (CFS) of LAB isolates without any treatment; B: cell-free supernatant (CFS) with pH neutralized to 6.0; C: cell-free supernatant (CFS) with pH neutralized to 6.0 and H_2O_2 eliminated; I, II, III, IV: classes of inhibition zones (I: 0–10 mm, II: 10.1–12 mm, III: 12.1–14 mm and IV: 14.1–18 mm).

Table 4. The effect of pepsin, trypsin, α-amylase and lipase treatment on the inhibitory activity of bacteriocin-like substances (BLS)produced by the LAB and *Bifidobacterium* isolates against indicators (*L. innocua* (ATCC 33090TM) and *E. coli* 0157:H7 (ATCC 69373) strains).

Isolated LAB and Bifidobacteria	Control *	*Listeria innocua* (ATCC 33090TM)				Control	*E. coli* 0157:H7 (ATCC 69373)			
		Pepsin	Trypsin	α-Amylase	Lipase		Pepsin	Trypsin	α-Amylase	Lipase
L. insectis 1	n.d.	n.d.	n.d.	n.d.	n.d.	n.d.	n.d.	n.d.	n.d.	n.d.
L. insectis 2	+**	−***	n.d.	n.d.	n.d.	n.d.	n.d.	n.d.	n.d.	n.d.
L. insectis 3	+	−	−	−	−	−	−	−	−	−
L. kunkeei 1	+	−	−	−	−	n.d.	n.d.	n.d.	n.d.	n.d.
L. kunkeei 2	n.d.	n.d.	n.d.	n.d.	n.d.	n.d.	n.d.	n.d.	n.d.	n.d.
L. casei 1	+		n.d.	n.d.		−	−	−	−	−
L. casei 2	+	−	−	−	−	−	−	−	−	−
L. casei 3	+	−	−	−	−	−	n.d.	n.d.	n.d.	n.d.
L. fermentum 1	+	−	−	−	−	−	−	−	−	−
L. fermentum 2	+	−	−	−	−	−	−	−	−	−
L. fermentum 3	+	−	−	−	−	n.d.	n.d.	n.d.	n.d.	n.d.
L. curvatus 1	n.d.	n.d.	n.d.	n.d.	n.d.	n.d.	n.d.	n.d.	n.d.	n.d.
L. curvatus 2	+	−	−	−	−	n.d.	n.d.	n.d.	n.d.	n.d.
L. paracasei subsp. *Paracasei* 1	+	−	−	−	−	−	−	−	−	−
L. paracasei subsp. *Paracasei* 2	+	−	−	−	−	−	−	−	−	−
L. paracasei subsp. *Paracasei* 3	+	−	+	−	−	−	−	−	−	−
L. plantarum 1	+	−	−	−	−	−	−	−	−	−
L. plantarum 2	+	−	−	−	−	−	−	−	−	−
L. reuteri 1	+	−	−	−	−	−	−	−	−	−
L. reuteri 2	+	−	−	−	−	−	−	−	−	−
L. reuteri 3	+	−	−	−	−	−	−	−	−	−
L. sakei 1	+	−	−	−	−	−	−	−	−	−
L. sakei 2	+	−	−	−	−	−	−	−	−	−
L. sakei 3	+	−	−	−	−	−	−	−	−	−
L. salivarius 1	+	−	−	−	−	−	−	−	−	−
L. salivarius 2	+	−	−	−	−	−	−	−	−	−
L. pentosus 1	+	−	−	−	−	−	−	−	−	−
L. pentosus 2	+	−	−	−	−	−	−	−	−	−
L. pentosus 3	+	−	−	−	−	−	−	−	−	−
L. kefiri 1	+	−	−	−	−	−	n.d.	n.d.	n.d.	n.d.
L. kefiri 2	+	n.d.	n.d.	n.d.	n.d.	−	−	−	−	−
L. kefiri 3	+	n.d.	n.d.	n.d.	n.d.	−	−	−	−	−
L. kefiri 4	+	−	−	−	−	−	−	−	−	−
L. kefiri 5	+	−	−	−	−	−	−	−	−	−

Table 4. *Cont.*

Isolated LAB and Bifidobacteria	Control *	*Listeria innocua* (ATCC 33090TM)				Control	*E. coli* 0157:H7 (ATCC 69373)			
		Pepsin	Trypsin	α-Amylase	Lipase		Pepsin	Trypsin	α-Amylase	Lipase
Lact. lactis 1	n.d.	n.d.	n.d.	n.d.	n.d.	n.d.	n.d.	n.d.	n.d.	n.d.
Lact. lactis 2	+	−	−	−	−	n.d.	n.d.	n.d.	n.d.	n.d.
Lact. lactis 3	+	−	−	−	−	n.d.	n.d.	n.d.	n.d.	n.d.
Lact. lactis subsp. *cremoris* 1	+	−	−	−	−	−	−	+	−	−
Lact. lactis subsp. *cremoris* 2	+	−	−	−	−	−	−	+	−	−
Lact. lactis subsp. *cremoris* 3	+	−	−	−	−	−	−	−	−	−
Leuconostoc spp. (1)	+	n.d.	n.d.	n.d.	n.d.	−	−	−	−	−
Leuconostoc spp. (2)	+	−	−	−	−	−	−	−	−	−
Leuconostoc spp. (3)	+	+	−	−	−	−	−	−	−	−
Leuconostoc spp. (4)	+	−	−	−	−	−	−	−	−	−
Leuconostoc spp. (5)	+	n.d.	n.d.	n.d.	n.d.	−	−	−	−	−
Bifidobacterium spp. (1)	+	n.d.	n.d.	n.d.	n.d.	−	−	−	−	−
Bifidobacterium spp. (2)	+	n.d.	n.d.	−	−	−	−	n.d.	n.d.	n.d.
Bifidobacterium spp. (3)	+	n.d.	n.d.	−	−	−	−	n.d.	n.d.	n.d.
Bifidobacterium spp. (4)	+	n.d.	n.d.	−	−	−	−	−	−	−

* Control = pH-neutralized and H_2O_2-eliminated cell-free supernatant (CFS) without addition of enzymes; ** (+): presence of inhibition zone; *** (−): absence of inhibition zone; n.d. = not determined.

4. Discussion

Fruits and vegetables are staples in human diets worldwide. In developed countries, there is a constantly increasing demand for these products, not to mention that social nutritional movements such as vegetarianism and veganism are gaining followers on a daily basis. From this perspective, the safety of fruits and vegetables is a public health issue of top priority.

In the present study, we purchased fruits (apples, pears and peaches) and vegetables (tomatoes, red peppers, white cabbage, carrots and RTU salad) from local open markets and retail stores. Table 2 shows the pathogens and spoilage bacteria isolated from these products, including *E. coli*, *L. monocytogenes* and *Salmonella* spp. These findings underline the danger that fruits and vegetables can carry a load of pathogens, which, under certain conditions, may initiate an outbreak. Twelve of these strains were selected and used as target cell indicators of the production of bacteriocins from LAB and *Bifidobacterium* strains.

Contamination with such pathogens can easily occur during preharvest (soil, manure, irrigation water), harvest (handling) and postharvest (transportation, storage, etc.) treatments. Because these products, and particularly the fruits, are eaten raw, they have been described as causative agents of outbreaks [5,50–53]. Although practices such as washing and sanitation reduce the microbial load to some extent, they do not eliminate the dangerous pathogens, e.g., *E. coli* O157:H7, *Salmonella* spp. and *L. monocytogenes* [54,55].

Furthermore, consumers expect and demand "natural and mild" means to preserve food safety from the industry—means that do not alter the sensorial characteristics of food and, of course, pose any health hazards to the consumer [52,56]. Bacteriocins, or BLS in general, seem to be an ideal solution to satisfy these demands since they are safe for consumers, natural in their origin and, by exerting antimicrobial activity, can extend the shelf life of fruits and vegetables [57,58]. Nature and natural products are excellent sources of BLS-producing bacteria. Our results revealed 101 strains of 16 species of LAB and Bifidobacteria (Table 1) isolated from oregano honeycombs, indicating an impressive biodiversity with the potential to harness BLS-producing microorganisms. This is a novel finding since there are no other reports in the literature concerning the microbiology of oregano honey. The wild types of *Lactobacilli* are stable and adapted to a variety of environments. They survive by (a) the production of organic acids, mainly lactic acid, as byproducts of the fermentation of carbohydrates, (b) the production of substances with antibacterial activity such as hydrogen peroxide and (c) the production of BLS.

In our study, and in order to analyze the antimicrobial effect of the LAB and *Bifidobacterium* strains isolated from oregano honeycombs to pathogens and spoilage bacteria isolated from fruits and vegetables, the cell-free supernatant (CFS) fluid of the cultures underwent three different categories and treatments: it was left untreated (Category A), modified to pH 6 by the addition of NaOH to eliminate organic acids (Treatment B) and, finally, in addition to NaOH, catalase was added to eliminate hydrogen peroxide (Treatment C). Our results (Table 3) show that in 10 out of 14 pathogens and spoilage bacteria, the distribution of the antibacterial potency of LAB and *Bifidobacterium* (expressed in millimeters of inhibition zones) was not affected by the different treatments of the CFS. This finding strongly suggests that, firstly, this antimicrobial activity was due to the BLS, and secondly, BLS' activity overshadows the antimicrobial effect of organic acids and hydrogen peroxide.

For example, in the case of *E. coli* strain No. 1, the three different treatments of CFS did not significantly change the distribution of LAB and *Bifidobacterium* to the different classes of inhibition zones (Table 3), which means that the potency of their antimicrobial action was not significantly modified when the organic acids and hydrogen peroxide were eliminated from the CFS. It follows that it was the action of BLS that exerted the inhibition zones observed and the contribution of organic acids and hydrogen peroxide to this effect was minimal, if not negligible. If the contribution of organic acids or activity of hydrogen peroxide was important, the distribution of the LAB and *Bifidobacterium* would have been significantly different. The exact same conclusions are also valid in the cases of the other two strains of *E. coli*, the two strains *of L. monocytogenes*, the two strains of *Salmonella* spp., the two strains of *B. cereus* and the reference strain *E. coli* O157:H7.

In the case of *L. innocua* (ATOC 33090TM), the picture is completely different (Table 3). Under Category A, 10 strains showed inhibition zones between 12.1 and 14.0 mm (Class III) and four strains showed inhibition zones between 14.1 and 18 mm (Class IV). When the organic acids were eliminated by the addition of NaOH (Treatment B), five strains showed inhibition zones between 12.1 and 14.0 mm (Class III) and one strain showed inhibition zones between 14.1 and 18 mm (Class IV). Finally, when hydrogen peroxide was eliminated by the addition of catalase (Treatment C), four strains showed inhibition zones between 12.1 and 14.0 mm (Class III) and no strains showed inhibition zones between 14.1 and 18 mm (Class IV). The entire Class IV antibacterial action was due to organic acids (in three out of four strains) and to a lesser extent to hydrogen peroxide (one strain out of four). In Class III, the antibacterial activity of 5 out of 10 strains was due to organic acid production, while the antibacterial activity of one strain was due to hydrogen peroxide. Similar findings are also valid in the case of the two *E. carotovora* strains and the *Xanthomonas* strain. A possible explanation is that these bacteria were, to some extent, resistant to the activity of the BLS produced by LAB and *Bifidocaterium*, thus leaving space for the organic acids and hydrogen peroxide to act.

Another indirect way to check the susceptibility of each pathogen and spoilage bacteria is to compare the bacterial target cells of each species under the same treatment. The three *E.coli* strains, isolated from fruits and vegetables, showed exactly the same sensitivity to the bacteriocins produced by the LAB and *Bifidobacterium*s trains under Category A (Table 3). A similar observation was valid for Treatments B and C. However, when compared with the reference *E. coli* O157:H7 strain, significant differences arose for all treatments, showing that the latter strain is more resistant to the BLS than the others. The same resistance to the BLS showed the reference *L. innocua* strain (a surrogate microorganism for *L. monocytogenes*) while the two strains of *L. monocytogenes*, when compared with each other under each treatment, showed no differences. It seems that the wild types are more sensitive to the action of BLIS than the standard registered strains. The two *Salmonella* spp. strains also showed no differences when compared under the same treatment. However, the two *B. cereus* strains showed significant differences under all treatments, implying that strain No. 1 was more susceptible than strain No. 2. Finally, the *E. carotovora* strains showed no significant differences under Category A and Treatment B, but showed significant differences under Treatment C. Since in Treatment C the organic acids and the hydrogen peroxide are eliminated, this finding suggests that *E. carotovora* strain No. 1 was more susceptible to the action of BLS than strain No. 2. These findings strongly point out that susceptibility to the BLS is not only species-specific but also, even within the same species, for some species, it is cell-specific. Of course, the possibility of the low concentration of BLIS in the CFS should be considered, as some LAB or *Bifidobacterium* strains could be poor producers of otherwise potent BLS. In such cases, the inhibitory effect of BLS to the target cells would be dose-responsive.

However, other studies in which BLS-producing LAB isolates were tested for their antimicrobial effects against Gram-negative and Gram-positive bacteria, as well as common spoilage fungi in fruits and vegetables, highlight that untreated CFS inhibited all tested bacteria and fungi except for *E. coli*, when, at the same time, after pH neutralization and H_2O_2 elimination, the CFS inhibited only *L. innocua* [17,59]. In a second statement, emphasizing the fact that the strong antimicrobial effects associated with *L. lactis* and *E. faecium* appeared to be a direct result of the organic acids and the H_2O_2 present in the CFS rather than the BLS [16], Hajikhani et al. [60] reported that antimicrobial activity compounds were produced by Enterococci against *Pseudomonas aeruginosa* and *Proteus vulgaris*, but left *E. coli* and *Yersinia enterocolitica* untouched.

Enzymatic treatment of the CFS strongly suggests that the inhibitory factor causing the inhibitory zones observed in our study is of protein origin (Table 4). Similar data were reported by Ghanbari et al. [61], who observed the inactivation of antimicrobial activity by the action of proteolytic enzymes and concluded that the particular finding was an indication of the proteinaceous nature of BLS. Additionally, the elimination of the inhibition zone promoted by α-amylase indicates that the BLS produced by LAB may be glycoproteins (carbohydrate moiety), which require both the glyco and the protein portion of the molecule in order to be active [62].

5. Conclusions

The isolation of LAB and *Bifidobacterium* strains originating from oregano honey, a very rare product, revealed the impressive biodiversity of these microorganisms. This finding is novel, and this study is the first report on the subject in the literature. Further screening of these microorganisms showed that they produced bacteriocin-like substances, which are effective not only against Gram-positive bacteria but also Gram-negative bacteria. These microorganisms are involved in the fruit or vegetable industry, either as human pathogen contamination or spoilage bacteria. This is an important finding since it is commonly accepted by the scientific community that these peptide-like substances are mainly active against Gram-positive bacteria.

Author Contributions: C.V., G.R., A.T. (Athina Tzora) and E.B.; methodology, G.R. and A.A.; software, C.V., A.T. (Athina Tzora), I.S., A.T. (Anastasios Tsinas). and E.B.; validation, A.A., A.T. (Anastasios Tsinas), G.R. and T.V.; formal analysis, C.V. and A.A.; investigation, C.V., A.T. (Athina Tzora), I.S. and E.B.; resources, C.V., A.A., G.R. and T.V.; data curation, C.V. and G.R.; writing—original draft preparation, C.V., A.T. (Athina Tzora), T.V. and E.B.; writing—review and editing, C.V., T.V. and E.B.; visualization, C.V., A.T. (Athina Tzora) and, E.B.; supervision, C.V., A.T. and E.B.; project administration. All authors have read and agreed to the published version of the manuscript.

Funding: This research received no external funding.

Conflicts of Interest: The authors declare no conflict of interest.

References

1.   Eurostat. Frequency of Fruit and Vegetables Consumption by Sex, Age and Degree of Urbanization. Available online: https://appsso.eurostat.ec.europa.eu/nui/show.do?dataset=hlth_ehis_fv1u&lang=en (accessed on 10 September 2020).

2.   Nishida, C.; Uauy, R.; Kumanyika, S.; Shetty, P. The joint WHO/FAO expert consultation on diet, nutrition and the prevention of chronic diseases: Process, product and policy implications. *Public Health Nutr.* **2004**, *7*, 245–250. [CrossRef] [PubMed]

3.   Garg, M.; Singh, V.K. Carbohydrate metabolism during fruit spoilage. *Biotechnol. Hortic.* **2016**, 178–197.

4.   Luna-Guevara, J.J.; Arenas-Hernandez, M.M.P.; De La Peña, C.M.; Silva, J.L.; Ramos-Cassellis, M.E. The role of pathogenic *E. coli* in fresh vegetables: Behavior, contamination factors, and preventive measures. *Int. J. Microbiol.* **2019**, *2019*, 1–10. [CrossRef] [PubMed]

5.   Beuchat, L.R. Ecological factors influencing survival and growth of human pathogens on raw fruits and vegetables. *Microbes Infect.* **2002**, *4*, 413–423. [CrossRef]

6.   Spoilage of vegetables and fruits. *Food Microbiol. Princ. Pract.* **2016**, 337–363. [CrossRef]

7.   European Data Journalism Network. The Antibiotic Resistance Crisis Deepens. Available online: https://www.europeandatajournalism.eu/eng/News/Data-news/The-antibiotic-resistance-crisis-deepens (accessed on 15 September 2020).

8.   Fair, R.J.; Tor, Y. Antibiotics and bacterial resistance in the 21st century. *Perspect. Med. Chem.* **2014**, *6*, PMC.S14459-64. [CrossRef] [PubMed]

9.   Morales, J.L.; Gutiérrez-Méndez, N.; Rivera-Chavira, B.E.; Pérez-Vega, S.B.; Nevárez-Moorillón, G.V. Biocontrol processes in fruits and fresh produce, the use of lactic acid bacteria as a sustainable option. *Front. Sustain. Food Syst.* **2018**, *2*, 1–13. [CrossRef]

10. Agriopoulou, S.; Stamatelopoulou, E.; Sachadyn-Król, M.; Varzakas, T. Lactic acid bacteria as antibacterial agents to extend the shelf life of fresh and minimally processed fruits and vegetables: Quality and safety aspects. *Microorganism* **2020**, *8*, 952. [CrossRef]

11. Fessard, A.; Remize, F. Genetic and technological characterization of lactic acid bacteria isolated from tropically grown fruits and vegetables. *Int. J. Food Microbiol.* **2019**, *301*, 61–72. [CrossRef]

12. Rodríguez, L.G.R.; Mohamed, F.; Bleckwedel, J.; Medina, R.; De Vuyst, L.; Hebert, E.M.; Mozzi, F. Diversity and functional properties of lactic acid bacteria isolated from wild fruits and flowers present in northern argentina. *Front. Microbiol.* **2019**, *10*, 1091. [CrossRef]

13. Thokchom, S.; Joshi, S.R. Probiotic and bacteriocin efficacy of lactic acid bacteria from traditionally fermented foods: A review. *Assam Univ. J. Sci. Technol.* **2012**, *10*, 142–155.

14. Buckenhüskes, H. Selection criteria for lactic acid bacteria to be used as starter cultures for various food commodities. *FEMS Microbiol. Rev.* **1993**, *12*, 253–271. [CrossRef]

15. Mokoena, M.P.; Mutanda, T.; Olaniran, A.O. Perspectives on the probiotic potential of lactic acid bacteria from African traditional fermented foods and beverages. *Food Nutr. Res.* **2016**, *60*, 1–12. [CrossRef]

16. Cleveland, J.; Montville, T.J.; Nes, I.F.; Chikindas, M.L. Bacteriocins: Safe, natural antimicrobials for food preservation. *Int. J. Food Microbiol.* **2001**, *71*, 1–20. [CrossRef]

17. Yang, E.; Fan, L.; Jiang, Y.; Doucette, C.; Fillmore, S. Antimicrobial activity of bacteriocin-producing lactic acid bacteria isolated from cheeses and yogurts. *AMB Express* **2012**, *2*, 48. [CrossRef]

18. Cotter, P.D.; Hill, C.; Ross, R.P. Bacteriocins: Developing innate immunity for food. *Nat. Rev. Genet.* **2005**, *3*, 777–788. [CrossRef] [PubMed]

19. Barbosa, A.A.T.; Mantovani, H.C.; Jain, S. Bacteriocins from lactic acid bacteria and their potential in the preservation of fruit products. *Crit. Rev. Biotechnol.* **2017**, *37*, 852–864. [CrossRef]

20. Silva, C.C.G.; Silva, S.P.M.; Ribeiro, S.C. Application of bacteriocins and protective cultures in dairy food preservation. *Front. Microbiol.* **2018**, *9*, 594. [CrossRef]

21. Bouchard, D.S.; Seridan, B.; Saraoui, T.; Rault, L.; Germon, P.; Gonzalez-Moreno, C.; Nader-Macias, F.M.E.; Baud, D.; François, P.; Chuat, V.; et al. Lactic Acid bacteria isolated from bovine mammary microbiota: Potential allies against bovine mastitis. *PLoS ONE* **2015**, *10*, e0144831. [CrossRef]

22. Dhundale, V.R.; Hemke, V.; Desai, D.; Dhundale, P. Evaluation and exploration of lactic acid bacteria for preservation and extending the shelf life of fruit. *Int. J. Fruit Sci.* **2018**, *18*, 1–14. [CrossRef]

23. Seow, J.; Agoston, R.; Phua, L.; Yuk, H.-G. Microbiological quality of fresh vegetables and fruits sold in Singapore. *Food Control* **2012**, *25*, 39–44. [CrossRef]

24. Ghaffar, N.A.; Hazman, Y.M.; Nabil, O.N.A. Rapid detection of pathogenic bacteria in vegetables and fruits in Egyptian Farms. *J. Am. Sci.* **2014**, *10*, 242–252.

25. FDA. Bacteriological Analytical Manual (BAM) Chapter 5: Salmonella. Available online: https://www.fda.gov/food/laboratory-methods-food/bam-chapter-5-salmonella (accessed on 10 September 2020).

26. Lennox, J.A.; Matthew, E.; Uwamere, E.; Chinyere, O.; Okpako, E.C. Incidence of Salmonella and Shigella species on some selected fruits and vegetables obtained from open area markets in Calabar Metropolis. *Int. J. Curr. Microbiol. App. Sci.* **2015**, *4*, 262–268.

27. Sospedra, I.; Rubert, J.; Soriano, J.; Mañes, J. Survey of microbial quality of plant-based foods served in restaurants. *Food Control* **2013**, *30*, 418–422. [CrossRef]

28. Ottaviani, F.; Ottaviani, M.; Agosti, M. Differential Agar Medium for Listeria monocytogenes. *Quimper Froid. Symp. Proc.* **1997**, *6*, 16–18.

29. Julien, C.-K.; Edith, A.A.; Thomas, A.D.; Mireille, D.; Coulibaly-Kalpy, J.; Agbo, E.A.; Dadie, T.A.; Dosso, M. Microbiological quality of raw vegetables and ready to eat products sold in Abidjan (CtedIvoire) markets. *Afr. J. Microbiol. Res.* **2017**, *11*, 204–210. [CrossRef]

30. Yu, P.; Yu, S.; Wang, J.; Guo, H.; Zhang, Y.; Liao, X.; Zhang, J.; Wu, S.; Gu, Q.; Xue, L.; et al. *Bacillus cereus* isolated from vegetables in China: Incidence, genetic diversity, virulence genes, and antimicrobial resistance. *Front. Microbiol.* **2019**, *10*, 948. [CrossRef]

31. Roach, R.; Mann, R.; Gambley, C.G.; Shivas, R.G.; Rodoni, B. Identification of Xanthomonas species associated with bacterial leaf spot of tomato, capsicum and chilli crops in eastern Australia. *Eur. J. Plant Pathol.* **2017**, *150*, 595–608. [CrossRef]

32. Matei, G.M.; Matei, S.; Matei, A.; Cornea, C.P.; Draghici, E.M.; Jerca, I.O. Bioprotection of fresh food products against blue mold using lactic acid bacteria with antifungal properties. *Rom. Biotechnol. Lett.* **2016**, *21*, 11201–11208.

33. Al-Hindi, R.R.; Alnajada, R.A.; Saleh, A.H.M. Isolation and identification of some fruit spoilage fungi: Screening of plant cell wall degrading enzymes. *Afr. J. Microbiol. Res.* **2011**, *5*, 443–448.

34. Cuppels, D. Evaluation of selective media for isolation of soft-rot bacteria from soil and plant tissue. *Phytopathology* **1974**, *64*, 468. [CrossRef]

35. Hadas, R.; Kritzman, G.; Gefen, T.; Manulis, S. Detection, quantification and characterization of Erwinia carotovora ssp. carotovora contaminating pepper seeds. *Plant Pathol.* **2001**, *50*, 117–123. [CrossRef]

36. Council of Europe. European pharmacopoeia suppl.4.2, chap.2.6.13. In *Test for Specified Micro-Organisms*, 4th ed.; Council of Europe: Strasbourg, France, 2002.

37. Savvaidis, T.K.I. Bacterial indicators and metal ions in high mountain lake waters. *Microb. Ecol. Health Dis.* **2001**, *13*, 147–152. [CrossRef]

38. Bezirtzoglou, E.; Romond, C. Occurrence of *Lactobacillus* sp. in newborns delivered by caesarean section. *Rev. Med. Microbiol.* **1997**, *8*, 101–103. [CrossRef]

39. Tajabadi, N. Comparison of lactic acid bacteria and bifidobacteria from honey stomachs and honeycombs of Giant honeybee (*Apis dorsata*) in Kedah and Terengganu. Master's Thesis, Universiti Putra Malaysia, Selangor, Malaysia, 2010.

40. Hwanhlem, N.; Buradaleng, S.; Wattanachant, S.; Benjakul, S.; Tani, A.; Maneerat, S. Isolation and screening of lactic acid bacteria from Thai traditional fermented fish (Plasom) and production of Plasom from selected strains. *Food Control* **2011**, *22*, 401–407. [CrossRef]

41. Tajabadi, N.; Mardan, M.; Manap, M.Y.A.; Shuhaimi, M.; Meimandipour, A.; Nateghi, L. Detection and identification of *Lactobacillus* bacteria found in the honey stomach of the giant honeybee *Apis dorsata*. *Apidologie* **2011**, *42*, 642–649. [CrossRef]

42. Feizabadi, F.; Sharifan, A.; Tajabadi, N. Isolation and identification of lactic acid bacteria from stored *Apis mellifera* honey. *J. Apic. Res.* **2020**, 1–6. [CrossRef]

43. Funke, G.; Monnet, D.; Debernardis, C.; Von Graevenitz, A.; Freney, J. Evaluation of the VITEK 2 system for rapid identification of medically relevant gram-negative rods. *J. Clin. Microbiol.* **1998**, *36*, 1948–1952. [CrossRef] [PubMed]

44. Karaduman, A.; Ozaslan, M.O.; Kilic, I.H.; Bayil-Oguzkan, S.; Kurt, B.S.; Erdogan, N. Identification by using MALDI-TOF mass spectrometry of lactic acid bacteria isolated from non-commercial yogurts in southern Anatolia, Turkey. *Int. Microbiol.* **2017**, *20*, 25–30.

45. Bungenstock, L.; Abdulmawjood, A.; Reich, F. Evaluation of antibacterial properties of lactic acid bacteria from traditionally and industrially produced fermented sausages from Germany. *PLoS ONE* **2020**, *15*, e0230345. [CrossRef]

46. Herreros, M.; Sandoval, H.; González, L.; Castro, J.M.; Fresno, J.M.; Tornadijo, M. Antimicrobial activity and antibiotic resistance of lactic acid bacteria isolated from Armada cheese (a Spanish goats' milk cheese). *Food Microbiol.* **2005**, *22*, 455–459. [CrossRef]

47. Djadouni, F.; Mebrouk, K.; Miloud, H. Control of E. coli and spoilage microorganisms in tomato sauce and paste using a synergistic antimicrobial formula. *J. Chem. Pharmac. Res.* **2015**, *7*, 1352–1360.

48. Smaoui, S.; Elleuch, L.; Bejar, W.; Karray-Rebai, I.; Ayadi, I.; Jaouadi, B.; Mathieu, F.; Chouayekh, H.; Bejar, S.; Mellouli, L. Inhibition of fungi and gram-negative bacteria by bacteriocin BacTN635 produced by *Lactobacillus plantarum* sp. TN635. *Appl. Biochem. Biotechnol.* **2009**, *162*, 1132–1146. [CrossRef] [PubMed]

49. Dawson, B.; Trapp, R.G. *Basic & Clinical Biostatistics*, 4th ed.; McGraw-Hill Education LLC: New York, NY, USA, 2004.

50. Bintsis, T. Foodborne pathogens. *AIMS Microbiol.* **2017**, *29*, 529–563. [CrossRef]

51. Artés, F.; Allende, A. Minimal fresh processing of vegetables, fruits and juices. In *Emerging Technologies for Food Processing*; Elsevier BV: Amsterdam, The Netherlands, 2005; pp. 677–716.

52. Allende, A.; Tomas-Barberan, F.; Gil, M.I. Minimal processing for healthy traditional foods. *Trends Food Sci. Technol.* **2006**, *17*, 513–519. [CrossRef]

53. Abadias, M.; Usall, J.; Anguera, M.; Solsona, C.; Viñas, I. Microbiological quality of fresh, minimally-processed fruit and vegetables, and sprouts from retail establishments. *Int. J. Food Microbiol.* **2008**, *123*, 121–129. [CrossRef]

54. Powell, D.; Luedtke, A. Fact Sheet: A Timeline of Fresh Juice Outbreaks. 2000. Available online: http://www.foodsafety.ksu.edu/en/article-details.php?a=2&c=6&sc=37&id=427 (accessed on 16 September 2020).

55. CDC. Multistate Outbreak of Human Salmonella Newport Infections Linked to Raw Alfalfa Sprouts (Final Update). Available online: https://www.cdc.gov/salmonella/2010/newport-alfalfa-sprout-6-29-10.html (accessed on 17 September 2020).

56. Xiao, Y.; Su, C.; Ouyang, Y.; Zhang, B. Trends of vegetables and fruits consumption among Chinese adults aged 18 to 44 years old from 1991 to 2011. *Zhonghualiuxingbingxue Za Zhi* **2015**, *36*, 232–236.

57. Barbosa, A.A.T.; De Araújo, H.G.S.; Matos, P.N.; Carnelossi, M.A.G.; De Castro, A.A. Effects of nisin-incorporated films on the microbiological and physicochemical quality of minimally processed mangoes. *Int. J. Food Microbiol.* **2013**, *164*, 135–140. [CrossRef]

58. Junior, A.A.D.O.; Couto, H.G.S.D.A.; Barbosa, A.A.T.; Carnelossi, M.A.G.; De Moura, T.R. Stability, antimicrobial activity, and effect of nisin on the physico-chemical properties of fruit juices. *Int. J. Food Microbiol.* **2015**, *211*, 38–43. [CrossRef]

59. Sharpe, V.D. Bioproservation of Fresh-Cut Salads Using Bacteriocinogenic Lactic Acid Bacteria Isolate from Commercial Produce. Master's Thesis, Dalhousie University, Halifax, NS, Canada, 2009.

60. Hajikhani, R.; Beyatli, Y.; Aslim, B. Antimicrobial activity of enterococci strains isolated from white cheese. *Int. J. Dairy Technol.* **2007**, *60*, 105–108. [CrossRef]

61. Ghanbari, M.; Jami, M.; Kneifel, W.; Domig, K.J. Antimicrobial activity and partial characterization of bacteriocins produced by lactobacilli isolated from Sturgeon fish. *Food Control* **2013**, *32*, 379–385. [CrossRef]

62. Heredia-Castro, P.Y.; Méndez-Romero, J.I.; Hernández-Mendoza, A.; Acedo-Félix, E.; González-Córdova, A.F.; Vallejo-Cordoba, B. Antimicrobial activity and partial characterization of bacteriocin-like inhibitory substances produced by *Lactobacillus* spp. isolated from artisanal Mexican cheese. *J. Dairy Sci.* **2015**, *98*, 8285–8293. [CrossRef]

Publisher's Note: MDPI stays neutral with regard to jurisdictional claims in published maps and institutional affiliations.

Article

Evaluation and Origin Discrimination of Two Monocultivar Extra Virgin Olive Oils, Cultivated in the Coastline Part of North-Western Greece

Vasiliki Skiada [1,2,*], Sofia Agriopoulou [1] , Panagiotis Tsarouhas [3], Panagiotis Katsaris [2], Eygenia Stamatelopoulou [1] and Theodoros Varzakas [1,*]

[1] Department of Food Science and Technology, University of the Peloponnese, 24100 Kalamata, Greece; sagriopoulou@gmail.com (S.A.); estamatel@gmail.com (E.S.)

[2] Department of Olive and Horticultural Plants, Hellenic Agricultural Organization-DEMETER, 24100 Kalamata, Greece; pankatsaris@yahoo.gr

[3] Department of Supply Chain Management (Logistics), International Hellenic University, 60100 Katerini, Greece; ptsarouhas@ihu.gr

* Correspondence: vpskiada@yahoo.gr (V.S.); tvarzakas@us.uop.gr (T.V.)

Received: 7 September 2020; Accepted: 24 September 2020; Published: 25 September 2020

Abstract: Extra virgin olive oil (EVOO) quality and authenticity are important and challenging factors nowadays for the assurance of consumers' protection, prevention of unfair competition, and disruption of the national economy by a false declaration of origin. Hence, the recognition of EVOO authenticity is of great interest in terms of commercial and quality aspects. The objective of this study was to evaluate and discriminate monovarietal extra virgin olive oils of the two dominant olive cultivars, Lianolia Kerkyras and Koroneiki, produced in the coastline part of Western Greece, based on their chemical characteristics, followed by statistical and chemometric analysis in order to profile for the first time the typical characteristics of Lianolia Kerkyras as well as to identify possible markers for authenticity purpose. A total of 104 olive oil samples were collected. Both cultivars had an overall high quality profile as far as their basic qualitative parameters (free fatty acid, peroxide value, and UV spectrometric indices) are concerned. A higher concentration in the mono-unsaturated oleic acid characterize olive oils of cv. Koroneiki compared to cv. Lianolia Kerkyras, while a clearly higher concentration in the poly-unsaturated linoleic acid was observed in olive oils of cv. Lianolia Kerkyras. In addition, olive oil samples of cv. Koroneiki showed a clear lower total sterols concentration with a percentage of 40.9% not surpassing the required EU Regulatory limit of 1000 mg/kg, an observation which strengthens previous published results of our research group and depicts an overall "intrinsic characteristic" of cv. Koroneiki. As far as the profile of the individual sterols is concerned, Lianolia Kerkyras samples exhibited higher mean value for the total sterol content as well as for β-sitosterol, the major phytosterol in olive oils, compared to the relative values of Koroneiki. Significant differences in the sterolic and fatty acid composition of the examined olive oil samples were shown by means of statistical analysis demonstrating a strong botanical effect and depicting that those compositional markers can be suggested as possible authenticity tools.

Keywords: olive oil; cv. Lianolia Kerkyras; cv. Koroneiki; fatty acid methyl esters; sterols; authenticity; quality

---

1. Introduction

Extra virgin olive oil (EVOO) is cherished as a fundamental ingredient not only in the Mediterranean diet but also internationally due to its proven health-promoting effects and nutritional properties [1–3]. This worldwide olive oil reputation is the major driving force for a continuously higher demand on the international olive oil consumption [4,5]. As a matter of fact, the exported olive oil quantities are continuously increasing not only from the main olive-oil producing counties (Spain, Italy and Greece) but also from non-traditional producing countries around the world such as Argentina, Australia, and China [6].

In this context, the demand of higher quality olive oils has led to the continuous appearance in the market of olive oils elaborated with specific and unique characteristics. Hence, the European Union (EU) has established a series of regulations for the certification, protection, and guarantee of the quality and authenticity of olive oils arising from unapproved and fraudulent activities [7–11]. Protected Designation of Origin (PDO) and Protected Geographical Indication (PGI) are two of the main systems created by EU in order to promote and protect food product's authenticity [12]. These denominations require precise definition of several parameters such as cultivar, geographical origin, agronomic practice, and production technology. From the above-mentioned parameters, the cultivar is of utmost importance since the olive cultivar and its characteristics are directly related to olive oil quality [13]. Therefore, identification of the botanical origin as well as adulteration of olive oils with lower quality or less costly cultivars are some of the goals of authenticity research [14,15].

Nowadays, investigation of one or more constituents (major and minor components) present in the olive oils is carried out addressing important information on olive cultivars and differentiating and comparing their botanical origin. For example, different studies from Italy, Greece, and Spain have been conducted to authenticate olive oils [16–19].

Although the number of Greek indigenous monocultivars is more than 40, with the most systematically cultivated variety being cv. Koroneiki, most of other local autochthonous monocultivars remain to be investigated in detail. In the coastline part of north-western Greece, and mainly in the regional units of Preveza, Parga, and Thesprotia, the dominant olive cultivar, yet poorly investigated, is a local variety named Lianolia Kerkyras. Lianolia Kekryras (referred to locally as Prevezana, Ntopia Pargas, Lianolia) is an indigenous olive cultivar, cultivated exclusively in the above-mentioned regions as well as in the island of Corfu (Kerkyra) in the Ionian sea [20,21]. It is a small-fruited cultivar with its fruits having an elongated cylindro-conical shape with a small nipple. Their weight ranges from 1.2–2.5 g and is normally harvested from late-November to late February. The oil content of Lianolia Kerkyras ranges from 18–20% and is considered as a variety of moderate productivity [20,21]. Respectively, Koroneiki is also a small-fruited cultivar with a cylindrical shape. The weight of Koroneiki olive fruit ranges from 0.6–1.5 g and its oil content from 22–25%. It is considered as a variety of excellent productivity and resistance to adverse weather conditions [20,21].

According to our knowledge, monovarietal olive oils of Lianolia Kerkyras have never before been characterized in depth. There are, in fact, a few publications concentrated on the classification of different Greek monocultivar olive oils, among which Lianolia Kekryras was one of the examined cultivars; however, the first study focuses on the classification of Western Greece's monocultivar olive oils based on their volatile profile [22] and the other two studies focused on their geographical discrimination [23,24].

Hence, the objective of this study was to evaluate and characterize monovarietal olive oils of cv. Lianolia Kerkyras produced in the coastline part of Western Greece and compare them with olive oils of Koroneiki variety produced in the same area. Free fatty acid, peroxide value, and UV absorption characteristics were analyzed as the quality parameters. Moreover, analysis of the sterolic and fatty acid profile were performed. In addition, emphasis was given on the potential of their discrimination for authenticity purpose in terms of their botanical origin. The identification of possible compositional markers was examined based on their chemical parameters.

2. Materials and Methods

2.1. Olive Oil Sampling

One hundred and four (N = 104) virgin olive oil samples were taken during the harvesting period of 2019–2020 from the coastline part of Western Greece. In more detail, sixty (60) olive oil samples of Lianolia Kerkyras and forty four (44) samples of Koroneiki cultivar were originated from the following regional units: Preveza area (at 38°57′33.35″ N and 20°25′06.18″ E)Parga area (at 39°17′07.20″ N and 20°24′01.80″ E) and Thesprotia area (at 39°29′36.60″ N and 20°22′42.76″ E) according to Google Earth version 7.0.1.8244-beta, Google, Inc., USA). Similar climatic conditions were evident at all regions, characterized mainly by mild climate with temperatures rarely falling below freezing. Summer months are hot, and the weather is the typical Mediterranean weather with not so much rain but lots of sunlight. Olive fruits were derived from conventional agriculture and harvested between mid-November to February according to specific technical details provided by the responsible agriculturists in the framework of the project. Olive fruits were transferred to local oil mills, equipped with 2-phase centrifugal systems (decanters) in all cases and the olive oil was extracted within 24 h by using the same protocol for all samples and as such they were not treated as variables. This involved washing and crushing of olive fruits followed by olive paste malaxation at 27–29 °C for 35–45 min and decanting. The resulting olive oil samples were transferred to 1 L air-tight dark-green glass bottles and stored at 4 °C in the fridge for analysis. All examined parameters were determined in duplicate.

2.2. Analysis of the Quality Parameters

Free fatty acid, peroxide value, and UV absorption characteristics (K_{232} and K_{268}) were determined as described by the analytical methods of Regulation EEC/2568/91 of the European Commission and later amendments [25].

2.3. Analysis of the Sterolic Profile

The oils, with the addition of α-cholestanol as internal standard (Sigma, St. Louis, MO, USA), were saponified with KOH in an ethanolic solution and the unsaponifiable matter was then extracted with diethylether. Separation of the different alcoholic compounds fractions took place by thin-layer chromatography (TLC) on a basic silica gel plate (Fluka, St. Louis, MO, USA). The addition of the appropriate silylation reagent (Alfa Aesar GmbH & Co., Karlsruhe, Germany) made possible the transformation of the recovered fractions into trimethyl-silyl-ethers. Analysis was performed by capillary column gas chromatography as described by EEC/2568/91 regulation, Annexes V [25].

2.4. Analysis of the Fatty Acid Profile

The principle of the method is based on the conversion through trans-esterification of the fatty acids into fatty acid methyl esters (FAME) and extraction with n-hexane. Analysis of the individual fatty acids was carried out according to Regulation EEC/2568/91, Annex IV with later amendments.

2.5. Statistical Analysis

The findings were presented as mean values ± standard deviation (SD). The use MINITAB 18 software (MINITAB, INC. State College, PA, USA) was applied to analyze the samples. In addition, minimum and maximum values of the samples, mean, and standard deviation (SD) were collected. In the theory of statistics, an indication of statistical dispersion is provided by the difference between the biggest and smallest values (range). Moreover, in this study, the statistical mean differences were evaluated based on the statistical tool analysis of variation (ANOVA). A statistical significance level was set at $p < 0.05$. The tests of normality and homogeneity for the variances were carried out, and we found that both conditions are satisfied. The principal component analysis (PCA) was also conducted to investigate the association of the two monocultivars with the chemical characteristics analyzed.

3. Results and Discussion

3.1. Quality Parameters of the Examined Olive Oils

Injuries caused to the olive fruits from insect attacks and fungal diseases, improper olive harvest practices, and extraction methods as well as poor storage conditions of the extracted olive oil are the main factors affecting the quality characteristics of the produced olive oil [26]. Hence, categorization of the collected monovarietal olive oils took place by means of determination of free fatty acids, peroxide value, and spectrophotometric absorption. Table 1 shows that all analyzed samples obtained from the two examined cultivars in the coastline part of Western Greece belong to the highest quality category of "extra virgin olive oil" since they satisfy the specifications set by EU Regulation 2568/91 [25]. More specifically, the mean free fatty acid was 0.24% and 0.27%, respectively for Koroneiki and Lianolia Kerkyras olive oils. Likewise, the mean peroxide value for cv. Koroneiki olive oils was 6.64 $meqO_2$ kg^{-1} whereas for cv. Lianolia Kerkyras the mean peroxide value was 5.21 $meqO_2$ kg^{-1}. Similarly, both monovarietal olive oils had K_{232} and K_{268} mean values quite below the limit set by the EU Regulation 2568/91. The results depict that both cultivars had an overall high quality profile in that crop year as far as their basic qualitative parameters is concerned.

Table 1. Quality indices for the examined Koroneiki and Lianolia Kerkyras olive oils from the coastline region of Western Greece.

Parameter	cv. Koroneiki (N = 44)		cv. Lianolia Kerkyras (N = 60)		EEC Limit for Extra Virgin Olive Oil (EVOO) Category
	Mean ± SD	Min–Max	Mean ± SD	Min–Max	
Free acidity (%)	0.24 ± 0.10	0.13–0.55	0.27 ± 0.12	0.12–0.75	≤0.80
Peroxide value ($meqO_2$/kg)	6.64 ± 1.26	3.81–9.66	5.21 ± 1.12	3.41–8.64	≤20
K_{232}	1.56 ± 0.14	1.39–2.04	1.61 ± 0.15	1.25–1.95	≤2.50
K_{268}	0.14 ± 0.01	0.11–0.19	0.14 ± 0.02	0.11–0.21	≤0.22

Values are expressed as means ± standard deviation (SD). N = 104.

3.2. Fatty Acid Profile of the Two Monocultivar Olive Oils

Fatty acid profile plays an important role in the quality and characterization of an olive oil as its composition reflects the nutritional properties of an olive oil [27]. Several researchers have reported that among other major components, fatty acids composition seems to represent a possible tool for varietal characterization and authentication [28–33]. Table 2 shows the mean fatty acid composition of the analyzed monovarietal olive oils. As it is shown, all fatty acids identified were found in the normal range expected for the extra virgin olive oil category for both monocultivars. With respect to the mono-unsaturated oleic acid (C18:1), olive oils of cv. Koroneiki presented a higher concentration with a mean value of 75.07% compared to cv. Lianolia Kerkyras (69.55%). Moreover, the saturated stearic acid (C18:0) concentration was higher for cv. Koroneiki with a mean value of 2.51% compared to the concentration of 2.04% for cv. Lianolia Kerkyras. On the other hand, olive oils of cv. Lianolia Kerkyras presented a clearly higher concentration of the poly-unsaturated linoleic acid (C18:2) with a mean value of 10.40% compared to the Koroneiki olive oils (6.43%) as well as a higher concentration in palmitic acid (14.46%). Regarding the comparison of the contents of the different lipids in the two varieties, the different flesh to stone ratio is characteristic of the olives themselves. The flesh to stone ratio is 3–4 to 1 in cv. Lianolia Kerkyras and 5 to 1 in cv. Koroneiki, and the stone is relatively large in the latter, according to the catalogue of "Apulian and Greek Olive Varieties". This anatomical characteristic combined with the fact that the olive seeds are richer in linoleic acid and beta-sitosterol than the olive pulp [34] could explain the higher content of these components in cv. Lianolia Kerkyras. Moreover, the discrimination of the two varieties can also be aided and explained by the 18:1/18:2 ratio which indicates that the ratio in cv. Koroneiki is double that of cv. Lianolia Kerkyras.

Table 2. Fatty acid profile of the examined monocultivar olive oils in the coastline region of Western Greece.

Parameter	cv. Koroneiki ($N = 44$)		cv. Lianolia Kerkyras ($N = 60$)		Calculated *p*-Value	EEC Limit for EVOO Category
	Mean ± SD	Min–Max	Mean ± SD	Min–Max		
Myristic C14:0 (%)	0.009 ± 0.002	0.006–0.018	0.008 ± 0.004	0.003–0.04	n.s	≤0.03
Palmitic C16:0 (%)	13.17 ± 1.01	11.16–17.59	14.76 ± 0.91	12.97–16.71	0.000	7.50–20.00
Palmitoleic C16:1 (%)	1.07 ± 0.17	0.83–1.69	1.47 ± 0.19	0.97–1.91	0.000	0.30–3.50
Heptadecanoic C17:0 (%)	0.04 ± 0.01	0.02–0.06	0.04 ± 0.01	0.02–0.07	n.s	≤0.40
Heptadecenoic C17:1 (%)	0.07 ± 0.01	0.05–0.12	0.08 ± 0.01	0.05–0.13	0.003	≤0.60
Stearic C18:0 (%)	2.51 ± 0.24	2.03–2.98	2.04 ± 0.15	1.78–2.64	0.000	0.50–5.00
Oleic C18:1 (%)	75.07 ± 1.71	69.76–77.96	69.55 ± 1.71	65.39–73.00	0.000	55.00–83.00
Linoleic C18:2 (%)	6.43 ± 1.27	4.21–9.55	10.40 ± 0.91	8.30–12.80	0.000	2.50–21.00
Linolenic C18:3 (%)	0.72 ± 0.07	0.63–0.88	0.79 ± 0.08	0.60–0.99	0.000	≤1.00
Arachidic C20:0 (%)	0.45 ± 0.03	0.34–0.53	0.40 ± 0.02	0.30–0.49	0.000	≤0.60
Eicosenoic C20:1 (%)	0.29 ± 0.04	0.23–0.37	0.28 ± 0.03	0.20–0.33	n.s	≤0.50
Behenic C22:0 (%)	0.13 ± 0.02	0.09–0.18	0.13 ± 0.02	0.09–0.18	n.s	≤0.20
Lignoceric C24:0 (%)	0.05 ± 0.02	0.01–0.10	0.05 ± 0.01	0.03–0.09	0.009	≤0.20

Values are expressed as means ± standard deviation (SD). n.s = not-significant. The $p < 0.05$ was set at the level of statistical significance.

Variability in fatty acid composition between the two monocultivar olive oil samples led to the performance of an analysis of variance (ANOVA) in order to assess their differences. Table 2 shows substantial statistical differences between Lianolia Kerkyras and Koroneiki samples in almost all the analyzed fatty acids ($p < 0.05$). The analysis of variance applied to the 13 GC analyzed variables allowed the variables with the highest discriminant power to be determined. The more discriminant variables are C18:2 (F = 343.19), C18:1 (F = 264.70), C18:0 (F = 149.88), C16:1 (F = 125.54), C16:0 (F = 71.21), C20:0 (F = 65.71), and C18:3 (F = 22.28).

The effect of cultivar in the fatty acid profile of the two monocultivar olive oils originating from the coastline region of north-western Greece is depicted from the results of this work proving the usefulness of fatty acid composition for this varietal discrimination. Those results are in agreement with our previously published data as well as other relevant studies, demonstrating that fatty acid profile plays a crucial role in the classification of virgin olive oils according to their cultivar [28–33]. Finally, according to our findings, this local Greek olive variety has the tendency of exhibiting higher concentrations in the poly-unsaturated omega-6 linoleic acid and lower concentration in the mono-unsaturated omega-9 oleic acid compared to the most systematically cultivated variety of cv. Koroneiki. Of course, further in-depth research for more crop years is necessary for the adequate characterization of the fatty acid profile of Lianolia Kerkyras olive oils.

3.3. Sterolic Profile of the Two Monocultivar Olive Oils

Phytosterols and triterpenic dialcohols belong to the unsaponifiable fraction of olive oil and constitute one of its minor components with an important health beneficial impact [35–37]. Many researchers have shown that each variety has a characteristic sterol "fingerprint", revealing that the sterolic profile can be used as a reliable indicator with a high discrimination potential for olive oil classification [38–43].

Taking into account the unexplored chemical characteristics of Lianolia Kerkyras, we employed the present study to determine and compare the sterolic profile of the Koroneiki and Lianolia Kekryras olive oils obtained from the coastline region of north-western Greece. The percentage of individual sterols as well as the concentrations of total sterols for the examined monovarietal olive oil samples are presented in Table 3. In general, total sterols concentration and individual sterols content of cv. Lianolia Kerkyras olive oil samples comply with the up to date EU legislation [18]. In contrast, olive oil samples of cv. Koroneiki showed lower concentration in total sterols with a mean value of 1020.8 mg/kg compared to olive oils of cv. Lianolia Kerkyras (1343.7 mg/kg). More precisely, 40.9% of the analyzed samples of cv. Koroneiki did not exceed the defined limit of 1000 mg/kg for total sterols (EEC Regulation 2568/91) as illustrated in Figure 1. This observation is in agreement with results of our previous publication regarding the tendency of low total sterol concentration in Koroneiki olive oils of the southern Peloponnese, enhancing and clearly depicting an "intrinsic characteristic" for Koroneiki cultivar [44].

The European Commission Regulation (EEC 2568/91) imposes limits or ranges for each type of sterol and total sterols based on the natural levels found in traditional olive oil types [25]. Sterol profiles outside these limits, in combination with other chemical parameters, could theoretically suggest that the oil is not authentic. However, a number of cases have found olive oils which naturally exceed or subceed the limits for sterols [45,46]. Analysis of virgin olive oils from cv Koroneiki in completely different geographical regions such as Crete and Australia has also shown a tendency of low total sterol concentration [47,48]. Thus, low mean value in the concentration of total sterols may depict a "special and intrinsic characteristic" for Koroneiki cultivar in general, which has to do with the cultivar itself [44].

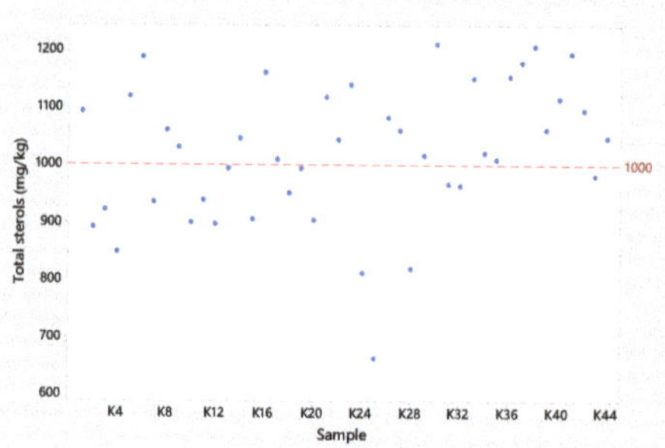

Figure 1. Diagram illustrating total sterols parameter (40.9% of Koroneiki olive oil samples are below the set limit of 1000 mg/kg). Dotted line represents the limit according to EEC/2568/91 for the EVOO category.

Table 3. Sterol profile of the examined monocultivar olive oils in the coastline region of Western Greece.

	cv. Koroneiki (N = 44)	cv. Lianolia Kerkyras (N = 60)	Calculating *p*-Value	EEC Limit for EVOO Category
Sterols and Triterpene Diols	**Mean ± SD**	**Mean ± SD**		
Cholesterol (%)	0.10 ± 0.08	0.12 ± 0.06	n.s	≤0.5
24-methylene-cholesterol %	0.23 ± 0.09	0.08 ± 0.04	0.000	
Campesterol %	3.82 ± 0.35	3.42 ± 0.17	0.000	≤4.0
Campestanol %	0.07 ± 0.03	0.04 ± 0.02	0.000	<campesterol
Stigmasterol %	0.63 ± 0.18	0.49 ± 0.15	0.000	
Chlerosterol %	0.81 ± 0.20	0.81 ± 0.16	n.s	
β-Sitosterol %	85.95 ± 2.68	89.21 ± 1.27	0.000	
Sitostanol %	0.48 ± 0.24	0.69 ± 0.17	0.000	
Δ-5-avenasterol %	6.93 ± 2.38	4.31 ± 1.27	0.001	
Δ-5, 24-stigm/dienol %	0.29 ± 0.14	0.22 ± 0.11	0.002	
Δ-7-stigmastenol %	0.32 ± 0.15	0.29 ± 0.11	n.s	≤0.5
Δ-7-avenasterol %	0.25 ± 0.16	0.26 ± 0.11	n.s	
Apparent b-Sitosterol %	94.63 ± 0.70	95.28 ± 0.35	0.000	≥93.0
Total Erythrodiol %	2.76 ± 1.07	1.43 ± 0.45	0.000	≤4.5
Total sterols (mg/kg)	1020.8 ± 120.7	1343.7 ± 115.1	0.000	≥1000

Values are expressed as means ± standard deviation (SD). n.s = not-significant. The *p* < 0.05 was set at the level of statistical significance.

With respect to the profile of the individual sterols, Lianolia Kerkyras olive oils samples showed a higher mean value for the major phytosterol β-sitosterol (89.21%) and for sitostanol (0.69%) compared to the relative values of Koroneiki olive oil samples (Table 3). Moreover, Lianolia Kerkyras exhibited lower mean values for the most abundant sterols, namely Δ-5-avenasterol (4.31%), campesterol (3.48%), stigmasterol (0.49%), as well as for total erythodiol content (1.43%). Although no previous reported data for the sterolic profile of cv. Lianolia Kerkyras is available, to the best of our knowledge, our results depict that this local Greek olive variety presents higher percentage mean values in β-sitosterol and total sterols compared to the most-known and exploited Koroneiki variety. The health-promoting aspects of β-sitosterol have been reported in several studies, recognizing its great hypocholesterolemic function in opposing the intestinal absorption of cholesterol and therefore contributing to the prevention of many diseases, including various types of cancer (colon, prostate, and breast) [49,50].

Comparison of the two monocultivar olive oils according to their sterolic profile, as shown in Table 3, by means of the calculated *p*-value, shows it to be in most cases close to 0.00 ($p \approx 0.00$), indicating a strong botanical effect. Thus, the dataset of individual and total sterols can enable the classification of the examined olive oils according to their cultivar and indicate them as a possible compositional marker for olive oil authentication. Future in-depth research by comparing the sterolic profile of olive oils derived from more olive cultivars would be advisable.

3.4. Chemometric Analysis

Principal component analysis (PCA) was carried out for the confirmation and strengthening of the classification of the examined monovarietal olive oils according to the cultivar. In order to restrict initial variables to a small number of new variables (known as principal components), the principal component analysis (PCA) is used to describe most of the original variations. The main goal of the key factor analysis is to identify the associated variables. The PCA score plot of Koroneiki versus Lianolia Kerkyras olive oils, according to their fatty acid and sterolic data set, is shown in Figure 2a,b respectively. In more detail, as shown in Figure 2a, the majority of K-points (corresponding to cv. Koroneiki) point to the left part of PC1, indicating that K has significant negative loads on component 2. Moreover, it can be observed that L-points (corresponding to cv. Lianolia Kerkyras) are shown on the right part of PC1, meaning that L has significant positive loads on component 1. Therefore, both K and L regions are independent and the data collection on fatty acids is not similar, impacting the two main regions. In accordance, the PCA score plot of Koroneiki vs. Lianolia Kerkyras olive oils' sterolic profile resulted in the creation of two separate clusters as shown in Figure 2b. In that case, we noted that L points are grouped very close to each other, depicting that the variability among Lianolia Kerkyras olive oils is very small compared to that of Koroneiki samples. A low variability for L points reveals that the samples appear to be very similar to the mean for the Lianolia Kerkyras, and therefore are not affected by any external factors.

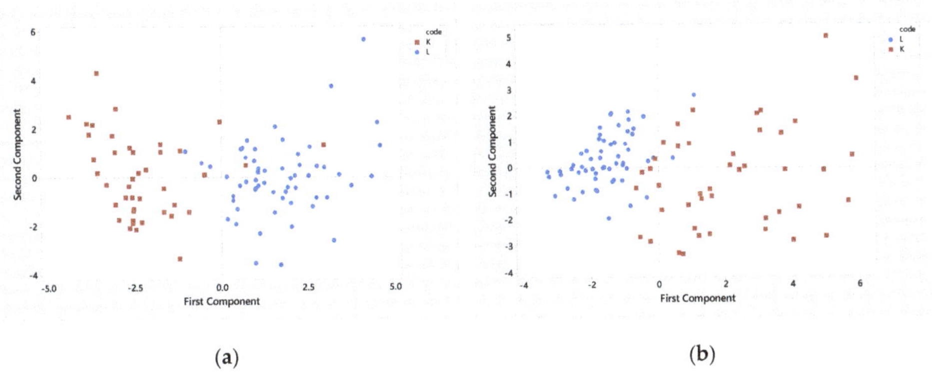

(a)   (b)

Figure 2. Score plot of principal component analysis (PCA) analysis based (**a**) on fatty acid (**b**) on the sterolic profiles of the examined monovarietal olive oils. K corresponds to Koroneiki olive oils (red spots) and L to Lianolia Kerkyras olive oils (blue spots).

In general, a discrete separation between the two cultivars was detected by applying the PCA algorithm to the data set of fatty acids and sterols. The obtained results are in accordance with other relevant studies concentrating on olive oil major and minor compounds as effective tools for studying olive oil authentication. For example, fatty acid and triglyceride composition data have shown significant potential for olive oil classification according to cultivar [29–32]. Likewise, clear differences have been observed in the content of the fatty acid as well as phenolic content of Tunisian olive varieties [33]. In addition, according to Lorenzo et al., botanical discrimination of olive oils can be achieved by examining the variables of stearic acid, campesterol, total sterols, and oxidative stability [42]. Another research group has recently used both sterol and phenolic fingerprints to discriminate Tunisian and Italian EVOOs, outlining their potential for authenticity evaluations [43].

Finally, a combined PCA was performed using the variables of the fatty acid compositional and sterolic data (a total of 28 variables in 104 observations). As shown in the score plot of Figure 3, a complete separation according to the cultivar was achieved. Hence, it can be concluded that the discrimination of cv. Koroneiki and Lianolia Kerkyras samples in terms of cultivar could occur based on both fatty acid and sterolic profile data.

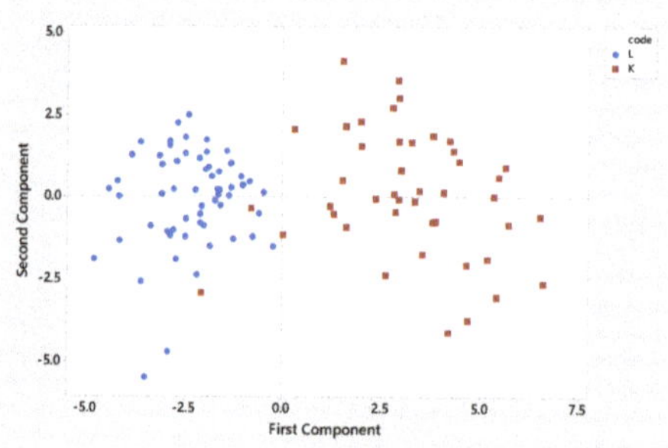

Figure 3. Score plot of PCA analysis based on 28 variables (combination of the sterolic and fatty acid data set) of the examined monovarietal olive oils. K corresponds to Koroneiki olive oils (red spots) and L to Lianolia Kerkyras olive oils (blue spots).

4. Conclusions

In the present study, we evaluated and profiled for the first time the chemical characteristics of Lianolia Kerkyras olive oils, a local Greek olive cultivar cultivated exclusively in the north-western coastline part of Greece as well as compared its chemical properties with those of the most well-known Greek olive cultivar, cv. Koroneiki. The high differentiation potential of sterols and fatty acid compositional data, as efficient authenticity tools for origin discrimination, was confirmed by employment of statistical analysis tools. The study and discrimination among local olive varieties is particularly important in order to preserve biodiversity and maintain the advantages of local varieties so as to promote and strengthen Greek olive sector. The obtained results not only lead to valuable information about the studied monocultivar olive oils, but can also contribute in the future to the establishment of a continuously enriched "Greek Authentic Olive Network" of indigenous, local, and less known and exploited monovarietal olive oils produced in Greece. Further in-depth research in combination with more examined parameters (e.g., sensory analysis, phenolic profile) and pioneering chemometric tools could secure the authenticity, traceability, and therefore higher commercial presence of Greek local olive varieties.

Author Contributions: Conceptualization, V.S., P.K., T.V.; methodology, V.S.; software, V.S.; validation, V.S., P.T.; formal analysis, V.S.; investigation, V.S., S.A., E.S.; resources, V.S.; data curation, P.T.; writing—original draft preparation, V.S., S.A., E.S., P.K., P.T., T.V.; writing—review and editing, V.S., A.S., E.S., P.K., P.T., T.V.; visualization, V.S., T.V.; supervision, T.V.; project administration, V.S., P.K.; funding acquisition, V.S., P.K., please turn to the CRediT taxonomy for the term explanation. All authors have read and agree to the published version of the manuscript.

Funding: This research received funding under the project "Authentic Olive Net" in the framework of the European program "Interreg Greece Italy" MIS Code 5003145.

Conflicts of Interest: The authors declare no conflict of interest.

References

1. Boskou, D. *Olive Oil: Chemistry and Technology*, 2nd ed.; AOCS Press: Champaign, IL, USA, 2006; pp. 45–93.
2. Preddy, V.; Watson, R. *Olives and Olive Oil in Health and Disease Prevention*, 1st ed.; Academic Press: Oxford, UK, 2010; pp. 1100–1220.
3. Roche, H.M.; Gibney, M.J.; Kafatos, A.; Zampelas, A.; Williams, C.M. Beneficial properties of olive oil. *Food Res. Int.* **2000**, *33*, 227–231. [CrossRef]
4. Official Website of the European Union. An Overview of the Production and Marketing of Olive Oil in the EU. Available online: https://ec.europa.eu/info/food-farming-fisheries/plants-and-plant-products/plant-products/olive-oil2020 (accessed on 20 July 2020).
5. International Olive Council Database. Available online: https://www.internationaloliveoil.org/what-we-do/economic-affairs-promotionunit/#exports (accessed on 10 July 2020).
6. Economic Affairs & Promotion Unit, International Olive Council. Available online: https://www.internationaloliveoil.org/what-we-do/economic-affairs-promotion-unit/ (accessed on 12 July 2020).
7. Council Regulation (EC). No. 2081/92 of 14 July 1992 on the protection of geographical indications and designations of origin for agricultural products and foodstuffs. *Off. J. Eur. Union* **1992**, *L208*, 1–8.
8. Council Regulation (EC). No. 2082/92 of 14 July 1992 on certificates of specific character for agricultural products and foodstuffs. *Off. J. Eur. Union* **1992**, *L208*, 9–14.
9. Council Regulation (EC). No. 510/2006 of 20 March 2006 on the protection of geographical indications and designations of origin for agricultural products and foodstuffs. *Off. J. Eur. Union* **2006**, *L93*, 12–25.
10. Council Regulation (EC). No. 1898/2006 of 14 December 2006 laying down detailed rules of implementation of Council Regulation (EC) no. 510/2006 on the protection of geographical indications and designations of origin for agricultural products and foodstuffs. *Off. J. Eur. Union* **2006**, *L369*, 1–23.
11. Council Regulation (EC). No. 182/2009 of 6 March 2009 amending Regulation (EC) no. 1019/2002 on marketing standards for olive oil. *Off. J. Eur. Union* **2009**, *L63*, 6–8.
12. Conte, L.; Bendini, A.; Valli, E.; Lucci, P.; Moret, S.; Maquet, A.; Lacoste, F.; Brereton, P.; García-González, D.L.; Moreda, W.; et al. Olive oil quality and authenticity: A review of current EU legislation, standards, relevant methods of analyses, their drawbacks and recommendations for the future. *Trends Food Sci. Technol.* **2019**, (in press). [CrossRef]
13. Aparicio, R.; Luna, G. Characterization of monovarietal virgin olive oil. *Eur. J. Lipid Sci. Technol.* **2002**, *104*, 614–627. [CrossRef]
14. Montet, D.; Ray, R. *Food Traceability and Authenticity*, 1st ed; CRC Press: Boca Raton, FL, USA, 2018.
15. Montealegre, C.; Alegre, M.; García-Ruiz, C. Traceability markers to the botanical origin in olive oils. *J. Agric. Food Chem.* **2009**, *58*, 28–38. [CrossRef]
16. Bajoub, A.; Medina-Rodríguez, S.; Gómez-Romero, M.; Ajal, E.A.; Bagur-González, M.G.; Fernández-Gutiérrez, A.; Carrasco-Pancorbo, A. Assessing the varietal origin of extra-virgin olive oil using liquid chromatography fingerprints of phenolic compound, data fusion and chemometrics. *Food Chem.* **2017**, *215*, 245–255. [CrossRef]
17. Alonso-Salces, R.M.; Moreno-Rojas, J.M.; Holland, M.V.; Reniero, F.; Guillou, C.; Héberger, K. Virgin olive oil authentication by multivariate analyses of ^{1}H NMR fingerprints and δ^{13}C and δ^2H data. *J. Agric. Food Chem.* **2010**, *58*, 5586–5596. [CrossRef] [PubMed]
18. Longobardi, F.; Ventrella, A.; Napoli, C.; Humpfer, E.; Schütz, B.; Schäfer, H.; Kontominas, M.G.; Sacco, D. Classification of olive oils according to geographical origin by using ^{1}H NMR fingerprinting combined with multivariate analysis. *Food Chem.* **2012**, *130*, 177–183. [CrossRef]
19. Becerra-Herrera, M.; Velez-Martín, A.; Ramos-Merchante, A.; Richter, P.; Beltrán, R.; Sayago, A. Characterization and evaluation of phenolic profiles and color as potential discriminating features among Spanish extra virgin olive oils with protected designation of origin. *Food Chem.* **2018**, *241*, 328–337. [CrossRef] [PubMed]
20. Kostelenos, G. *Olive Fruit Data, History, Description and Geographical Distribution of Olive Varieties in Greece*, 1st ed.; Kostelenos Georgios: Poros island, Greece, 2011; pp. 1–436. (In Greek)

21. Kostelenos, G. Greek olive varieties. In *Elaiokomia Egkiklopedia-The Olive Oil*, 1st ed.; Zambounis, V., Ed.; Axion Ekdotiki: Athens, Greece, 2017; Volume 2, pp. 39–53. (In Greek)

22. Pouliarekou, E.; Badeka, A.; Tasioula-Margaria, M.; Kontakos, S.; Longobardi, F.; Kontominas, M.G. Characterization and classification of Western Greek olive oils according to cultivar and geographical origin based on volatile compounds. *J. Chromatogr. A* **2011**, *1218*, 7534–7542. [CrossRef]

23. Longobardi, F.; Ventrella, A.; Casielloa, G.; Sacco, D.; Tasioula-Margari, A.K.; Kiritsakis, A.K.; Kontominas, M.G. Characterisation of the geographical origin of Western Greek virgin olive oils based on instrumental and multivariate statistical analysis. *Food Chem.* **2012**, *133*, 1169–1175. [CrossRef]

24. Karabagias, I.; Michos, C.; Badeka, A.; Kontakos, S.; Stratis, I.; Kontominas, M.G. Classification of Western Greek virgin olive oils according to geographical origin based on chromatographic, spectroscopic, conventional and chemometric analyses. *Food Res. Int.* **2013**, *54*, 1950–1958. [CrossRef]

25. Commission Regulation (EEC). No. 2568/91 of 14 July 1991 on the characteristics of olive oil and olive-residue oil and on the relevant methods of analysis. *Off. J. Eur. Union* **1991**, *L208*, 1–8.

26. Kiritsakis, A.; Christie, W. Analysis of edible oil. In *Handbook of Olive Oil*; Harwood, J., Aparicio, R., Eds.; Springer: Gaithersburg, MD, USA, 2000; pp. 129–158.

27. Boskou, D. Olive oil. In *More on Mediterranean Diets*; Simopoulos, A.P., Visioli, F., Eds.; Karger: Basel, Switzerland, 2007; Volume 97, pp. 180–210.

28. Di Bella, G.; Maisano, R.; La Pera, L.; Lo Turco, V.; Salvo, F.; Dugo, G. Statistical characterization of Sicilian olive oils from the Peloritana and Maghrebian zones according to the fatty acid profile. *J. Agric. Food Chem.* **2007**, *55*, 6568–6574. [CrossRef]

29. Ollivier, D.; Artaud, J.; Pinatel, C.; Durbec, J.P.; Guerere, M. Triacylglycerol and fatty acid compositions of French virgin olive oils. Characterization by chemometrics. *J. Agric. Food Chem.* **2003**, *51*, 5723–5731. [CrossRef]

30. D'Imperio, M.; Dugo, G.; Alfa, M.; Mannina, L.; Segre, A.L. Statistical analysis on Sicilian olive oils. *Food Chem.* **2007**, *102*, 956–965. [CrossRef]

31. Skiada, V.; Tsarouhas, P.; Varzakas, T. Comparison and discrimination of two major monocultivar extra virgin olive oils in the southern region of Peloponnese, according to specific compositional/traceability markers. *Foods* **2020**, *9*, 155. [CrossRef] [PubMed]

32. Stefanoudaki, E.; Kotsifaki, F.; Koutsaftakis, A. Classification of virgin olive oils of the two major Cretan cultivars based on their fatty acid composition. *J. Am. Oil Chem. Soc.* **1999**, *76*, 623–626. [CrossRef]

33. Krichene, D.; Taamalli, W.; Daoud, D.; Salvador, M.; Fregapane, G.; Zarrouk, M. Phenolic compounds, tocopherols and other minor components in virgin olive oils of some Tunisian varieties. *J. Food Biochem.* **2007**, *31*, 179–194. [CrossRef]

34. Alves, E.; Simoes, A.; Rosário Domingues, M. Fruit seeds and their oils as promising sources of value-added lipids from agro-industrial byproducts: Oil content, lipid composition, lipid analysis, biological activity and potential biotechnological applications. *Crit. Rev. Food Sci. Nutr.* 2020. [CrossRef]

35. Moreau, R.A.; Norton, R.A.; Hicks, K.B. Phytosterols and phytostanols lower cholesterol. *Inform (Champaign)* **1999**, *10*, 572–577.

36. Boskou, D.; Tsimidou, M.; Blekas, D. *Olive Oil Chemistry and Technology*; AOCS Press: Champaign, IL, USA, 2006; pp. 41–72.

37. López-Miranda, J.; Pérez-Jiménez, F.; Ros, E.; De Caterina, R.; Badimón, L.; Covas, M.I.; Escrich, E.; Ordovás, J.M.; Soriguer, F.; Abiá, R.; et al. Olive oil and health: In Proceedings of the II international conference on olive oil and health consensus report, Jaén and Córdoba, Spain. *Nutr. Metab. Cardiovasc. Dis.* **2010**, *20*, 284–294. [CrossRef]

38. Lukic, M.; Lukic, I.; Krapac, M.; Sladonja, B.; Pilizota, V. Sterols and triterpene diols in olive oil as indicators of variety and degree of ripening. *Food Chem.* **2013**, *136*, 251–258. [CrossRef]

39. Bucci, R.; Magrı', A.D.; Magrı, A.L.; Marini, D.; Marini, F. Chemical authentication of extra virgin olive oil varieties by supervised chemometric procedures. *J. Agric. Food Chem.* **2002**, *50*, 413–418. [CrossRef]

40. Matos, L.C.; Cunha, S.C.; Amaral, J.S.; Pereira, J.A.; Andrade, P.B.; Seabra, R.M.; Oliveira, B. Chemometric characterization of three varietal olive oils (cvs. Cobrancosa, Madural and Verdeal Transmontana) extracted from olives with different maturation indices. *Food Chem.* **2007**, *102*, 406–414. [CrossRef]

41. Brescia, M.A.; Alviti, G.; Liuzzi, V.; Sacco, A. Chemometrics classification of olive cultivars based on compositional data of oils. *J. Am. Oil. Chem. Soc.* **2003**, *80*, 945–950. [CrossRef]

42. Lorenzo, I.M.; Perez Pavon, J.L.; Fernandez Laespada, M.E.; García Pinto, C.; Moreno Cordero, B.; Henriques, L.R.; Peres, M.F.; Simoes, M.P.; Lopes, P.S. Application of headspace-mass spectrometry for differentiating sources of olive oils. *Anal. Bioanal. Chem.* **2002**, *374*, 1205–1211. [CrossRef] [PubMed]

43. Mohamed, M.B.; Rocchetti, G.; Montesano, D.; Ali, S.B.; Guasmi, F.; Grati Kamoun, N.; Lucini, L. Discrimination of Tunisian and Italian extra-virgin olive oils according to their phenolic and sterolic fingerprints. *Food Res. Int.* **2018**, *106*, 920–927. [CrossRef] [PubMed]

44. Skiada, V.; Tsarouhas, P.; Varzakas, T. Preliminary study and observation of "Kalamata PDO" extra virgin olive oil, in the Messinia region, southwest of Peloponnese, Greece. *Foods* **2019**, *8*, 610. [CrossRef] [PubMed]

45. Casas, J.S.; Bueno, E.O.; Garcia, A.M.M.; Cano, M.M. Sterol and erythrodiol uvaol content of virgin olive oils from cultivars of Extramadura (Spain). *Food Chem.* **2004**, *87*, 225–230.

46. Rivera del Alamo, R.M.; Fregapane, G.; Aranda, F.; Gomez-Alonso, S.; Salvador, M.D. Sterol and alcohol composition of Cornicabra virgin olive oil: The campesterol content exceeds the upper limit of 4% established by EU regulations. *Food Chem.* **2004**, *84*, 533–537. [CrossRef]

47. Mailer, R.J.; Ayton, J.; Graham, K. The Influence of growing region, cultivar and harvest timing on the diversity of Australian olive oil. *J. Am. Oil. Chem. Soc.* **2010**, *87*, 877–884. [CrossRef]

48. Stefanoudaki, E.; Chartzoulakis, K.; Koutsaftakis, A.; Kotsifaki, F. Effect of drought stress on qualitative characteristics of olive oil of cv Koroneiki. *Grasas Aceites* **2001**, *52*, 202–206. [CrossRef]

49. Awad, A.B.; Chan, K.C.; Downie, A.C.; Fink, C.S. Peanuts as a source of β-sitosterol, a sterol with anticancer properties. *Nutr. Cancer* **2000**, *36*, 238–241. [CrossRef]

50. Awad, A.B.; Chen, Y.C.; Fink, C.S.; Hennessey, T. Beta-Sitosterol inhibits HT-29 human colon cancer cell growth and alters membrane lipids. *Anticancer Res.* **1996**, *16*, 2797–2804.

 *applied sciences*

Article

The Effect of Eugenol and Chitosan Concentration on the Encapsulation of Eugenol Using Whey Protein–Maltodextrin Conjugates

Iceu Agustinisari [1,2], **Kamarza Mulia** [1] and **Mohammad Nasikin** [1,*]

1   Department of Chemical Engineering, Universitas Indonesia, Depok 16424, Indonesia;
    iceu.agustinisari@ui.ac.id (I.A.); kmulia@che.ui.ac.id (K.M.)
2   Indonesian Agency for Agricultural Research and Development, Jl. Ragunan 29, Pasar Minggu,
    Jakarta 12540, Indonesia
*   Correspondence: mnasikin@che.ui.ac.id

Received: 8 April 2020; Accepted: 29 April 2020; Published: 4 May 2020

Featured Application: The spray-dried eugenol formulation can be used as a food preservative preparation or for pharmaceutical purposes. The encapsulation formulation and method can be applied to other essential oils.

Abstract: Eugenol has many functional properties for food and pharmaceutical purposes, especially as an antimicrobial agent. However, its use is constrained by its volatility and shelf life because it is easily degraded due to temperature, oxidation, and light. Research on encapsulation technology using biopolymers is still required to obtain the appropriate formulation in a eugenol delivery system. The aims of this research were to develop a new formulation of protein and polysaccharides in eugenol encapsulation and to determine the effect of eugenol and chitosan concentration on the characteristics of the emulsions and spray-dried powder produced. In this study, eugenol was encapsulated in whey protein–maltodextrin conjugates and chitosan through the double layer encapsulation method. The emulsions which were prepared with 2.0% eugenol were relatively more stable than those of 1.0% eugenol based on the polydispersity index and zeta potential values. Spray-dried powder which was prepared using an emulsion of 2.0% w/w eugenol and 0.33% w/w chitosan had the highest eugenol loading. The presence of chitosan resulted in more stable emulsions based on their zeta potential values, improved thermal stability of eugenol, increased eugenol loading to become twice as much as the loading obtained without chitosan, and modified release profile of eugenol from the spray-dried powders.

Keywords: eugenol; encapsulation; whey protein–maltodextrin conjugates; chitosan

---

1. Introduction

Eugenol ($C_{10}H_{12}O_2$) which is a phenylpropanoid group consisting of an allyl chain-substituted by guaiacol is a major bioactive compound that has a concentration of 45–90% in clove. This essential oil has shown many pharmacological uses due to its antibacterial, antifungal, antiplasmodial, antivirus, anthelmintic, anti-inflammatory, analgesic, and antioxidant activities [1]. However, eugenol has limited applications, because of its volatility, slight solubility in water, and ease of damage when exposed to high temperatures, air, and light. Nevertheless, encapsulation technology has been used to overcome these problems and to improve the utilization of essential oils as pharmaceuticals and food ingredients. Some benefits of encapsulating essential oils include increasing stability, protecting active compounds from interaction with other ingredients, increasing the activity or functional properties, and decreasing volatility and toxicity [2,3]. In addition, encapsulation protects active compounds from oxidation, masks flavor, allows controlled release, and increases bioavailability and efficacy [4–6].

The benefit of encapsulation is achieved by using an appropriate polymer matrix as a wall material. Some factors affecting the retention and release of active compounds from the film include the type of polymer, the preparation method, the interaction between polymeric and active compounds, and the environment [7,8]. Generally, protein and carbohydrates are good combinations for wall materials for encapsulation [9]. Whey protein (WP) as a protein and maltodextrin (MD) as a polysaccharide are frequently used as a matrix in essential oil encapsulation, emulsifying agents, and stabilizers [10,11]. Moreover, protein-conjugated polysaccharides are better emulsifiers than proteins or polysaccharides alone. The difference between the present work and previous studies lies in the type of whey protein used. The whey protein used in this study was derived from bovine milk and had a protein content of 7–11%, whereas the isolates had a protein content of more than 90%.

No research has been conducted on double-layer eugenol encapsulation using a whey protein–maltodextrin (WPMD) conjugate as the first layer and the effect of chitosan as the second layer. Generally, double-layer encapsulation is formed using layer-by-layer electrostatic deposition of polyelectrolytes on oppositely charged surfaces and is processed by emulsification and homogenization [12,13]. The advantages of a double-layer system are better stability against environmental stress than a single layer [14] and the controlled release of active compounds [15].

Chitosan was chosen as the second layer in consideration of its characteristics and the findings of some related research. Previous studies have confirmed that a multilayer oil-in-water emulsion containing lecithin, chitosan, and pectin made through the layer-by-layer method exhibited good stability to aggregation [16]. Several studies on the effectiveness of chitosan as an encapsulating agent and its profile release have been carried out by some researchers [17–19]. Furthermore, according to Estevinho et al. [20], chitosan has the ability to form ionic or covalent bonds with crosslinking agents and form a network in which the active compounds can be maintained. These properties are highly advantageous and necessary for controlling the release of active compounds.

The objectives of this study were (1) to develop a new formulation of biopolymers using the double-layer method for eugenol encapsulation and (2) to investigate the effect of different eugenol and chitosan concentrations on emulsification, encapsulation performance, and the physical properties of spray-dried powders. In addition, the kinetics and mechanism of eugenol release from microcapsules were also evaluated.

2. Materials and Methods

2.1. Materials

WP from bovine milk (W1500) was obtained from Sigma-Aldrich, and technical MD was obtained from PT Sarana Mitra Anugrah (Bogor, Indonesia). Food-grade chitosan (88.59% degree of diacetylation) was purchased from Biotech Surindo (Cirebon, West Java, Indonesia), sodium acetate was purchased from Merck (127-09-4|106268), and eugenol (W246700) was obtained from Sigma-Aldrich.

2.2. Preparation of Whey Protein–Maltodextrin (WPMD) Conjugates

The preparation of WPMD conjugates was adopted from Shah et al. and Akhtar and Dickinson [11,21] with modifications. Whey protein was dissolved at 40 g/1000 g in distilled water and stirred for 15 min to homogenize the solution. The maltodextrin (80 g) was dissolved in the whey protein solution, which was, then, hydrated up to 18 h at room temperature (28–30 °C). The samples were spray-dried at 150 °C inlet temperature, 0.4 kPa compressed air, and 52 m^3/h air flow rate using a spray dryer (Lab Plant Spray Dryer SD05). The outlet temperature was recorded at 70–80 °C, then, the spray-dried powders were heated at 90 °C in a cabinet oven for 1.5 h, which was then continued at 110 °C for 1.5 h.

2.3. Encapsulation of Eugenol by Emulsification and Spray Drying

Emulsification to encapsulate eugenol in the WPMD conjugate and chitosan as the wall material was adopted from the method reported by Shah et al. and Preetz et al. [11,12] with modifications. The WPMD conjugate product (5% *w/w*) was dissolved in distilled water and stirred using a magnetic stirrer at 700 rpm for 15 min to homogenize the solution. The conjugate solution had two functions, i.e., an emulsifier to emulsify the essential oil and a first layer in the double-layer system. In the first step, the emulsification was prepared using a high-shear homogenizer (Ultraturrax homogenizer, IKA T25, Germany). The conjugate solution was homogenized at 15,000 rpm for 1 min to produce the primary emulsion. This was achieved by adding eugenol of 2.0% or 1.0% *w/w* of the WPMD conjugate solution mass into the solution of WPMD conjugates under stirring at 15,000 rpm. Chitosan solutions (0.0%, 0.067%, 0.2%, and 1.0% in 0.1 M acetate buffer, pH 4.5) were gradually added, separately, to the primary emulsion according to the treatments that were determined. The addition was carried out under stirring at 15,000 rpm for 3 min to obtain the secondary emulsion. The mass ratio between the eugenol primary emulsion and chitosan solution was 2:1. The secondary emulsion was treated with a high-pressure homogenizer (Panda 2000, Gea Niro Soavi) for five cycles at 400–450 bar. There were eight emulsion samples consisting of four chitosan concentrations in each of two eugenol concentrations. The final emulsions had eugenol concentrations of 2.0% and 1.0% and were coded as F1 and F2, respectively. The chitosan concentrations in the final emulsion were 0.0%, 0.067%, 0.20%, and 0.33%. All emulsion samples were spray-dried at 150 °C inlet temperature to obtain dried powder samples. The formulation and the name of the samples is described in Table 1.

Table 1. The formulation of emulsion and name of the samples obtained.

Samples	Eugenol Concentration in Primary Emulsion (%)	Chitosan Concentration in Chitosan Solution (%)	Eugenol Concentration in Secondary Emulsion (%)	Chitosan Concentration in Secondary Emulsion (%)
F1-chi 0.0%	3.0	0.0	2.0	0.0
F1-chi 0.067%	3.0	0.2	2.0	0.067
F1-chi 0.2%	3.0	0.6	2.0	0.2
F1-chi 0.33%	3.0	1.0	2.0	0.33
F2- chi 0.0%	1.5	0.0	1.0	0.0
F2-chi 0.067%	1.5	0.2	1.0	0.067
F2-chi 0.2%	1.5	0.6	1.0	0.2
F2-chi 0.33%	1.5	1.0	1.0	0.33

2.4. Encapsulation Performance

Total eugenol was determined using the method of Shah et al. [11] with modifications. The spray-dried powder (40 mg) was dissolved in 10 mL of 60% methanol by sonication for 30 min. The samples were filtered through a membrane with a 0.45 μm pore size and the injection volume was 20 μL. Analyses were performed using HPLC equipped with an RS diode array detector at 280 nm. The mobile phase consisted of 60 mL/100 mL aqueous methanol using an isocratic mode at a flow rate of 1 mL/min in a C-18 column (Agilent, Zorbax Eclipse Plus; 4.6 mm × 25 cm, 5 μm, 100 Å). The eugenol concentration was determined from the sample peak area and used in the calculation of the loading capacity (LC). Encapsulation efficiency (EE) is defined as the percentage of total eugenol mass in spray-dried product with reference to the corresponding mass of eugenol in feed. The calculation of mass of eugenol in feed involved non-solvent mass in emulsion and mass of collected product.

$$\text{Loading (g/100g)} = \frac{Mass\ of\ eugenol}{Mass\ of\ collected\ product} \times 100 \tag{1}$$

$$\text{Mass of eugenol in feed (g)} = \frac{mass\ of\ eugenol\ in\ emulsion}{non\ solvent\ mass\ in\ emulsion} \times \text{mass of collected product} \tag{2}$$

$$\text{Encapsulation efficiency (g/100g)} = \frac{Mass\ of\ eugenol}{mass\ of\ eugenol\ in\ feed} \times 100 \tag{3}$$

2.5. Particle Size and Zeta Potential Measurement of Emulsion and Spray-Dried Powder Dispersion

The particle size and zeta potential of the emulsion and the dispersion of dried powder were measured using a Zetasizer Nano ZS (Malvern Instrument, Malvern, UK). The measurement principle for particle size was based on dynamic light scattering, while zeta potential was determined using laser doppler micro electrophoresis. The light source is He-Ne laser and a detector angle of 173°. Three drops (0.07 g) of emulsion samples were diluted in 20 mL of distillation water. The particle dimension of the spray-dried powder was measured by dispersing the powder (1.33 mg) in distilled water (20 mL), followed by sonication (Branson 3510) for 5 min. The dispersion was poured into the disposable plastic micro cuvette for particle size and disposable folded capillary cell (DTS 1070), for zeta potential measurement. The particle size measurement results showed the mean of a droplet diameter (z-average) and polydispersity index (PdI) value. Each data value was an average of three measurements.

2.6. Differential Scanning Calorimetry (DSC)

The thermal denaturation properties of eugenol-encapsulated microparticles using the WPMD conjugate and chitosan as wall materials were analyzed with a DSC 8000 Perkin Elmer. The samples were weighed to be approximately 7–10 mg and prepared in aluminum pans. The samples were heated from 40 to 300 °C at a rate of 10.00 °C/min. The thermal transitions were evaluated in terms of peak transition temperature (Tp) and enthalpy (ΔH). On the basis of the International Confederation for Thermal Analysis and Calorimetry (ICTAC) standard, the peak value in DSC thermograms is known as the melting point in polymer samples https://www.perkinelmer.com/CMSResources/Images/44-74542GDE_DSCBeginnersGuide.pdf.

2.7. Fourier Transform Infrared (FTIR)

All infrared spectra of eugenol microcapsules were obtained using a Nicolet iS50 FTIR spectrometer (Thermo Scientific). The instrument was equipped with a KBr beam splitter and a DTGS KBr detector. The scanning process was carried out at a resolution of 2 cm^{-1} with a frequency range of 400 and 4000 cm^{-1}.

2.8. Study of In Vitro Release of Eugenol and Release Kinetic

The in vitro release test of eugenol followed the method used by Hosseini et al. [22] and Chen et al. [23]. The sample (40 mg) was placed in an Erlenmeyer flask containing 10 mL of 60% phosphate-buffered saline (pH 7.4) and 40% ethanol. Agitation was carried out using a shaker incubator at a temperature of 27–30 °C with a speed of 75 rpm. At the appointed time intervals, 5 mL of the sample was sucked out for analysis and replaced with 5 mL of the buffer phosphate solution. The sample was transferred into a tube for further filtration using 0.45 μm PVDF and put into a 2 mL vial for the measurement process by HPLC. The cumulative percentage of the amount of active compound/essential oil released from the spray-dried powder was obtained by dividing the cumulative amount of essential oil released at any given time interval (Mt) by the initial weight of the encapsulated essential oil (M0) (Equation (3)):

$$\text{Release Cumulative Percentage} = \sum_{t=0}^{t} \frac{Mt}{M0} \times 100\% \tag{4}$$

$$\frac{Mt}{M0} = kt^{n} \tag{5}$$

The release mechanism and release kinetics of eugenol from the spray-dried powder were investigated using the Korsmeyer–Peppas kinetic model [24] (Equation (4)). Mt is the amount of eugenol released at a given time, M0 is the initial amount of eugenol contained in the spray-dried powder or the maximum amount of eugenol that can be released from the spray-dried powder, t is the release time, k is the kinetic constant for the system, and n is the release characteristic and determines the release mechanism.

2.9. Statistical Analysis

The data analyses were carried out with the MINITAB 19 software using analysis of variance (ANOVA), and differences were considered significant at $\alpha < 0.05$.

3. Results and Discussion

3.1. Effect of Chitosan and Eugenol Concentration on Eugenol Emulsions

The data showed that the droplet size of F1 emulsions tended to increase along with the increasing concentration of chitosan solution as the second layer in the eugenol encapsulation (Table 2). This result is in line with that of Lertsurthiwong and Rojsitthisak [25], who investigated oil encapsulation using chitosan and alginate as a biopolymer. The increased droplet size indicated polymer attachment on the surface of the oil core. However, the eugenol emulsion F2 did not show the same phenomenon. The results of the ANOVA test indicated that the increase in droplet size, due to the chitosan concentration, was not significantly different. Furthermore, the Tukey's test results showed that there were significant differences in droplet size between F2-chi 0.0% and F2-chi 0.067%.

Table 2. Droplet size, polydispersity index, and zeta potential of emulsions.

Treatments (*)	Droplet Size (nm)	Polydispersity Index	Zeta Potential (mV)
F1 emulsion-chi 0.0%	243.6 ± 18.3 [bcd]	0.239 ± 0.020 [bcd]	−16.4 ± 1.1 [c]
F1 emulsion-chi 0.067%	224.4 ± 10.5 [cd]	0.202 ± 0.009 [d]	20.5 ± 0.3 [ab]
F1 emulsion-chi 0.2%	288.7 ± 5.2 [bc]	0.213 ± 0.006 [cd]	24.8 ± 2.8 [a]
F1 emulsion-chi 0.33%	311.5 ± 4.9 [b]	0.204 ± 0.016 [cd]	24.9 ± 2.9 [a]
F2 emulsion-chi 0.0%	200.5 ± 12.7 [d]	0.264 ± 0.013 [bc]	−10.5 ± 4.2 [c]
F2 emulsion-chi 0.067%	417.4 ± 61.3 [a]	0.486 ± 0.042 [a]	14.4 ± 5.4 [b]
F2 emulsion-chi 0.2%	301.1 ± 24.5 [b]	0.250 ± 0.024 [bcd]	19.1 ± 5.5 [ab]
F2 emulsion-chi 0.33%	286.9 ± 15.0 [bc]	0.299 ± 0.020 [b]	15.9 ± 1.3 [ab]

(*) F1 and F2 contained 2.0% and 1.0% eugenol, respectively. Data represent mean ± standard deviation. Different letters indicate significant differences among samples ($p < 0.05$).

The eugenol emulsion of F2-chi 0.067% had the largest particle size and was significantly different from the other samples. This was probably due to the instability of the emulsion, which eased the aggregation. The instability of the emulsion was also shown by its PdI and zeta potential values. The PdI of F2-chi 0.067% was the highest among all the samples, whereas the zeta potential value was the lowest (Table 2). A high PdI value indicates instability of an emulsion [26].

Overall, the data showed that F1 emulsions had PdI values closer to 0.2, which represented the homogeneity of the emulsions. The F1 emulsions also showed a higher zeta potential than F2, but they were not significantly different. The F1 emulsion-chi 0.0% and the F2 emulsion-chi 0.0% showed negative values of zeta potential due to the absence of chitosan. The values were close to 30 mV, which revealed the stability of the emulsions. Zeta potential is an indicator of the stability of an emulsion and is influenced by the electric charge of the interface [27,28]. This phenomenon indicated that the formula F1 emulsion was more stable than the formula F2 emulsion. The difference between the F1 and F2 emulsion formulas was in the eugenol concentration. The ratios of the mass emulsifier (WPMD conjugates) and eugenol in F1 and F2 emulsions were 5:3 and 5:1.5, respectively. The higher eugenol concentration in this study produced a better emulsion. The effect of the clove oil content on droplet size and stability in nanoemulsions was investigated by Shahavi et al. [29].

3.2. Performance of Spray-Dried Powder

The spray-dried samples redispersed in water had a larger mean droplet size of emulsion before spray-drying. This phenomenon denoted that some of the nanoparticles formed aggregations during the spray-drying process. Similar studies on the change of particle size during drying have shown that the particle size of essential oils encapsulated in zein nanoparticle samples also increased after lyophilizing [30]. Increased particle size was probably caused by the imperfections of the hydration or the occurrence of structural changes within atomized droplets during spray-drying. From the research by Chen et al. [23], structural changes can occur due to the differences in the distribution of essential oils that evaporated and condensed during spray-drying which later cooled. The difference in the concentration of eugenol used in the F1 and F2 treatment samples perhaps caused differences in the particles' structural changes during drying. Therefore, F1 samples (2.0% eugenol) had a larger dispersed particle size than that of the F2 samples (1.0% eugenol). Moreover, increasing the chitosan concentration also increased the size of the dried particles. The chitosan content of 0.33% in the emulsion significantly increased the size of the spray-dried powder particles. The F1 powder-chi 0.33% had the largest particle size (1243 nm), as well as PdI value (0.71) (Table 3).

Table 3. Particle size, polydispersity index, and zeta potential of powders.

Treatments (*)	Particle Size (nm)	Polydispersity Index	Zeta Potential (mV)
F1-chi 0.0%	403.2 ± 17.2 [e]	0.639 ± 0.048 [abc]	−26.9 ± 0.8 [d]
F1-chi 0.067%	679.2 ± 55.4 [bcd]	0.511 ± 0.041 [cd]	−2.2 ± 0.2 [c]
F1-chi 0.2%	834.6 ± 66.3 [b]	0.669 ± 0.016 [ab]	10.3 ±1.5 [ab]
F1-chi 0.33%	1243 ± 72.6 [a]	0.705 ± 0.085 [a]	11.4 ±1.0 [ab]
F2-chi 0.0%	404.3 ± 83.6 [e]	0.468 ± 0.016 [d]	18.7 ± 1.2 [d]
F2-chi 0.067%	498.0 ± 25.1 [de]	0.495 ± 0.067 [cd]	3.0 ± 8.4 [bc]
F2-chi 0.2%	543.9 ± 109 [cde]	0.540 ± 0.068 [bcd]	16.1 ± 3.2 [a]
F2-chi 0.33%	702.6 ± 34.5 [bc]	0.559 ± 0.043 [abcd]	19.2 ±0.3 [a]

(*) F1 and F2 contained 2.0% and 1.0% eugenol, respectively, in emulsion. Data represent mean ± standard deviation. Different letters indicate significant differences among samples ($p < 0.05$).

In contrast to the emulsions, the F1 spray-dried samples had smaller zeta potential values than that of the F2 spray-dried samples. However, the trend remained the same, where F1-chi 1.0% had a higher zeta potential value than that of the F1-chi 0.2% and F1-chi 0.067%, likewise for the F2 samples.

Encapsulation performance of eugenol using WPMD conjugates and chitosan was determined by the percentage of encapsulation efficiency (EE) and loading capacity (LC). The percentage of EE and LC increased with increasing levels of chitosan concentration of 0.067%, 0.2%, and 0.33% (Table 4). Chitosan content of 0.33% in formula F1 significantly increased the LC and EE, however, formula F2 showed a different phenomenon. This finding is similar to that of Dima et al. [31], who concluded that coriander essential oil-loaded microspheres prepared in an alginate/chitosan system with a 1:2 ratio had a higher value of entrapment efficiency than microspheres prepared in an alginate/chitosan system with a 1:1 ratio.

Table 4. Eugenol feed and encapsulation performance.

Treatments (*)	Eugenol in Feed (%)	Mass of Collected Product (g)	Loading Capacity (%)	Efficiency Encapsulation (%)
F1-chi 0.0%	37.50	3.91 ± 0.08	6.6 ± 1.2 [b]	17.6 ± 3.9 [c]
F1-chi 0.067%	37.13	2.19 ± 0.19	4.7 ± 1.1 [bc]	12.6 ± 2.8 [c]
F1-chi 0.2%	36.41	2.58 ± 0.22	6.3 ± 1.2 [bc]	17.4 ± 3.2 [c]
F1-chi 0.33%	35.71	2.83 ± 0.40	16.8 ± 0.9 [a]	47.1 ± 2.5 [a]
F2-chi 0.0%	23.08	3.00 ± 0.06	6.5 ± 1.0 [b]	28.1 ± 3.5 [b]
F2-chi 0.067%	22.80	2.29 ± 0.36	3.6 ± 0.8 [c]	15.8 ± 2.6 [c]
F2-chi 0.2%	22.26	2.54 ± 0.33	6.7 ± 0.7 [b]	30.2 ± 2.4 [b]
F2-chi 0.33%	21.74	3.23 ± 0.59	7.4 ± 1.0 [b]	34.2 ± 3.6 [b]

(*) F1 and F2 contained 2.0% and 1.0% eugenol, respectively, in emulsion. Data represent mean ± standard deviation. Different letters indicate significant differences among samples ($p < 0.05$).

The F1 spray-dried sample with 0.33% chitosan solution (F1 powder-chi 0.33%) had the highest loading capacity and was significantly different from the other samples (Table 4). The spray-dried samples with 0.067% chitosan solution for both formulas F1 and F2 showed the lowest percentage of EE and LC. These values were lower than the EE and LC of formulas F1 and F2 without chitosan. This could be attributed to the instability of the emulsion and the droplet size. Chen et al. [23] stated that large particle sizes at low pH caused a decrease in total oil content and encapsulation efficiency in spray-dried powder. Another researcher stated that during the spray-drying process, smaller particles had a higher mass transfer rate and more quickly formed semipermeable membranes around the atomized droplets, which reduced the loss of volatile compounds [32]. This phenomenon could have happened to the F2-chi 0.067% spray-dried powder sample. This sample came from an emulsion that had the largest emulsion droplet size, the highest PdI value, and a low zeta potential value. However, the spray-dried powder of F1-chi 0.067% had a small particle size and PdI value, and therefore the opinion of Jafari et al. [32] is not appropriate for this case. There could be differences in interactions between the WPMD conjugate, eugenol, and chitosan solutions in the acetate buffer and the interactions between the WPMD conjugates, eugenol, and acetate buffer without chitosan. The difference in interaction caused a higher percentage of LC and EE of eugenol in the spray-dried powder F1-chi 0.0% than in F1-chi 0.067%, as well as between F2-chi 0.0% and F2-chi 0.067%.

The eugenol spray-dried powder produced from this research had higher loading than some similar research results. The eugenol loading of the F1 sample powder-chi 0.33% (16.8%) was higher than the result of Woranuch and Yoksan [33]. They stated that eugenol loading in their eugenol-loaded chitosan nanoparticle study ranged from 0.85% to 12.80%. Research on eugenol loaded in zein/casein nanocomplexes resulted in a loading of 5.5% [23]. The encapsulation with the WPMD conjugate produced a loading ranging from 5.0% to 7.9% [11]. Generally, the encapsulation efficiency of the F2 powders was higher than that of the F1 powders, except for F1-chi 0.33%. We found that a higher eugenol concentration lowered the encapsulation efficiency. This result is in line with that of Hosseini et al. [22], who reported that EE decreased with an increase in the initial essential oil content. This could be due to the saturation of essential oil loading into the wall material during the encapsulation process and also the effect of chitosan concentration.

The level of saturation of eugenol was known from experiments encapsulating eugenol with four levels of eugenol concentration. The result showed that the percentage of EE increased at eugenol concentrations up to 2%, and then decreased when loaded with 4.0% eugenol. The LC increased until 4.0% eugenol concentration, and then decreased at 8% eugenol concentration (Figure S1). Volatile compound retention can be influenced by an optimum percentage of infeed solid that varies for each carrier or encapsulating agent. However, in this research, all formulas used the same mass and type of infeed solid (WPMD conjugate and chitosan). Each formulation is distinguished by the percentage of eugenol. In general, the encapsulation process with spray drying usually used 20% flavor load based on carrier solids. The higher percentage of that flavor load usually causes unacceptably high losses of flavors during spray drying [34]. The formula with a concentration of eugenol 2% (F-2%) has a maximum percentage of encapsulation efficiency (EE). The EE is influenced by the mass of eugenol in feed, which also involves a solid carrier in its calculation. The percentage of eugenol in feed for formula F-2% is 35.71% (data not shown). This value was greater than the formula with 1% eugenol (21.74%) and smaller than the eugenol formula 4% (52.63%), and the eugenol formula 8% (68.97%). The percentage of eugenol in feed for F-2% was probably the optimum value in the formulation of this study because it produced the maximum percentage of EE, although the maximum loading was at F-4%. At F-4% the percentage of EE decreased. The ANOVA test results showed that the loading percentage between F-2% and F-4% was not significantly different, while the percentage of EE was significantly different. Therefore, it can be concluded that the 2.0% eugenol concentration was the best at obtaining the optimal percentages of loading and encapsulation efficiency in the encapsulation of double-layer eugenol using WPMD conjugates and chitosan.

3.3. Differential Scanning Calorimetry

All spray-dried powders containing chitosan, except the microcapsule of F2 powder-chi 0.067%, showed endothermic processes of temperature peak at 74.8 °C until 78.6 °C (Figure 1A,B). Chitosan had two endothermic peaks at 85.88 and 275.26 °C (Figure S2). The endothermic peak for eugenol is 258.81 °C [35]. The disappearance of the peaks located at these temperatures indicated that there was an interaction between the WPMD conjugates and eugenol, as well as with chitosan. Previous studies reported that the individual characteristics of the graph polymer wall material and the active compounds are not visible in the DSC graphs if the interaction and encapsulation efficiency are good [36–38]. The interaction between core and the wall material determined the retention of the active compound. According to Hill et al. [39], the varied ratio of essential oil to the wall material resulted in the differences in the interaction of essential with the wall material which, then, influenced the encapsulation efficiency.

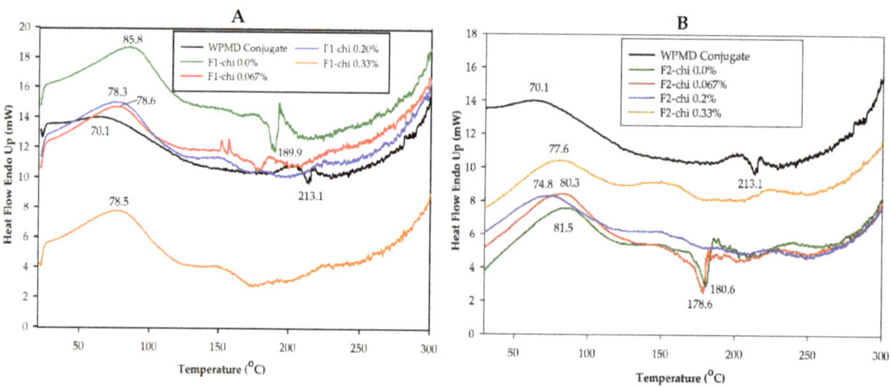

Figure 1. Differential scanning calorimetry (DSC) thermogram of F1 (**A**) and F2 (**B**) spray-dried powders.

According to the thermograms, the spray-dried powders with 0.067%, 0.2%, and 0.33% chitosan concentrations for both eugenol concentrations (F1 and F2) appeared to have adjacent endothermic peak temperatures (Figure 1A,B). Therefore, the chitosan concentration did not have a different effect on the thermal behavior of the eugenol spray-dried powder. However, the addition of chitosan as the second layer in eugenol encapsulation can protect the sample from denaturation and decomposition. This was indicated by the absence of the exothermic peak and only one endothermic peak with temperatures below 100 °C. The endothermic peak in the range of 100 °C can be related to the gelatinization process. There was a decomposition of carbohydrates and similar structures in the presence of water during heating. This process was complex because, at the same time, water vaporization also occurred. Meanwhile, the endothermic peak in the range of 200 °C is related to the melting materials and the possibility of partial and liquid evaporation [40].

The thermograms showed that the formulations without chitosan (F1 powder-chi 0.0% and F2 powder-chi 0.0%) had endothermic and exothermic peak temperatures. The F1 powder-chi 0.0% had 85.75 and 189.97 °C endothermic and exothermic peaks, respectively. The temperature peaks of F2 powder-chi 0.0% at 81.49 and 180.61 °C, respectively, were lower than those of F1 powder-chi 0.0%. These two types of peak temperatures were also observed in the thermograms of the WPMD conjugate powder and the F2 powder-chi 0.067% at 70.14 and 213.09 °C and at 80.32 and 178.62 °C, respectively. The exothermic peak was attributed to the occurrence of decomposition.

According to Ronkart et al. [41] and Beirão-da-Costa et al. [42], the material reorganization during the spray-drying process could affect the changes in microcapsule melting temperatures. In their investigation, reorganization occurred when the polymer was heated above its transition glass temperature (Tg). In this study, the spray-drying process was held at 150 °C. The DSC measurements showed that the temperature of Tg maltodextrin and chitosan was not detected, while the Tg temperature of whey protein was 119.47 °C (Figure not shown). Therefore, the occurrence of material reorganization in eugenol encapsulation cannot be ascertained.

3.4. Fourier Transform Infrared Spectroscopy (FTIR)

The eugenol spray-dried powder of formula F1 with different levels of chitosan concentrations had similar spectra profiles (Figure 2A). The presence of eugenol and the formation of encapsulation could be detected from the appearance of peaks at wavenumbers in the range of 1517–1561 cm^{-1} in the F1 spray-dried powder samples. The encapsulation process caused a shift in wavenumbers of the spray-dried powder sample, however, the wavenumbers were in the range of functional groups that represent C = C aromatic ring and secondary amines, NH bend. This wavenumbers' shift probably indicated physical interactions among eugenol, WPMD, and chitosan. The results of Piletti [43] explained that the physical interaction between eugenol with the β-cyclodextrin molecules caused the modification of the O-H functional group. The IR bands of eugenol, WPMD conjugates, and chitosan are presented in Table S1. The existence of a functional group in the sample could also be shown by the percentage of transmittance. In general, the percentage of transmittance in the wavenumber range of 1550–1650 cm^{-1} for F1-chi 0.0% was lower than for F1-chi 0.067% and F1-chi 0.33%. This proved that F1-chi 0.33% had more chitosan than other F1 samples.

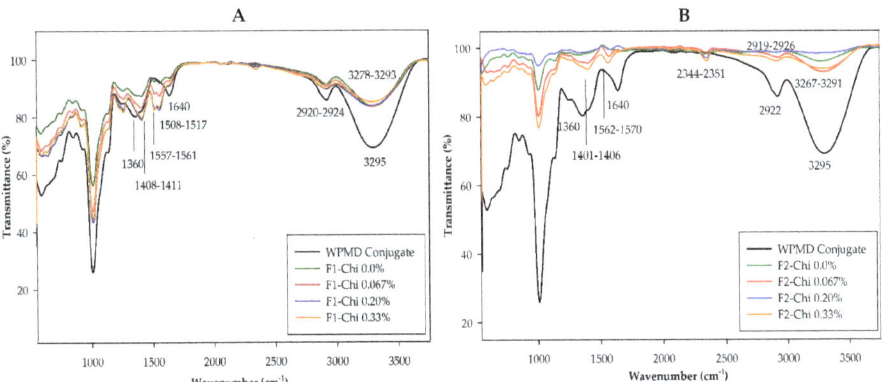

Figure 2. Infrared spectra of F1 (**A**) and F2 (**B**) spray-dried powders.

The existence of a C = C aromatic group in all F1 powder samples was seen from peaks in the 1508 cm^{-1} region. These results are in line with those of Sajomsang et al. [44] and Woranuch and Yoksan [33]. Strong absorption was detected in F1 powder-chi 0.33%, indicating the higher eugenol content (Table S2). The data in Table 2 show that the F1 powder-chi 0.33% had the highest percentage loading capacity of eugenol. This was also reinforced by the detection of the OH bend at 1408–1411 cm^{-1} and a C–H aromatic ring at 559–998 cm^{-1}. The wavenumber range of 1400–1419 cm^{-1} represents phenol and OH bend groups [45]. The intensity of these peaks in F1 powder-chi 0.33% was slightly higher than in F1 powder-chi 0.0%, 0.067%, and 0.2% (Figure 2A). The peak shifts that occurred were also likely related to the physical interaction between eugenol and the wall materials in the encapsulation process as explained above.

The profiles of FTIR spectra of F2 spray-dried powder samples were generally different from F1 powders (Figure 2B). Compared with the spectra of F1 powders, the percentage of transmittance of spectra of F2 powders was higher, meaning that the absorption was lower, which indicates the weak intensity of certain functional groups (Table S3). Peaks in the range of 3271–3295 cm^{-1} experienced an increase in transmittance. This indicated the intensity of the hydrogen bonds between eugenol and the WPMD conjugate or between eugenol and free maltodextrin in F2 powders was lower than in F1 powders. Peaks in the range of 1563–1577 cm^{-1} in a spray-dried powder containing chitosan represented NH functional groups that showed the presence of chitosan. The existence of eugenol was detected at peaks of 1401–1402 cm^{-1}, which could have been shifted from specific wavenumbers of eugenol (Table S1). However, the intensity of these wavenumbers was lower than that of F1 powders, as seen from the percentage of the transmittance (Figure 2A,B). This was because the eugenol loading of F2 powders was lower than that of F1 powders (Table 4).

3.5. In Vitro Release Characteristic of Eugenol

A release study of eugenol from the spray-dried powder was carried out to determine the effect of chitosan concentration on the eugenol released from microcapsules at a predetermined time. The results showed that F1-chi 0.0% powder had the lowest cumulative percentage of eugenol release (41.35%) among other F1 spray-dried powders at the end of the observation time (Figure 3A). However, among the samples containing chitosan, F1-chi 0.33% powder showed the lowest eugenol cumulative release (57.22%) as compared with F1-chi 0.067% (69.78%) and F1-chi 0.2% (59.22%). Chitosan concentration seemed to have the effect of inhibiting the release of eugenol in F1 encapsulation formulations.

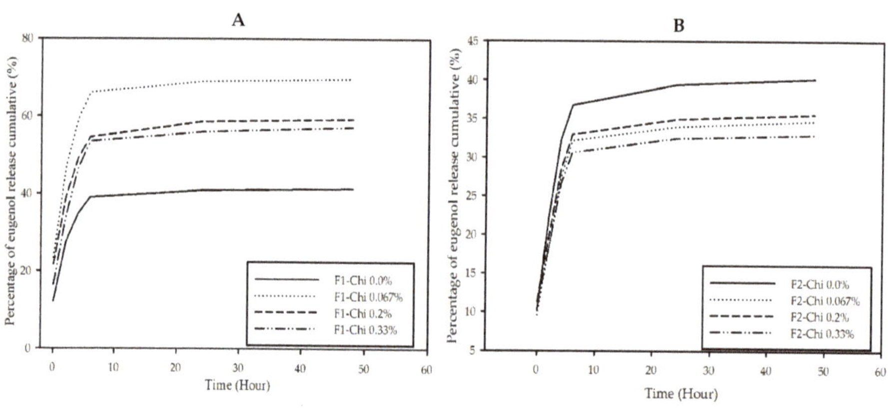

Figure 3. Release profile of eugenol from F1 (**A**) and F2 (**B**) spray-dried powders.

The difference in eugenol concentration affected its interaction with the WPMD conjugates and chitosan, which was thought to influence the release profile of eugenol from F2 microcapsules. The spray-dried powder of F2-chi 0.0% (40.16%) had a greater percentage of eugenol cumulative release than the F2 powder containing chitosan at the end of the observation time (Figure 3B). The F2-chi 0.067% (34.68%), 0.6% (35.55%), and 1.0% (32.89%) had adjacent release graphs; however, the F2-chi 0.33% appeared to have the lowest cumulative release percentage at 48 h. Thus, it can be concluded that the presence of chitosan retarded the release of eugenol from microcapsules.

The percentage of eugenol release increased for 6 h and began to slow down after that until 48 h (Figure 3A,B). The retardation of eugenol release was probably related to the diffusion of active compounds contained in the particles [24]. Furthermore, Agnihotri et al. [46] explained that drug release through a diffusion mechanism began with penetration of the medium into the particle system, which caused the particles to swell, and then the drug was released from the swollen matrix.

The mechanism and eugenol release kinetic model of the spray-dried powders were also studied. The calculation to determine the mechanism and release kinetics used the Korsmeyer–Peppas equation (Equation (5)). The results of the calculation of the release kinetics showed that the values of the diffusion exponential (n) of all samples were less than 0.43 (Table 5). This indicates that the release mechanism in the eugenol microcapsules followed Fickian release (case I transport).

Release of eugenol occurred through a diffusion process. Ritger and Peppas [47] explained the criteria of the n value as follows: n ≤ 0.43 indicates Fickian release, n = 0.85 indicates case II transport, and 0.43 < n < 0.85 indicates non-Fickian release (an anomalous behavior). The results also revealed the diffusion constant value (k) and Pearson coefficient (R^2). The k value indicates the kinetics of the release. The greater the value of k, the faster the release of an active compound from microcapsules [48].

Table 5. Kinetic data of eugenol released from microcapsules using the Korsmeyer–Peppas kinetic model.

Samples	n	k	R^2
F1 powder-Chi 0.0%	0.2350	0.2061	0.8243
F1 powder-Chi 0.067%	0.2061	0.3729	0.8379
F1 powder-Chi 0.2%	0.1926	0.3204	0.8683
F1 powder-Chi 0.33%	0.2479	0.2688	0.8538
F2 powder-Chi 0.0%	0.2667	0.1756	0.8619
F2 powder-Chi 0.067%	0.2519	0.1590	0.8696
F2 powder-Chi 0.2%	0.2264	0.1756	0.8765
F2 powder-Chi 0.33%	0.2263	0.1635	0.8720

F1 and F2 contained 2.0% and 1.0% eugenol, respectively, in emulsion, n = diffusion exponential, k = diffusion constant, R^2 = Pearson coefficient.

Regarding some of the phenomena encountered in this experiment, the F1 formula samples had a cumulative release percentage greater than the formula F2 samples. The result was supported by the k value of the samples. The k value of F1 samples was higher than that of the F2 samples, indicating that the release of eugenol from F1 samples was faster than that from the F2 samples. The F1-chi 0.0% had the lowest cumulative release percentage and k value among other F1 microcapsules, while F2-chi 0.0% had the highest cumulative release percentage among other F2 microcapsules. According to Keawchaoon and Yoksan [24], the amount and release rate of a component were influenced by the pH of the media. However, the media used in this release trial had the same pH (7.4) for all samples. The differences between each sample were the combination of eugenol concentration and chitosan concentration. The eugenol spray-dried powders were obtained from eugenol emulsions with different emulsion pH values. The addition of different concentrations of chitosan solution to the first eugenol emulsion (primary emulsion) produced a different pH value in the second eugenol emulsion (secondary emulsion). The F1-chi 0.0% and F2-chi 0.0% emulsions had a lower pH than F1 and F2 containing chitosan. The pH value of the emulsion formula F1 was lower than the pH value of the emulsion formula F2. The pH difference could have influenced the electrostatic interaction among eugenol, WPMD conjugates, and chitosan which, then, influenced the release of eugenol. The mechanism was explained by Combrinck et al. [48]. Another possible reason for the release phenomenon of F1 and F2 powders is the difference in eugenol loading (Table 4). An increase in the rate of release of essential oils from microcapsules containing more essential oils is due to a decrease in the thickness of the microcapsule wall because insufficient wall material encapsulates the oil completely [49].

4. Conclusions

In this study, WPMD conjugates acted as an emulsifier, as well as a coating agent, in eugenol encapsulation. The concentration of eugenol and chitosan had an effect on the properties of the emulsions and spray-dried powders, including particle size, zeta potential, encapsulation efficiency, and loading capacity. The spray-dried powder prepared using an emulsion of 2.0% *w/w* eugenol and 0.33% *w/w* chitosan had the highest eugenol loading of 16.8%. The eugenol loading of this formulation with chitosan was twice as much as the loading obtained without chitosan reported previously. The eugenol encapsulation using chitosan as the second layer turned out to be effective at stabilizing the emulsion, as revealed by the zeta potential value. The presence of chitosan also improved the thermal stability and prevented the decomposition of the encapsulation product. The differences in eugenol concentration affected the release profile, whereas chitosan concentration had a role in slowing eugenol release. Generally, the results of the study indicate that the eugenol-loaded formulation yielded a biocompatible product with potential applications as food additives and pharmaceuticals.

Supplementary Materials: The following are available online at http://www.mdpi.com/2076-3417/10/9/3205/s1, Figure S1: Encapsulation efficiency and loading capacity of eugenol microcapsules loading with various concentration of eugenol, Figure S2: DSC thermogram of chitosan powder, Table S1: IR band WPMD conjugate, eugenol, and chitosan obtained from the FTIR measurement of the research, Table S2: IR band and % transmittance of F1 samples, Table S3: IR band and % transmittance of F2 samples.

Author Contributions: Conceptualization, I.A., K.M. and M.N.; methodology, I.A. and K.M.; formal analysis, I.A.; investigation, I.A.; data curation, I.A. and K.M.; writing—original draft preparation, I.A.; writing—review and editing, I.A., K.M. and M.N.; visualization, I.A.; supervision, K.M. and M.N.; project administration, I.A. All authors have read and agreed to the published version of the manuscript.

Funding: This research received funding for the experimental work (2017–2020) and the APC (2020) from the Indonesian Agency for Agricultural Research and Development.

Acknowledgments: The authors thank the Indonesia Agency for Agricultural Research and Development, Ministry of Agricultural and Universitas Indonesia for the research funding.

Conflicts of Interest: The authors declare no conflict of interest.

References

1. Kamatou, G.; Vermaak, I.; Viljoen, A. Eugenol—From the Remote Maluku Islands to the International Market Place: A Review of a Remarkable and Versatile Molecule. *Molecules* **2012**, *17*, 6953–6981. [CrossRef] [PubMed]
2. Donsi', F.; Annunziata, M.; Sessa, M.; Ferrari, G. Nanoencapsulation of essential oils to enhance their antimicrobial activity in foods. *LWT* **2011**, *44*, 1908–1914. [CrossRef]
3. Bilia, A.R.; Guccione, C.; Isacchi, B.; Righeschi, C.; Firenzuoli, F.; Bergonzi, M.C. Essential Oils Loaded in Nanosystems: A Developing Strategy for a Successful Therapeutic Approach. *Evid. Based Complement. Altern. Med.* **2014**, *2014*, 1–14. [CrossRef] [PubMed]
4. Nedović, V.; Kalusevic, A.; Manojlovic, V.; Levic, S.; Bugarski, B. An overview of encapsulation technologies for food applications. *Procedia Food Sci.* **2011**, *1*, 1806–1815. [CrossRef]
5. Neethirajan, S.; Jayas, D.S. Nanotechnology for the Food and Bioprocessing Industries. *Food Bioprocess Technol.* **2010**, *4*, 39–47. [CrossRef] [PubMed]
6. Sowasod, N.; Nakagawa, K.; Tanthapanichakoon, W.; Charinpanitkul, T. Development of encapsulation technique for curcumin loaded O/W emulsion using chitosan based cryotropic gelation. *Mater. Sci. Eng. C* **2012**, *32*, 790–798. [CrossRef]
7. Cagri, A.; Ustunol, Z.; Ryser, E.T. Antimicrobial Edible Films and Coatings. *J. Food Prot.* **2004**, *67*, 833–848. [CrossRef]
8. Cha, D.; Cooksey, K.; Chinnan, M.; Park, H. Release of nisin from various heat-pressed and cast films. *LWT* **2003**, *36*, 209–213. [CrossRef]
9. Young, S.; Sarda, X.; Rosenberg, M. Microencapsulating Properties of Whey Proteins. 2. Combination of Whey Proteins with Carbohydrates. *J. Dairy Sci.* **1993**, *76*, 2878–2885. [CrossRef]
10. Rodea-González, D.A.; Cruz-Olivares, J.; Román-Guerrero, A.; Rodríguez-Huezo, M.E.; Vernon-Carter, E.J.; Pérez-Alonso, C. Spray-dried encapsulation of chia essential oil (*Salvia hispanica L.*) in whey protein concentrate-polysaccharide matrices. *J. Food Eng.* **2012**, *111*, 102–109. [CrossRef]
11. Shah, B.; Davidson, P.M.; Zhong, Q. Encapsulation of eugenol using Maillard-type conjugates to form transparent and heat stable nanoscale dispersions. *LWT* **2012**, *49*, 139–148. [CrossRef]
12. Preetz, C.; Rübe, A.; Reiche, I.; Hause, G.; Mäder, K.; Mäder, K. Preparation and characterization of biocompatible oil-loaded polyelectrolyte nanocapsules. *Nanomed. Nanotechnol. Boil. Med.* **2008**, *4*, 106–114. [CrossRef]
13. Calero, N.; Muñoz, J.; Cox, P.; Heuer, A.; Guerrero, A. Influence of chitosan concentration on the stability, microstructure and rheological properties of O/W emulsions formulated with high-oleic sunflower oil and potato protein. *Food Hydrocoll.* **2013**, *30*, 152–162. [CrossRef]
14. Zhang, Y.; Niu, Y.; Luo, Y.; Ge, M.; Yang, T.; Yu, L.; Wang, Q. Fabrication, characterization and antimicrobial activities of thymol-loaded zein nanoparticles stabilized by sodium caseinate—Chitosan hydrochloride double layers. *Food Chem.* **2014**, *142*, 269–275. [CrossRef] [PubMed]
15. Weiss, J.; Gaysinsky, S.; Davidson, M.; McClements, J. CHAPTER 24—Nanostructured Encapsulation Systems: Food Antimicrobials. In *Global Issues in Food Science and Technology*; Barbosa-Cánovas, G., Mortimer, A., Lineback, D., Spiess, W., Buckle, K., Colonna, P., Eds.; Academic Press: San Diego, CA, USA, 2009; pp. 425–479. [CrossRef]
16. Ogawa, S.; Decker, E.A.; McClements, D.J. Production and Characterization of O/W Emulsions Containing Droplets Stabilized by Lecithin–Chitosan–Pectin Mutilayered Membranes. *J. Agric. Food Chem.* **2004**, *52*, 3595–3600. [CrossRef]

17. Hsieh, W.-C.; Chang, C.-P.; Gao, Y.-L. Controlled release properties of Chitosan encapsulated volatile Citronella Oil microcapsules by thermal treatments. *Colloids Surf. B Biointerfaces* **2006**, *53*, 209–214. [CrossRef]

18. Peng, H.; Xiong, H.; Li, J.; Xie, M.; Liu, Y.; Bai, C.; Chen, L. Vanillin cross-linked chitosan microspheres for controlled release of resveratrol. *Food Chem.* **2010**, *121*, 23–28. [CrossRef]

19. Souza, J.M.; Caldas, A.L.; Tohidi, S.D.; Molina, J.; Souto, A.P.; Fangueiro, R.; Zille, A. Properties and controlled release of chitosan microencapsulated limonene oil. *Rev. Bras. Farm.* **2014**, *24*, 691–698. [CrossRef]

20. Estevinho, B.N.; Rocha, F.; Santos, L.; Alves, A. Microencapsulation with chitosan by spray drying for industry applications—A review. *Trends Food Sci. Technol.* **2013**, *31*, 138–155. [CrossRef]

21. Akhtar, M.; Dickinson, E. Whey protein—Maltodextrin conjugates as emulsifying agents: An alternative to gum arabic. *Food Hydrocoll.* **2007**, *21*, 607–616. [CrossRef]

22. Hosseini, S.F.; Zandi, M.; Rezaei, M.; Farahmandghavi, F. Two-step method for encapsulation of oregano essential oil in chitosan nanoparticles: Preparation, characterization and in vitro release study. *Carbohydr. Polym.* **2013**, *95*, 50–56. [CrossRef] [PubMed]

23. Chen, H.; Zhong, Q. A novel method of preparing stable zein nanoparticle dispersions for encapsulation of peppermint oil. *Food Hydrocoll.* **2015**, *43*, 593–602. [CrossRef]

24. Keawchaoon, L.; Yoksan, R. Preparation, characterization and in vitro release study of carvacrol-loaded chitosan nanoparticles. *Colloids Surf. B Biointerfaces* **2011**, *84*, 163–171. [CrossRef] [PubMed]

25. Rojsitthisak, P.; Rojsitthisak, P.; Nimmannit, U. Preparation of turmeric oil-loaded chitosan-alginate biopolymeric nanocapsules. *Mater. Sci. Eng. C* **2009**, *29*, 856–860. [CrossRef]

26. Eid, A.; Elmarzugi, N.; El Enshasy, H. Preparation and evaluation of olive oil nanoemulsion using sucrose monoester. *Int. J. Pharm. Pharm. Sci.* **2013**, *5*, 434–440.

27. Gharsallaoui, A.; Saurel, R.; Chambin, O.; Cases, E.; Voilley, A.; Cayot, P. Utilisation of pectin coating to enhance spray-dry stability of pea protein-stabilised oil-in-water emulsions. *Food Chem.* **2010**, *122*, 447–454. [CrossRef]

28. Bouyer, E.; Mekhloufi, G.; Rosilio, V.; Grossiord, J.-L.; Agnely, F. Proteins, polysaccharides, and their complexes used as stabilizers for emulsions: Alternatives to synthetic surfactants in the pharmaceutical field? *Int. J. Pharm.* **2012**, *436*, 359–378. [CrossRef]

29. Shahavi, M.H.; Hosseini, M.; Jahanshahi, M.; Meyer, R.L.; Darzi, G.N. Evaluation of critical parameters for preparation of stable clove oil nanoemulsion. *Arab. J. Chem.* **2019**, *12*, 3225–3230. [CrossRef]

30. Wu, Y.; Luo, Y.; Wang, Q. Antioxidant and antimicrobial properties of essential oils encapsulated in zein nanoparticles prepared by liquid–liquid dispersion method. *LWT* **2012**, *48*, 283–290. [CrossRef]

31. Dima, C.; Cotarlet, M.; Balaes, T.; Bahrim, G.; Alexe, P.; Dima, S. Encapsulation of Coriander Essential Oil in Beta-Cyclodextrin: Antioxidant and Antimicrobial Properties Evaluation. *Rom. Biotechnol. Lett.* **2014**, *19*, 9128–9140.

32. Jafari, S.M.; He, Y.; Bhandari, B. Effectiveness of encapsulating biopolymers to produce sub-micron emulsions by high energy emulsification techniques. *Food Res. Int.* **2007**, *40*, 862–873. [CrossRef]

33. Woranuch, S.; Yoksan, R. Eugenol-loaded chitosan nanoparticles: I. Thermal stability improvement of eugenol through encapsulation. *Carbohydr. Polym.* **2013**, *96*, 578–585. [CrossRef] [PubMed]

34. Reineccius, G.A. Flavor encapsulation. *Food Rev. Int.* **1989**, *5*, 147–176. [CrossRef]

35. Pramod, K.; Suneesh, C.V.; Shanavas, S.; Ansari, S.H.; Ali, J. Unveiling the compatibility of eugenol with formulation excipients by systematic drug-excipient compatibility studies. *J. Anal. Sci. Technol.* **2015**, *6*, 265. [CrossRef]

36. Cocero, M.J.; Martín, Á.; Mattea, F.; Varona, S.; Alonso, M.J.C. Encapsulation and co-precipitation processes with supercritical fluids: Fundamentals and applications. *J. Supercrit. Fluids* **2009**, *47*, 546–555. [CrossRef]

37. Feyzioglu, G.C.; Tornuk, F. Development of chitosan nanoparticles loaded with summer savory (*Satureja hortensis* L.) essential oil for antimicrobial and antioxidant delivery applications. *LWT* **2016**, *70*, 104–110. [CrossRef]

38. Souza, A.C.P.; Gurak, P.D.; Marczak, L.D.F. Maltodextrin, pectin and soy protein isolate as carrier agents in the encapsulation of anthocyanins-rich extract from jaboticaba pomace. *Food Bioprod. Process.* **2017**, *102*, 186–194. [CrossRef]

39. Hill, L.E.; Gomes, C.; Taylor, T.M. Characterization of beta-cyclodextrin inclusion complexes containing essential oils (trans-cinnamaldehyde, eugenol, cinnamon bark, and clove bud extracts) for antimicrobial delivery applications. *LWT* **2013**, *51*, 86–93. [CrossRef]

40. Otálora, M.C.; Carriazo, J.G.; Iturriaga, L.; Nazareno, M.A.; Osorio, C. Microencapsulation of betalains obtained from cactus fruit (*Opuntia ficus-indica*) by spray drying using cactus cladode mucilage and maltodextrin as encapsulating agents. *Food Chem.* **2015**, *187*, 174–181. [CrossRef]

41. Ronkart, S.N.; Paquot, M.; Blecker, C.S.; Fougnies, C.; Doran, L.; Lambrechts, J.C.; Norberg, B.; Deroanne, C. Impact of the Crystallinity on the Physical Properties of Inulin during Water Sorption. *Food Biophys.* **2008**, *4*, 49–58. [CrossRef]

42. Beirão-Da-Costa, S.; Duarte, C.; Bourbon, A.I.; Pinheiro, A.C.; Januário, I.; Vicente, A.A.; Da Costa, M.L.B.; Delgadillo, I. Inulin potential for encapsulation and controlled delivery of Oregano essential oil. *Food Hydrocoll.* **2013**, *33*, 199–206. [CrossRef]

43. Piletti, R.; Bugiereck, A.; Pereira, A.; Gussati, E.; Magro, J.D.; Mello, J.; Dalcanton, F.; Ternus, R.; Soares, C.; Riella, H.G.; et al. Microencapsulation of eugenol molecules by β-cyclodextrine as a thermal protection method of antibacterial action. *Mater. Sci. Eng. C* **2017**, *75*, 259–271. [CrossRef] [PubMed]

44. Sajomsang, W.; Nuchuchua, O.; Gonil, P.; Saesoo, S.; Sramala, I.; Soottitantawat, A.; Puttipipatkhachorn, S.; Ruktanonchai, U.R. Water-soluble β-cyclodextrin grafted with chitosan and its inclusion complex as a mucoadhesive eugenol carrier. *Carbohydr. Polym.* **2012**, *89*, 623–631. [CrossRef] [PubMed]

45. Coates, J. Interpretation of Infrared Spectra, a Practical Approach. *Encycl. Anal. Chem.* **2006**, *12*, 10815–10837. [CrossRef]

46. Agnihotri, S.A.; Mallikarjuna, N.N.; Aminabhavi, T.M. Recent advances on chitosan-based micro- and nanoparticles in drug delivery. *J. Control. Release* **2004**, *100*, 5–28. [CrossRef]

47. Ritger, P.L.; Peppas, N.A. A simple equation for description of solute release II. Fickian and anomalous release from swellable devices. *J. Control. Release* **1987**, *5*, 37–42. [CrossRef]

48. Combrinck, J.; Otto, A.; Du Plessis, J. Whey Protein/Polysaccharide-Stabilized Emulsions: Effect of Polymer Type and pH on Release and Topical Delivery of Salicylic Acid. *AAPS PharmSciTech* **2014**, *15*, 588–600. [CrossRef]

49. Patil, D.K.; Agrawal, D.S.; Mahire, R.R.; More, D.H. Synthesis, characterization and controlled release studies of ethyl cellulose microcapsules incorporating essential oil using an emulsion solvent evaporation method. *Am. J. Essent. Oils Nat. Prod.* **2016**, *4*, 9.

Article

Relationship between Color and Redox Potential (E_h) in Beef Meat Juice. Validation on Beef Meat

Paolo Cucci, Aimeric C. K. N'Gatta, Supakakul Sanguansuk, André Lebert and Fabrice Audonnet *

Institut Pascal, Université Clermont Auvergne, CNRS, SIGMA Clermont, F-63000 Clermont-Ferrand, France; paolo.cucci@orange.fr (P.C.); charlesaimeric@gmail.com (A.C.K.N.); supakakul1@gmail.com (S.S.); andre.lebert@uca.fr (A.L.)
* Correspondence: fabrice.audonnet@uca.fr; Tel.: +33-4-73-40-78-17

Received: 20 March 2020; Accepted: 29 April 2020; Published: 1 May 2020

Abstract: In France, around 3.5 million cattle are slaughtered each year, which represents 1.3 million tons of beef carcasses. However, waste due essentially to organoleptic defects is estimated at 3.4% of the production or 45,000 tons of beef carcasses. Microbiological contamination and color are the two major causes of defect. In order to prevent color defect, a study was performed to develop a new method for measuring rapidly and instantly the redox potential as an indicator of color changes in carcasses without slowing down the slaughter line. This measurement would allow to classify them upstream according to their time of colors changes in order to sort them and to avoid food waste in the future. Meat juice has been shown to be a good mimetic medium for the study of color changes. The effect of different parameters was studied in order to fix experimental conditions. Color change is faster in the juice than in the meat and faster at 20 °C than at 4 °C. Redox potential allows following color changes and a symmetry has been highlighted between this thermodynamic measure and color changes.

Keywords: redox potential; color transfer; beef juice; beef meat

---

1. Introduction

In France, each year, around 3.5 million cattle are slaughtered for meat production, which represents 1.3 million tons of beef carcasses [1]. Most of this meat is distributed in large and medium-sized stores and alone accounts for 75% of purchases of fresh red meat in France [2]. One in two consumers buy fresh red meat based on color [3]. However, if the red meat does not meet the consumer's requirements (a beautiful color and a good smell), the latter does not buy it, which leads to food waste. In 2016, waste due to organoleptic defects in beef was estimated at 3.4% of production or 45,000 tons of beef carcasses [4]. Microbiological contamination and color are the two major causes of observed defect in beef meat.

The color of the meat is due to the presence of myoglobin. Myoglobin is a water-soluble intramuscular protein that binds to oxygen and allows cellular respiration in the muscle [5]. It is made up of globin and a prosthetic group called heme. At the center of the heme is an iron atom with its six coordination sites. The sixth coordination site makes it possible to bind a ligand such as dioxygen or even water [6]. Depending on the oxidation state of iron (ferrous Fe^{2+} or ferric Fe^{3+}) and the bound ligand, myoglobin is commonly found in 3 forms: deoxymyoglobin, oxymyoglobin, or metmyoglobin [7]. Each redox form of myoglobin, when it is mainly present in meat, generates different colors. Deoxymyoglobin is purplish in color and is combined with a vacuum product or immediately cut. Oxymyoglobin is bright red in color and is associated with good quality red meat in trays, while metmyoglobin, brown in color on meat, and results in consumer rejection of the product. Indeed, there is a dynamic of interconversions (redox reactions) between these three forms of myoglobin [7]. Initially, myoglobin or deoxymyoglobin undergoes oxygenation or blooming by binding to oxygen and becomes red-colored oxymyoglobin. Under low partial oxygen pressure such as during cellular respiration by muscle tissue, oxymyoglobin can revert to deoxymyoglobin, which is sensitive to oxidation. This oxidation at the level of the iron atom in the center of the heme causes the passage of the deoxymyoglobin of purplish color in metmyoglobin of brown color.

The aim of this study is to carry out a rapid and instant measurement of the redox potential, E_h, directly on a carcass (without slowing down the slaughter line), at the separation between forequarter and hindquarter in order to evaluate the time necessary for meat color to change from red to brown. To the best of the authors' knowledge, there is no technology to date to measure color and/or redox potential directly on slaughter lines. This would allow carcasses to be classified upstream in order to sort them. Indeed, a carcass whose E_h value would indicate an early transfer would then be sent to a line of processed products, while a carcass whose E_h value would indicate a later transfer would be distributed in large and medium-sized stores and offered in cut and packaged meat. However, since meat is a complex medium and redox probes for semi-solid media do not exist, the use of a mimetic medium such as meat juice (i.e., meat exsudate) has been proposed as a first approach.

The strategy applied in this work consisted in simultaneously measuring the color of a beef meat juice and the E_h in order to search for a relationship between these two variables. The main sources of variability have been identified in order to limit their effect. A validation of the relationship on beef meat was also performed.

2. Materials and Methods

2.1. Biological Products

Rib steaks from young cattle (Puigrenier slaughterhouse, Montluçon, France) were semi-dressed, vacuum-packed, and labeled for all the experiments. After reception, the meat is stored in the refrigerator at (4 ± 2) °C. Rib steak was selected because it is easily and quickly accessible during the cutting of the carcass in the forequarter and hindquarter.

2.1.1. Meat Juice

The meat juice preparation is based on the study by Kim and Jeong [8]. After reception, muscles were cut into pieces 3 cm thick with a weight of about 80 g and packaged in Polyamide/Polyethylene (PA/PE) storage bags impermeable to gases (3 welds, 200×300 mm, Sovapack, Cuiseaux, France). The bags were hermetically sealed either under half vacuum (500 mbar) or under vacuum with a double-chamber machine (Multivac, Lagny Sur Marne, France) for packaging in bags. The samples were then frozen at (-20 ± 2) °C for 24 h and then thawed at (4 ± 2) °C for 72 to 96 h depending on the size of the samples. The obtained juices after freezing and thawing of the samples were collected in a sterile manner under a microbiological safety station. The juices recovered have undergone different preparatory stages or not depending on the objectives of the study:

1. centrifugation at 4 °C at 10,000 g for 5 min;
2. filtration through a filter with a pore diameter of 0.22 μm;
3. dilution according to the desired volume and rate;
4. use of milliQ water or physiological water for dilution.

2.1.2. Meat Decontamination

Rib steaks were decontaminated with peracetic acid following the study of Lebert et al. [9]. Then, they are cut into slices about 2 to 3 cm thick for a weight of 350 to 400 g. The slices are then placed in sealable trays in PS/EVOH/PE (Form'plast, Chantrans, France). They were finally placed in a modified atmosphere using a T200 semi-automatic sealer (Multivac, Lagny Sur Marne, France). OPP/T504 film (Soussanna, Thiais, France) was used to seal the trays.

2.2. Color Monitoring

2.2.1. Meat Juice

A climatic chamber (Binder KMF 240, VWR, Fontenay-sous-Bois, France) was used to reproduce the conditions of an industrial storage fridge for meat. Indeed, it allows temperature between 2 and 20 °C and relative humidity between 95 and 99%. In order to light up the meat and follow the color changes inside the climatic chamber, two strips with fluorescent tubes (T5, cool white color, 27.7 cm, 6 W, Diall, Paris, France) spaced apart 30 cm have been installed. The lighting delivers about 360 lux on the samples (dual-display traceable luxmeter, VWR Collection, France). In addition, a balance (Kern SXS-6K-3M, Timber Production, Esmans, France) was used in situ to monitor the weight over time to ensure that the samples did not dry out during the experiment. A camera (Logitech C270 720p, Logitech, Paris, France) was installed vertically approximately 20 cm above the samples to follow meat juice color.

2.2.2. Meat

An industrial system (ADIV, Clermont-Ferrand, France) was used to follow the color changes of red meat. In a cold room (Dagard, Boussac, France) supplied by a cold group (Arcos, Gorrevod, France), strips with fluorescent tubes (T8, cool white color, 60 cm, 18 W, Sylvania, Saint-Etienne, France) were fixed on both sides others of the device with a distance of 45 cm between each strip. In addition, they were placed 30 cm above the meat trays just like the cameras (Logitech C270 720p, Logitech, Paris, France). Lighting at this facility delivers approximately 1700 lux on meat samples.

2.3. Development of a Redox Probe for Solid Media

Since commercial redox probes for semi-solid media do not exist, a built in-house redox electrode has been developed. It is composed of two "working" electrodes of 1 mm in diameter, one of which consists of a platinum rod (99.95% purity, Surepure Chemetals, New-Jersey, United States) and the other of an oxidized iron rod. Each rod is connected by electric cables in order to recover the electrical signals (measured voltages) at the level of the computer (the whole is molded in a resin).

2.4. Physico-Chemical Measurements

For meat juices, pH probes (HI11310, Hanna Instruments, Lingolsheim, France) and E_h probes (HI36180, Hanna Instruments, Lingolsheim, France) were used. The measurements were taken and recorded automatically every 15 min using tablets (pH-/mV-meter Edge HI2002-02, Hanna Instruments, Lingolsheim, France). For meat, the built in-house probes were used. The data was measured and recorded using our internally developed software. The probes for liquid and semi-solid media were sterilized with 70° alcohol before being used.

Since at 4 °C, the time of color transfer is quite long (up to 400 h), and based on the fact that all the studied reactions are physico-chemical reactions, it has been decided to speed up these reactions by performing the experiment at 20 °C.

2.5. Color Monitoring and Images Analysis

Images of the samples (juice and meat) were taken every 15 min by using a software developed with LabVIEW 2018 and the Vision Development module (National Instruments Corporation, USA). These images in bmp format were saved on a computer before being processed.

An image processing process was performed for each saved image. Using ImageJ software (National Institutes of Health, Bethesda, Maryland, USA), the values of Lightness (L*), Red/Green index (a*), and Blue/Yellow index (b*) were calculated automatically on the selected area (use of macros developed in-house). The target values were calculated to verify that the camera calibration did not drift over time. In this case, the difference between the initial target value and the target value after derivation has been subtracted or added to the measured value of the sample. This procedure gave the kinetics of a*.

2.6. Data Normalization

The normalization allows adjusting of a series of values according to a transformation function in order to make them comparable with a few specific reference points. It is necessary when the results and their interpretations can be affected by the incompatibility of the units or scales of measurement between variables. This data normalization method was used for the comparison of the red index (a*) and the redox potential (E_h) variation curves. The transformation function (Equation (1)) used is:

$$f(V_i) = (V_i - V_{max})/(V_{max} - V_{min}) \tag{1}$$

where V_i = value to normalize, V_{max} = maximum value, V_{min} = minimum value. This function allows to normalize (and then compare) the red index a* and E_h between 0 and −1.

3. Results

3.1. Operating Conditions

In the next subsections, different operating conditions applied on the juices were tested: volume (20 mL, 35 mL and 50 mL), dilution (1/5, 1/10 and 1/15), water used for the dilution (milliQ or physiological), packaging (air or vacuum), and centrifugation (or not). Even if it should be interesting *per se*, it is worth mentioning that juice composition and visible spectrophotometry were not used in this study. Firstly, concerning the juice composition, we have considered the composition of the juice as a pre-slaughter factor. It is part of the initial conditions and is dependent on the animal such as the type of animal diet, the stress of animal experienced before slaughter, or also the type of muscle as mentioned by Bekhit et al. [10]. So, it was not relevant to master it for all the different experiments since each animal is different and it is impossible to perform this analysis on the slaughter line. Secondly, monitoring the evolution of visible spectrophotometry peaks of the meat juice in order to quantify the relative percentages of deoxymyoglobin, oxymyoglobin, or metmyoglobin over time using empirical equations as proposed by Tang et al. [11], were not undertaken because theses analyzes are not there either feasible on the slaughter line. Lastly, one has to note that the pH evolution of the different juices was also studied, but in all cases, its evolution was stable (between 5.7 and 5.5). These slight variations are in the uncertainties of measurement of the probe (± 0.1 pH unit, corresponding to two times the standard deviation) and do not explain color changes. Consequently, it has been decided to not represent it in the following parts.

3.1.1. Influence of the Samples Volumes and Dilutions

The purpose of this experiment is to show the importance of standardizing samples when measuring color, since the volume and the dilution have an impact on the initial and final color (after transfer) of a meat juice. Samples with different volumes (20 mL, 35 mL, and 50 mL) and different dilutions (1/5, 1/10, and 1/15) were made in order to obtain the best possible compromise (Figure 1).

The juice samples were obtained following the freezing/thawing of a rib steak cut into small pieces (3 cm thick with a weight of about 80 g) packed in air; the juices were centrifuged (at 4 °C, 10,000 g for 5 min), filtered (through a filter with a porosity of 0.22 μm) and diluted with milliQ water. Each sample was transferred to a crystallizer of the same format and monitoring of the color change was carried out for 24 h at 20 °C in the climatic chamber, the choice of 20 °C being motivated by the acceleration of the reaction kinetics. At t = 0 h, the juices of 20 mL, 35 mL, and 50 mL diluted 1/5, 1/10, and 1/15, respectively, showed an intermediate red color. After 24 h, all the juices were found to have turned from a red to brown color where the smaller the volume and dilution, the darker the color of the juice, rending the color change difficult to analyze. Moreover, in order to measure E_h reliably, the largest volume (50 mL) is better to properly immerse the redox probes in these crystallizers. Finally, in order to modify the properties of the juice initially recovered as little as possible, the smallest dilution (1/5) was used. Following this analysis, a volume of 50 mL and a 1/5 dilution were used in identical crystallizers for all the experiments.

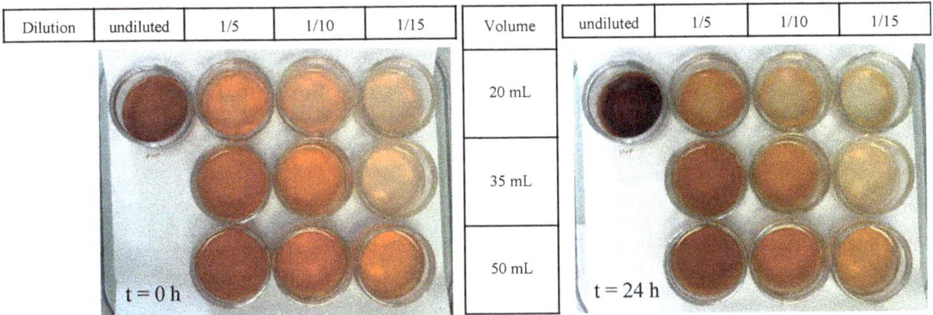

Figure 1. Visualization of the color of the same rib steak juice with different volumes (20 mL, 35 mL, and 50 mL) and different dilutions (1/5, 1/10, and 1/15) at t = 0 h and t = 24 h, at 20 °C. Pure meat juice (undiluted) is also presented for comparison to diluted juices.

3.1.2. Influence of the Water Type Used for Dilution

In this experiment, two types of water were tested, milliQ water and physiological water. These two waters were chosen because they are the most used in the world of the food industry (except running water). Each meat juice from the same whole rib steak, vacuum-packed before freezing and thawing, was centrifuged, filtered, and then diluted 1/5 in 50 mL milliQ water or 50 mL physiological water previously sterilized. A monitoring of the transfer of duplicate juices was implemented at 20 °C in the climatic chamber. The results showed that the two diluted meat juices had the same initial a* value (Figure 2a). After the normalization of the data, it was found that the meat juice diluted with water milliQ turned faster than meat juice diluted with physiological water. The juice diluted in milliQ water turned completely yellow after 24 h while the juice diluted in physiological water changed completely in color (also yellow) after 48 h (Figure 2b). In fact, in order to not add ionic species in the juices, it was decided to use milliQ water for all the other experiments.

3.1.3. Influence of the Packaging and Centrifugation Conditions

It is known that the oxygen has a role on the oxidation of myoglobin. Thus, it is necessary to determine whether the packaging conditions of the meat, in air or under vacuum, before freezing/thawing, have an influence on the color change of the juice from which it comes. Two rib steak muscles were packed whole under two conditions, one in air (21% of O_2) and the other in vacuum (0% of O_2) before being frozen/thawed. The juices obtained were diluted 1/5 in 50 mL of milliQ water, centrifuged, filtered, and then placed at 20 °C in the climatic chamber. It was found that the meat packaged in air gives more volume of juice (Factor 2) than that packed in vacuum. Figure 3a shows that the normalized values of a* are identical.

When the juice has been collected after the freeze/thaw step, it may contain residues of connective tissue and adipose tissue. In order not to affect the color change due to the presence of residual, easily oxidizable fats, and so that the juice passes more easily through the filter during the filtration stage, the juice has been centrifuged and therefore clarified. The aim of this experiment was to verify the possible impact of centrifugation on the color change of the meat juice. The juice samples from a vacuum-packed whole rib steak were centrifuged or not centrifuged then filtered (through a 0.22 μm porosity filter) and diluted 1/5 in 50 mL with milliQ water. Monitoring of the color change of duplicate juices was implemented at 4 °C in the climatic chamber. This temperature of 4 °C was chosen compared to previous experiments at 20 °C in order to have a first approximation of the transfer time of a juice at 4 °C. After normalization, it was found that the curves of the a* values of the centrifuged and non-centrifuged juices are identical (Figure 3b).

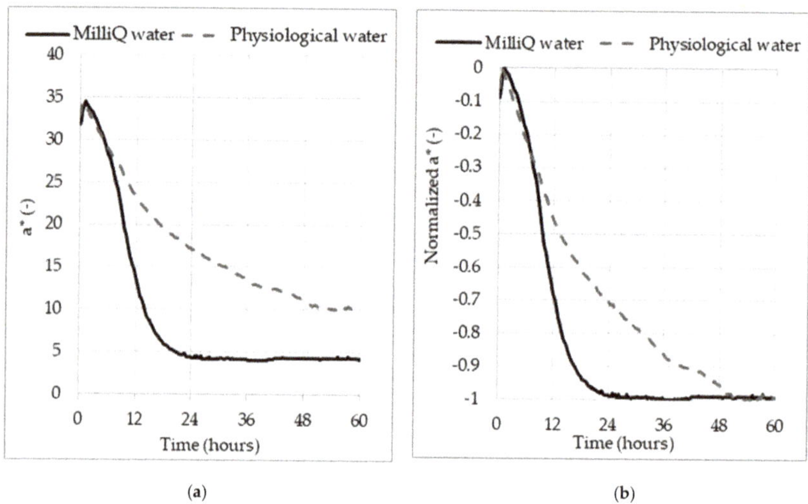

Figure 2. Monitoring of the color change (a*) as a function of time of (**a**) a rib steak juice (50 mL of centrifuged juice, filtered and 1/5 diluted) at 20 °C according to the type of water (milliQ water, black solid line, or physiological water, grey dashed line) used to dilute the meat juice as a function of time and (**b**) the same data in normalized values, using Equation (1).

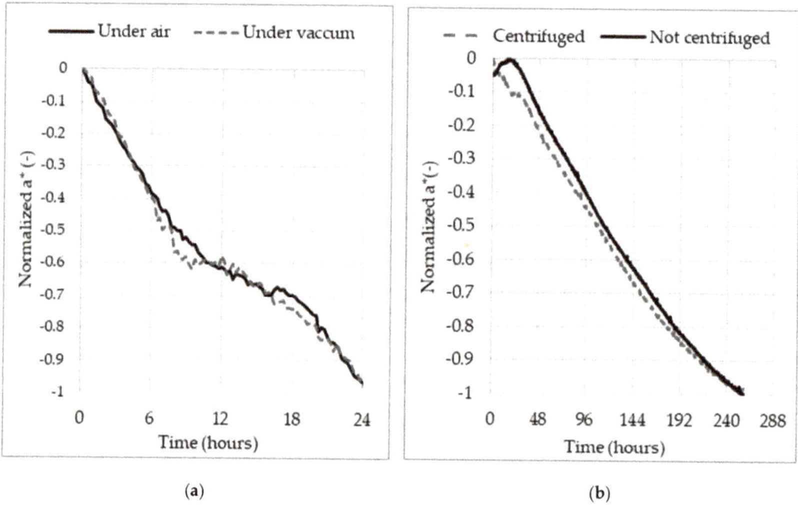

Figure 3. Monitoring of the color change (normalized values of a*) as a function of time for (**a**) a rib steak juice (50 mL of centrifuged juice, filtered and 1/5 diluted) at 20 °C depending on the type of packaging (air, in black solid line), or under vacuum, in grey dashed line) and (**b**) a rib steak juice (50 mL of filtered juice and diluted 1/5) at 4 °C with, in grey dashed line, or without, black solid line, the step of centrifugation (centrifuged or not centrifuged).

3.2. Correlation between Color Transfers and E_h in the Juice

In order to be as close as possible to industrial conditions, it was decided to carry out this experiment with no filtering of the juice. Indeed, in the meat industry, carcasses, muscles, and meat do not undergo any decontamination step between the start and the end of the process. The goal is to check if E_h is really a good indicator of the color change in this case. The parameters of the protocol for obtaining the juice were as follows: freezing/thawing of a whole rib steak muscle under vacuum, centrifugation, and 1/5 dilution of the juice in 50 mL of sterile milliQ water. The same juice was separated in four crystallizers (two duplicates without probes and two duplicates with probes), which were then placed at 4 °C in the climatic chamber. The results of the normalized values of a* and of E_h are represented in Figure 4. The first observation which can be made is that a certain "symmetry" is observed between a* and E_h (when a* decreases, E_h increases). First, the values of a* start with a phase of slight decrease, up to 105 h where the juices are still red. Then, for the next 45 h, a sudden decrease phase is observed; at this end point, the juices have turned and are yellow/greenish in color. Finally, a new plateau phase is observed. Concerning the E_h evolution, a slight increase is observed during the first 96 h, then, an exponential increase is measured before reaching a plateau after 140 h. One has to note that a crossover is observed at 125 h, when the color transfer begins to be well established.

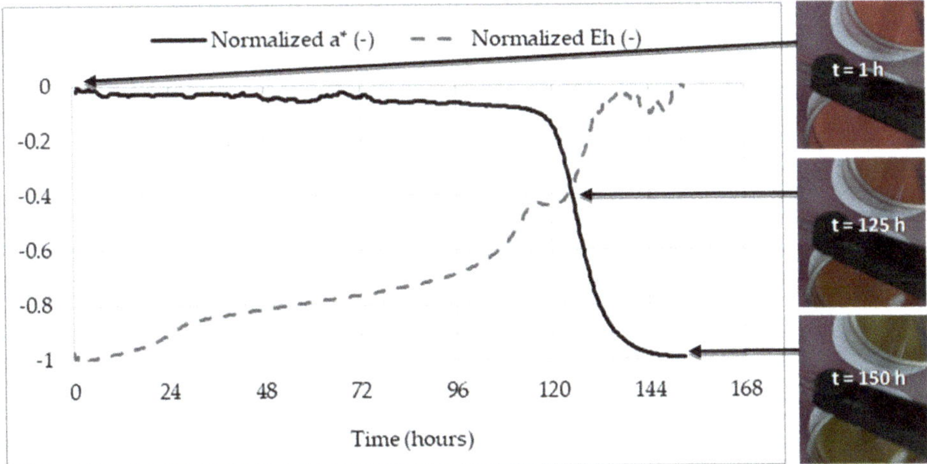

Figure 4. Monitoring of the color transfer (normalized values of a*, in black solid line) of a non-decontaminated rib steak juice (50 mL of juice diluted 1/5) at 4 °C and of the redox potential (normalized values of E_h, in grey dashed line) as a function of time and associated photos showing the color transfer of non-decontaminated juices (duplicated on the right) at t = 1 h, t = 125 h, and t = 150 h.

3.3. From Beef Juice to Beef Meat Measurements

3.3.1. Detection Limit for Color Changes in Meat

As the color transfers in the juice samples are homogeneous, the image processing process made it possible to detect the color transfers using the whole surface of the juice. However, for meat, a detection limit had to be determined. Indeed, if the values of a* are determined by taking into account the entire surface of the meat, the tacking time cannot be detected on the curve (only some pixels are concerned by a color transfer) as shown in Figure 5a. To solve this problem, different sizes of meat sample area were used as the area for measuring the values of a*. These different areas were calculated by dividing the total number of pixels of the area of the meat sample by 1, 2, 4, 16, 32, 64, and 128 (Figure 5a). The results presented in Figure 5b show that from the 1/64 cut, the values of a* obtained no longer vary for the same sample. Indeed, the precision is such that the transfer is then detectable on the curve of values of a*. Therefore for all meat samples, a measurement of the values of L*, a* and b* was carried out on 1/64 of the total area at the place where the tack was observed.

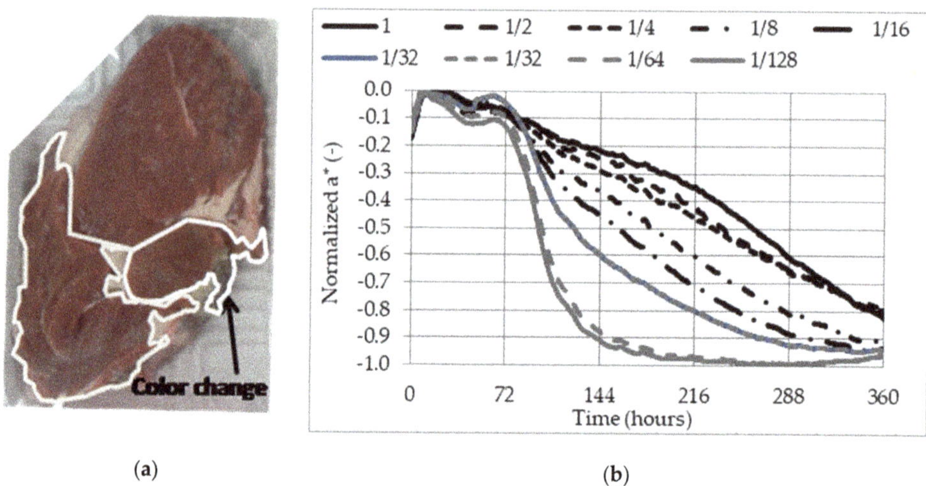

(a) (b)

Figure 5. (**a**) Example of a limit detection of the color transfer, using a surface of 1/2 (white separation); (**b**) normalized a* values for different areas of meat measured for the same sample (area divided by 1, 2, 4, 16, 32, 64, and 128) over time.

3.3.2. Comparison between Beef Juice and Beef Meat

The aim of this experiment is to compare the color changes of beef meat and its associated juice in order to know if the juice is really a mimetic medium of the meat. For this experiment, a rib steak and its associated juice were collected and placed at 4 °C in the climatic chamber. The juice underwent the same protocol as for the other experiments and results are shown in Figure 6, with top left, pure juice and top right, 1/5 diluted juice. The results represent the normalized values of a* of rib steaks treated in 1/64 and its 1/5 diluted juice. With regard to the juice, the values of a* have the same tendency as previously observed: a slight decrease up to about 96 h, followed by a more pronounced slope up to 240 h. For meat, the measured a* values follow exactly the same trends as the juice, i.e., first of all a slight decrease (t < 80 h) then a more significant fall (t < 240 h). A color change from red to greenish was identified. The normalized values shown in Figure 6 are pretty close, even if the fall was slightly more rapid for the a* values of the juice than for those of the meat. The juice is therefore a good mimetic medium with regard to color transfers although the kinetics of the transfer is faster in the juice than in the meat (which is consistent because the diffusion coefficients are different).

3.3.3. Calibration of the Built in-House Redox Probes

Meat juice being a good mimetic medium for meat to study color changes, it was decided to use the built in-house redox probe in the juice to perform their calibration. Indeed, the probes were built with two "working" electrodes, as shown in Figure 7a (see Material and Methods section for details), so that there is no longer a reference electrode and the values cannot be compared with values of commercial redox electrode. To verify the validity of our built in-house electrode, it was introduced into an unfiltered juice, diluted to 1/5 with a commercial redox probe (Hanna Instrument). The juice recovery protocol is the same as for the previous experiment. The results of this experiment, in normalized values, are shown in Figure 7b. Even if there is no reference electrode on our built in-house electrode, the normalized E_h curves follow the trends. Following these results in the juice, the built in-house electrode developed for semi-solid media seems to be very promising for tests on meat.

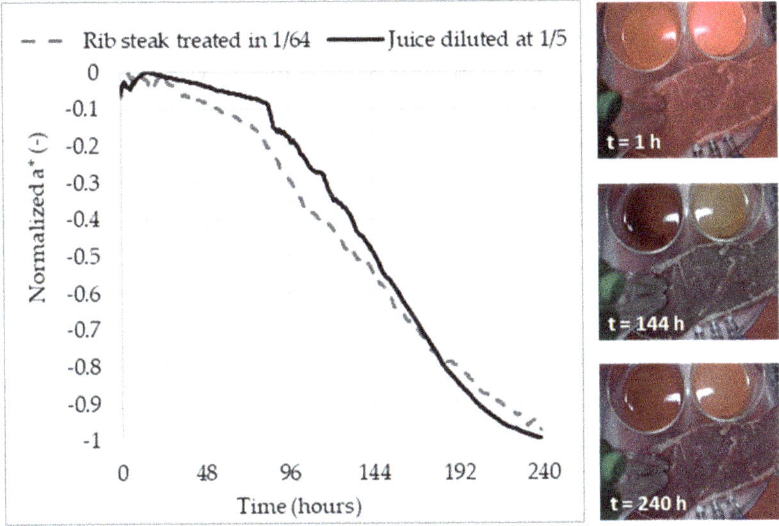

Figure 6. Monitoring of the color transfer (normalized values of a*) of a rib steak, in grey dashed line, and its associated juice (50 mL pure on the left and 50 mL diluted 1/5 on the right), in black solid line, at 4 °C as a function of time, and the associated photos showing their color transfers at t = 1 h, t = 144 h, t = 240 h.

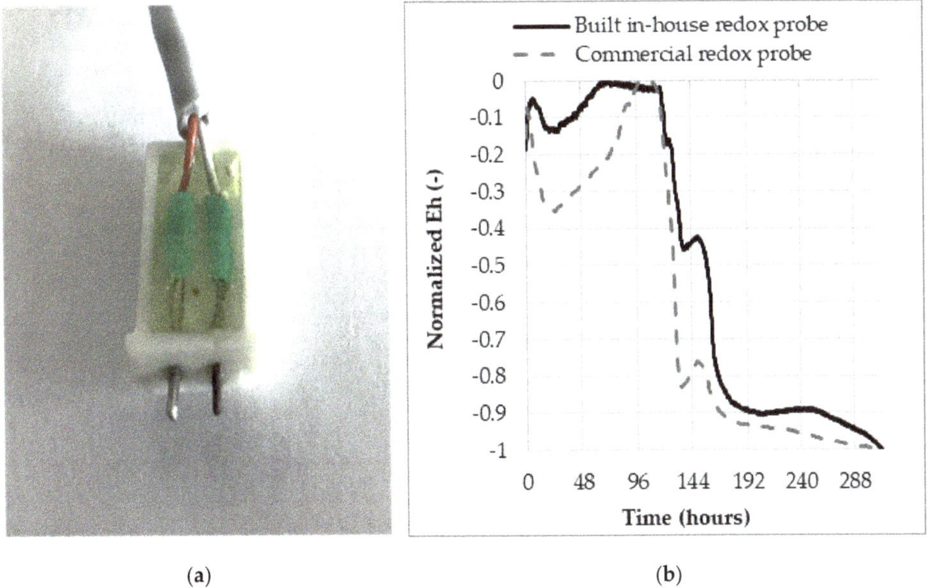

(a)   (b)

Figure 7. (a) Built in-house redox electrode composed of two "working" electrodes of 1 mm in diameter, one of which consists of a platinum rod (left) and the other of an oxidized iron rod (right) molded in a resin; (b) Monitoring of normalized E_h measured with a commercial redox probe (Hanna Instrument, in grey dashed line) and our built in-house redox probe (in black solid line) in a steak juice diluted 1/5 unfiltered as a function of time.

3.4. Validation on Beef Meat

This experiment was done under semi-industrial conditions. The rib steaks were placed in a heat-sealed tray and the built in-house probes were placed on the meat before air sealing (21% of oxygen). The experiment was carried out with 6 samples (6 successive slices) from the same muscle. As for the location of the probe on the meat, it was positioned in the muscle making up the rib steak which turned most often according to other experiments carried out during this study. At this stage, the important point is to test the probe on the meat medium to see if it works. The values of a* were measured as close as possible to the probe (a few mm) and represent 1/64 of the slice of meat. The results of this experiment are shown in Figure 8. At the start of the experiment, the value of a* is 20 and the meat is red. A sharp drop was measured after about 5 h. A color transfer was detected at the redox probe after 20 h of experience. A greenish color can then be observed at the level of the redox probe. After 96 h, the values of a* reach a plateau corresponding to a marked green color on the beef meat. For E_h, the trends are the same as what has just been presented, the values ranging from 460 mV to 520 mV during the increase of the redox potential, then a plateau of a few hours was measured and finally a start of fall occurs, ending at 485 mV after 144 h. Normalized values are shown in Figure 8 on the right. This standardization highlights that E_h follows the color change (change from red to green) and that there is a crossover after 20 h which corresponds to the start of the color change. The built in-house redox probe therefore makes it possible to follow the color change in the semi-solid medium that is beef meat.

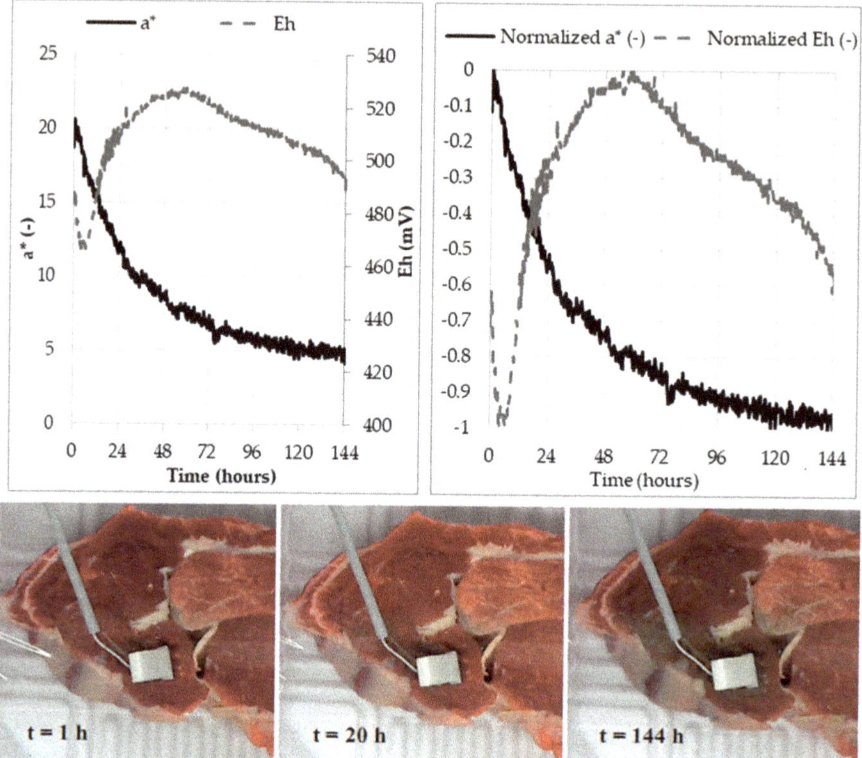

Figure 8. Monitoring the color transfer a* of a rib steak in a heat-sealed tray in air and the redox potential, E_h, measured with a built in-house redox probe at 4 °C as a function of time (**left**), the values normalized as a function of time (at **right**) and the associated photos showing their color transfers at t = 1 h, t = 20 h, and t = 144 h.

4. Discussion

The aim of this study is to carry out a measurement of the E_h on a carcass, at the separation between forequarter and hindquarter in order to evaluate, by the measurement of the color transfer a*, the time necessary for meat color to change from red to brown. Since meat is a complex media, linked to the fact no commercial redox probes exist for semi-solid media, a preliminary study was the use of meat juice as a meat mimetic medium. For that, the volume which has been retained is 50 mL so that the redox probes used can be completely immersed. Regarding the dilution, it was decided to modify the juice as little as possible to stay as close as possible to the pure juice, the 1/5 dilution in milliQ water was therefore selected. The more the juice is diluted the more the amount of myoglobin (Mb) present decreases, which can influence the results. In fact, the muscles containing high concentrations of Mb appear darker than those containing less. These differences in myoglobin concentrations, oxygen consumption rate and autoxidation rate may explain the variations in color stability between different muscles [12]. The juice preparation steps are just as important and should be monitored to ensure that they have the least impact on color changes. Pure juice is not a good choice to follow color changes because it is too dark. A 1/5 dilution was applied with two different waters. The results showed that the color change was slowed down with the use of physiological water. Indeed, the use of NaCl is well known to act as a preservative and this can explain this slowing down although the study done by Maruitti and Bragagnolo [13] has highlighted the pro-oxidant effect of NaCl on lipid oxidation and so, on color transfers. In contrast, milliQ water has the same behavior as distilled water and according to the experience of Thiem et al. [14] on the preparation of meat juices, distilled water was used and the components in the meat juices were not changed after dilution. This thus validates the method used in this study.

Experience on the packaging conditions of meat before freezing has highlighted the fact that the juice extracted from meat frozen in air has a value of a* higher than that of meat frozen in vacuum. This can be explained by the fact that the meat packaged in air was able to oxygenate during the freezing and thawing stages, resulting thus in a higher beginning value. However, no difference was observed on the color transfer itself. According to Baran et al. [15], when fresh meat is packaged with films having a high impermeability to oxygen (under vacuum), anaerobic growth of native microorganisms is favored and the growth of aerobic native microorganisms is limited unlike packaging under oxygen (under air). Moreover, Daniloski et al. [16] also showed similar results with these two types of packaging. The different types of packaging can influence the color change due to the presence of microorganisms. However the filtration step done in this study has removed all the microorganisms from the juices. This could explain why no difference was observed in the color transfer between the two studied packaging conditions.

The centrifugation step did not affect the color transfer but only the initial a* value. Indeed, the non-centrifuged juice has a value of a* higher than the centrifuged juice. This can be explained by the fact that the centrifuged juice has been clarified and is therefore clearer than the non-centrifuged juice. As a result, only the value of a* was impacted. The purpose of the centrifugation step is thus to standardize the juice samples by removing any debris recovered from the juice after thawing and to facilitate filtration.

During the experiments on the unfiltered beef juices at 4 °C and on beef meat at 4 °C, it was observed that the redox potential and color changes are linked; it is thus possible to follow the color changes using the redox potential and vice versa. Indeed, a symmetry was highlighted between this thermodynamic quantity (E_h) and the color changes (a*). During the color transfer, heme iron oxidizes and changes from ferrous to ferric [17]. The increase in the redox potential would therefore be linked to the loss of the red color of the meat which is closely linked to the oxidation of heme iron. In the work of Ke et al. [18], the redox potential of the *Psoas major* muscle (PM) increased significantly between 0 and 7 days and was associated with an increase of the percentage of metmyoglobin and a decrease in color stability (values of a*). Additionally, the redox potential of the *Longissimus lumborum* (LL) muscle did not show any significant variation over time, and was associated with a stability of the color of the muscle. We thus confirmed the observations of Ke et al. [18] in meat juice at 4 °C by the E_h increase, reflecting the oxidation of the medium during the loss of red color of the juice. Moreover, the experience comparing the color transfer between beef meat and its juice has shown that the juice and the meat have a simultaneous color transfer but that the kinetics of the transfer are faster in the juice. This latter observation is supported by the fact that the juice being a liquid medium, the physico-chemical reactions has to be faster than in a semi-solid media, where the species diffusion coefficients are greater.

The temperature has also an influence on the kinetics of the color change. The color transfer curves for juice or meat at 4 °C are generally composed of three phases in our study. First, a plateau which corresponds to the latency time, that is to say the time necessary for the oxidation to set up, then a fairly brutal fall corresponding to the oxidation (change from red to yellow/brown) in itself and finally a plateau phase once the transfer is complete. On the other hand, in experiments at 20 °C, this first stage of the plateau is shorter, even sometimes nonexistent and the transfer time is twice as short. Indeed, at 4 °C, transfers started between 96 and 144 h for juice and meat and ended after 144 and 240 h for juice and meat, respectively, while at 20 °C transfers started quickly between 1 and 10 h depending on the juices and ended after 24 h. This is explained by the oxidation of myoglobin which is kinetically favored and faster at high temperatures [17,19]. However, for the experiment on the color transfer of meat with the monitoring of E_h with the built in-house redox probe at 4 °C, the color transfer was earlier than for the experiment of comparison of the color transfer meat and its juice. Two hypotheses can be put forward. The first would be that this meat is more sensitive to the color change than another because of its internal variability. The second would be that the redox probe favored the color change. Indeed, although the built in-house redox probe correctly follows the color change, with an almost constant delta compared to the values measured with the commercial redox probe, the oxidized iron that composes it may have initiated the color change at the location where it was inserted. Moreover, one has to note that this observation was exactely the same for the five other slices of meat under study, indicating that the probes helped trigger the color transfer on the six different samples. For exemple, Warner et al. [20] have shown that the iron concentration has an impact on the color change in mutton, where the color stability is reduced when the muscle had a high iron content. This argument would explain the systematic faster color transfer observed using the built in-house redox probe since, an electrode is made using an oxidized iron (Fe(III) state) rod. This iron rod was used since the aim was to not introduce other reducing metal species which would have produced metal ions which could have caused complexation reactions, which could themselves influence the color transfer of the meat. It is clear that the use of these probes is ultimately problematic in an industrial point of view, paricularly in slaughterhouses. However, the clear demonstrated relationship between the redox potential E_h and the color transfer a* seems to open the way to color measurement directly in contact with the meat. This would seem all the more interesting as this measurement is faster than that of the redox potential, for which probe development would also have to be carried out. Future work will then to model and predict these color transfers according to different factors, such as oxygen partial pressure, temperature or maturation time; this would allow carcasses to be classified upstream in order to sort them.

Author Contributions: Conceptualization, A.L. and F.A.; methodology, P.C.; investigation, P.C., A.C.K.N. and S.S.; writing—original draft preparation, P.C.; writing—review and editing, A.L. and F.A.; project administration, A.L. and F.A.; funding acquisition, A.L. and F.A. All authors have read and agreed to the published version of the manuscript.

Funding: This research was conducted with the support of the European Regional Development Fund (FEDER) program of 2014–2020, the Regional Council of Auvergne and the ADIV (Association pour le Développement de l'Institut de la Viande, Clermont-Ferrand, France).

Acknowledgments: Authors thank Pascal LAFON for wise advice for the choice and the implementation of cameras, David DUCHEZ for his help to take control of the software for image acquisition, Emilie PARAFITA and Valérie SCISLOWSKI for the relations with the slaughterhouse and the handling of the industrial installation of ADIV.

Conflicts of Interest: The authors declare no conflict of interest.

References

1. FranceAgriMer. Rapport D'activité. 2018, p. 60. Available online: https://www.franceagrimer.fr/FranceAgriMer2/Vie-de-l-etablissement/Rapports-d-activites (accessed on 10 March 2020).
2. FranceAgriMer. Les Unités de Vente Consommateurs dans les Filières Carnées et Aquatiques. 2012, p. 8. Available online: https://www.franceagrimer.fr/fam/content/download/19198/document/SYN-MUL-2012-UVC%20poisson%20viande.pdf (accessed on 10 March 2020).

3. Syndicat National de l'Industrie des Viandes (SNIV) et Syndicat National du Commerce du Porc (SCNP). Dossier de Presse: Palmarès des Critères D'achat de Viande des Français. 2014, p. 19. Available online: http://www.cultureviande.fr/wp-content/uploads/2014/06/dossier-de-presse-2014.pdf (accessed on 10 March 2020).

4. ADEME. Pertes et Gaspillages Alimentaires. L'état des Lieux et Leur Gestion par étapes de la Chaîne Alimentaire. 2016, p. 165. Available online: https://www.ademe.fr/sites/default/files/assets/documents/pertes-et-gaspillages-alimentaires-201605-rapport.pdf (accessed on 10 March 2020).

5. Wittenberg, J.B.; Wittenberg, B.A. Myoglobin function reassessed. *J. Exp. Biol.* **2003**, *206*, 2011–2020. [CrossRef] [PubMed]

6. Cornforth, D.; Jayasingh, P. Chemical and physical characteristics of meat color and pigment. In *Encyclopedia of Meat Sciences*, 1st ed.; Jensen, W.K., Devine, C., Dikeman, M., Eds.; Elsevier Sci. Ltd.: Oxford, UK, 2004; Volume 1, pp. 249–256.

7. American Meat Science Association (AMSA). Meat Color Measurement Guidelines. Revised 21 December 2012. Available online: http://www.meatscience.org/publications-resources/printed-publications/amsa-meat-color-measurement-guidelines (accessed on 10 March 2020).

8. Kim, G.D.; Jeong, T.C. Proteomic analysis of meat exudates to discriminate fresh and freeze-thawed porcine Longissimus thoracis muscle. *LWT Food Sci. Technol.* **2015**, *62*, 1235–1238. [CrossRef]

9. Lebert, I.; Robles-Olvera, V.; Lebert, A. Application of polynomial models to predict growth of mixed cultures of *Pseudomonas* spp. and *Listeria* in Meat. *Int. J. Food Microbiol.* **2000**, *61*, 27–39. [CrossRef]

10. Bekhit, A.E.D.; Morton, J.D.; Bhat, Z.F.; Kong, L. Meat color: Factors affecting color stability. In *Encyclopedia of Food Chemistry*, 1st ed.; Varelis, P., Melton, L., Shahidi, F., Eds.; Elsevier Inc.: Amsterdam, The Netherlands, 2019; Volume 2, pp. 202–210.

11. Tang, J.; Faustman, C.; Hoagland, T.A. Krzywicki Revisited: Equations for Spectrophotometric Determination of Myoglobin Redox Forms in Aqueous Meat Extracts. *J. Food Sci.* **2004**, *69*, 717–720. [CrossRef]

12. Renerre, M.; Labas, R. Biochemical factors influencing metmyoglobin formation in beef muscles. *Meat Sci.* **1987**, *19*, 151–165. [CrossRef]

13. Maruitti, L.R.B.; Bragagnolo, N. Influence of salt on lipid oxidation in meat and seafood products: A review. *Food Res. Int.* **2017**, *94*, 90–100. [CrossRef] [PubMed]

14. Thiem, I.; Lüpke, M.; Seifert, H. Extraction of meat juices for isotopic analysis. *Meat Sci.* **2005**, *71*, 334–341. [CrossRef] [PubMed]

15. Baran, W.I.; Kraft, A.; Walker, H.W. Effects of carbon dioxide and vacuum packaging on color and bacterial count of meat. *J. Milk Food Technol.* **1970**, *33*, 77–82. [CrossRef]

16. Daniloski, D.; Petkoska, A.T.; Galic, K.; Scetar, M.; Kurek, M.; Kalevska, T.; Nedelkoska, D.N. The effect of barrier properties of polymeric films on the shelf-life of vacuum packaged fresh pork meat. *Meat Sci.* **2019**, *158*, 107880. [CrossRef] [PubMed]

17. Faustman, C.; Cassens, R.G. Strategies for improving fresh meat color. In Proceedings of the 35th ICoMST, Danish Meat Research Institute, Copenhagen, Denmark, 20–25 August 1989; pp. 446–453.

18. Ke, Y.; Mitacek, R.M.; Abraham, A.; Mafi, G.G.; VanOverbeke, D.L.; DeSilva, U.; Ramanathan, R. Effects of Muscle-Specific Oxidative Stress on Cytochrome c Release and Oxidation–Reduction Potential Properties. *J. Agri. Food Chem.* **2017**, *65*, 7749–7755. [CrossRef] [PubMed]

19. Gutzke, D.; Trout, G.R. Temperature and pH Dependence of the Autoxidation Rate of Bovine, Ovine, Porcine, and Cervine Oxymyoglobin Isolated from Three Different Muscles *Longissimus dorsi*, *Gluteus medius*, and *Biceps femoris*. *J. Agric. Food Chem.* **2002**, *50*, 2673–2678. [CrossRef] [PubMed]

20. Warner, R.D.; Kearney, G.; Hopkins, D.L.; Jacob, R.H. Retail colour stability of lamb meat is influenced by breed type, muscle, packaging and iron concentration. *Meat Sci.* **2017**, *129*, 28–37. [CrossRef] [PubMed]

MDPI

St. Alban-Anlage 66

4052 Basel

Switzerland

Tel. +41 61 683 77 34

Fax +41 61 302 89 18

www.mdpi.com

Applied Sciences Editorial Office

E-mail: applsci@mdpi.com

www.mdpi.com/journal/applsci